JAPAN ANNUAL REVIEWS IN ELECTRONICS,
COMPUTERS & TELECOMMUNICATIONS
Vol. 16
AMORPHOUS SEMICONDUCTOR
TECHNOLOGIES & DEVICES

Editor-in-chief

Yoshihiro Hamakawa
Osaka University

Editorial Board

Hiromu Haruki
Fuji Electric Corporate Research and Development, Ltd.
Eiichi Maruyama
Hitachi, Ltd.
Yoshihiko Mizushima
Hamamatsu Photonics Co., Ltd.
Kazunobu Tanaka
Electrotechnical Laboratory
Fumiko Yonezawa
Keio University

▲Still expanding consumer electronics applications of a-Si solar cells. (presented by Sanyo Electric Co.)

▲Role to role mass production in line system of a-Si solar cells. (presented by Sharp-ECD Solar Inc.)

▲NEDO size (40×120 cm^2) amorphous silicon solar cell module produced by continuous mass production line. (presented by Fuji Electric Co., Ltd.)

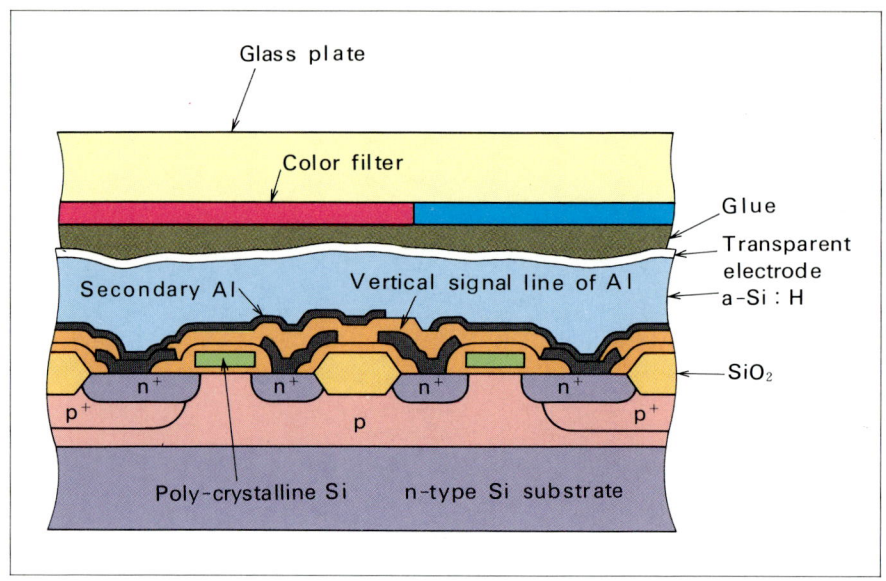

(a)

(b)
▲Monolithic tri-color a-Si:H image sensor stacked on silicon MOS (a) and its reproduced picture (b).
(presented by Hitachi, Ltd.)

JAPAN ANNUAL REVIEWS IN ELECTRONICS, COMPUTERS & TELECOMMUNICATIONS
Vol. 16

AMORPHOUS SEMICONDUCTOR
TECHNOLOGIES & DEVICES

1984

Editor Y. HAMAKAWA

OHM
Tokyo・Osaka・Kyoto

NORTH-HOLLAND
Amsterdam・New York・Oxford

OHMSHA, LTD. and NORTH-HOLLAND PUBLISHING COMPANY

Copyright © 1984 by OHMSHA, LTD.

ISBN 4-274-03035-0 (OHMSHA)
ISBN 0-444-87584-0 (NORTH-HOLLAND)
ISSN 0167-5036

Co-published by
OHMSHA, LTD.
1-3 Kanda Nishiki-cho, Chiyoda-ku, Tokyo 101, Japan
Sole distributors for Japan
and
ELSEVIER SCIENCE PUBLISHERS B.V. (NORTH-HOLLAND)
P.O. Box 1991, 1000 BZ Amsterdam, the Netherlands
Sole distributors for outside Japan

All rights reserved. No part of this publication may be reproduced, stored in a retrieval system, or transmitted in any form or by any means, electronic or mechanical, photocopy, recording, or otherwise, without the prior written permission of the publishers.

Printed in Japan

Preface

The recent discovery of an existence of valency controllability in hydrogen passivated amorphous silicon strongly promotes the evaluation of amorphous semiconductor as a new electronics material. This new amorphous material is able to form both p-n and p-i-n junctions and it has excellent photoconductivity with a considerably high absorption coefficient. The foregoing characteristics coupled with massproduceability of large area non-epitaxial growth on any substrate material match very timely with the strong current need for the development of a low cost solar cell as a new energy resource. With the aid of the national project for renewable energy development, substantial progress in the amorphous silicon field has been seen in recent years in both basic physics and technology. These integrated new knowledges here opened some other new application fields such as TFT, electrophotography, three-dimensional integrated devices, and quantum well devices, etc., and has triggered to start many related R & D efforts in the broad areas of optoelectronics and electronics.

Recently, there have been a number of books published on the subject of the physics of amorphous semiconductors. However, there are still very few books on amorphous devices and technologies. The purpose of publishing this volume is to sum up the present status of the Japanese activities in the device field and their technological accomplishments and to stimulate scientists and engineers toward the advancement of this newly born electronics area, even though the basic features of this material have not yet been well identified.

It is our editors earnest hope that this annual review will be helpful to all researchers and engineers in this area, and that the contents will lead to further accelerative advances in this rapidly expanding technological field. In selection of the contents of this volume, many people have assisted the editorial board and have offered their support to the domestic amorphous semiconductor scientists. First we would like to express our sincere appreciation to the Ministry of Education, Special Research Project Office on "Amorphous Material & Physics", and also the Sunshine Project Headquarters Office for releasing their supported research data to the authors. At the publishers, we want to acknowledge Mr. S. Sato, Director of the publishing office, who encouraged us to undertake this series of editions and Mr. M. Mori, Mr. Y. Moriwaki and Mr. K. Watanabe who handled the prodution of this book.

(Yoshihiro Hamakawa)

Toyonaka, Osaka, Japan
Full Cherry Blossom Time, Spring, 1984

Contents

PREFACE

Chapter 1 INTRODUCTION 1
    Y. Hamakawa ; Osaka University
  1.1 Remarkable Progress of Amorphous Silicon Solar Cell Technology·· 2
  1.2 Organization of the Chapters 8

Chapter 2 SOME TOPICS IN THE BASIC PHYSICS 11
  2.1 Optical Absorption Properties in Amorphous Silicon Alloys 12
    F. Yonezawa ; Keio University
  2.2 Defects in Si-Based Films——ESR and NMR Studies—— 21
    T. Shimizu ; Kanazawa University
  2.3 Raman Study on Bonding Structure of a-Si and Related Materials ·· 33
    T. Shimada and Y. Katayama ; Hitachi, Ltd.
  2.4 Photoinduced Effect in Hydrogenated Amorphous Silicon 42
    K. Morigaki ; University of Tokyo

Chapter 3 NEW ASPECTS OF PREPARATIONS AND GROWTH
           KINETICS 51
  3.1 Plasma Spectroscopies and Film Deposition 52
    *K. Tanaka, N. Hata and A. Matsuda ; Electrotechnical
    Laboratory*
  3.2 a-Si : H Deposition from Higher Silanes 67
    M. Hirose ; Hiroshima University
  3.3 Microcrystalline Silicon (μc-Si) Prepared by Plasma-Chemical
    Techniques 80
    Y. Osaka and T. Imura ; Hiroshima University
  3.4 Photo-CVD of a-Si : H 98
    *K. Aota and Y. Tarui ; Tokyo University of Agriculture
    and Technology*
    T. Saitoh ; Hitachi Ltd.
  3.5 Tetrahedral Alloys 108
    Y. Kuwano and S. Tsuda ; Sanyo Electric Co., Ltd.

CONTENTS

Chapter 4 TOPICS IN DEVICE PHYSICS FIELD ················ 119
    4.1 Photogenerated Carrier Transport ············· 120
        H. Okamoto ; Osaka University
    4.2 Impurity Effects ············· 134
        A. Hiraki ; Osaka University
    4.3 Laser Scribing Lithography ············· 149
        S. Yamazaki, S. Watabe and K. Itoh ; Semiconductor Energy Laboratory Co., Ltd.
    4.4 Dry Etching Processing ············· 168
        A. Yoshikawa and Y. Utsugi ; Atsugi Electrical Communication Laboratory, NTT
    4.5 Quantum Wells in Amorphous Semiconductors ············· 173
        T. Ogino ; Atsugi Electrical Communication Laboratory, NTT
        Y. Mizushima ; Hamamatsu Photonics Co., Ltd.

Chapter 5 AMORPHOUS SILICON SOLAR CELLS ············· 179
    5.1 p-i-n & n-i-p Basis Solar Cells ············· 180
        Y. Uchida ; Fuji Electric Corporate Research and Development Ltd.
    5.2 Heterojunction Stacked Solar Cells ············· 200
        Y. Hamakawa and H. Okamoto ; Osaka University
    5.3 Mass Production Technology in a Roll-to-Roll Process ············· 212
        H. Morimoto ; Sharp Corp.
        M. Izu ; Energy Conversion Devices, Inc.
    5.4 a-Si Solar Cell Industrialization and Systems ············· 222
        Y. Kuwano and S. Nakano ; Sanyo Electric Co., Ltd.

Chapter 6 AMORPHOUS SEMICONDUCTOR ELECTRONIC DEVICES ············· 235
    6.1 Thin Film Diodes and Transistors ············· 236
        M. Matsumura ; Tokyo Institute of Technology
    6.2 Microfabrication Technology Using Multicomponent Amorphous Alloy ············· 242
        K. Murase ; Atsugi Electrical Communication Laboratory, NTT
        Y. Mizushima ; Hamamatsu Photonics Co., Ltd.
    6.3 Anodic Oxidation and Its Device Applications ············· 252
        H. Hasegawa, H. Yamamoto, S. Arimoto and H. Ohno ; Hokkaido University

Chapter 7 OPTO-ELECTRONIC APPLICATIONS OF AMORPHOUS SEMICONDUCTORS ············· 265
    7.1 Optical Memories ············· 266
        M. Takenaga and M. Mikoda ; Matsushita Electric Industrial

Co., Ltd.

7.2 A CCD Imager Overlaid with An a-Si : H Layer ················ 283
   N. Harada ; Toshiba Corp.

7.3 Amorphous Silicon Linear Image Sensors ···················· 290
   T. Tsukada ; Hitachi, Ltd.

7.4 Enhancement of Long Wavelength Sensitivity ················ 300
   I. Shimizu ; Tokyo Institute of Technology

Authors' Profile ·· 311

CHAPTER 1

INTRODUCTION

Yoshihiro HAMAKAWA*

As the introduction of this volume, the recent significant progress in the field of amorphous silicon and its alloys are examined with demonstrating the strong potential needs for a low cost solar cell in the photovoltaic project. Characteristic features as new electronic material are analyzed, and prospect of remarkable growth and application fields are also discussed together with the topics compiled in this edition. In the final section, brief remarks are made on the efforts paid by the editorial board for the seletion of topics in this rapidly expanding field. The organization and structure of the contents in this volume are also introduced.

---

* Faculty of Engineering Science, Osaka University, Toyonaka, Osaka 560.

INTRODUCTION

1.1 Remarkable Progress of Amorphous Silicon Technology

This year marks 10th anniversary of the sunshine project which has been organized in 1974 for the new energy resources to supply considerable portion of Japanese total energy demands by the year of 2000. Among a wide variety of the renewable energy technologies in the sunshine project such as solar energy, geothermal energy, coal gasification and liquefaction, hydrogen energy etc., the solar photovoltaics is growing to become the most promised technology for future energy resource, which is clean in nature, pollution free and abundantly available anywhere in the world or even in space as everybody knows. A significant evidence of this remarkable progress in the photovoltaic technology is more than one order of magnitude reduction of the solar cell price accomplished during these ten years. In fact, it was more than 100 $/Wp of the module cost in 1974, and now has come down to 6~8 $/Wp as shown in Fig. 1.1. The main reason of this cost down is due to a

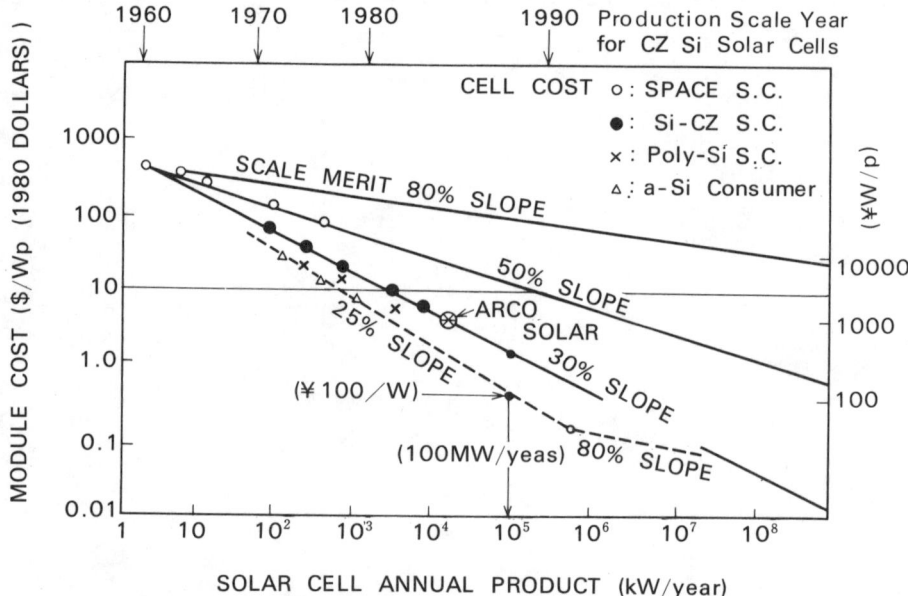

Fig. 1.1 Solar cell module cost vs mass-production scale, 30% slope lines of the scale merit match with Si-CZ solar cell and 25% for a-Si solar cell.

big scale merit of solar cell production with increasing about two order of magnitude of the annual productions in recent few years as shown in Fig. 1.2.[1]

As it has been pointed out elsewhere[2] that cost reduction of the solar cell module is prime importance to succeed in the photovoltaic project. Therefore tremendous amount of R & D efforts have been paid in the wide areas of technical fields from solar cell material, cell structure, mass production processes to photovoltaic systems. As the results, it could be expected that there are roughly two steps of technological innovations for further solar cell cost reductions as shown in Fig. 1.3. One is low cost substrate preparation technologies

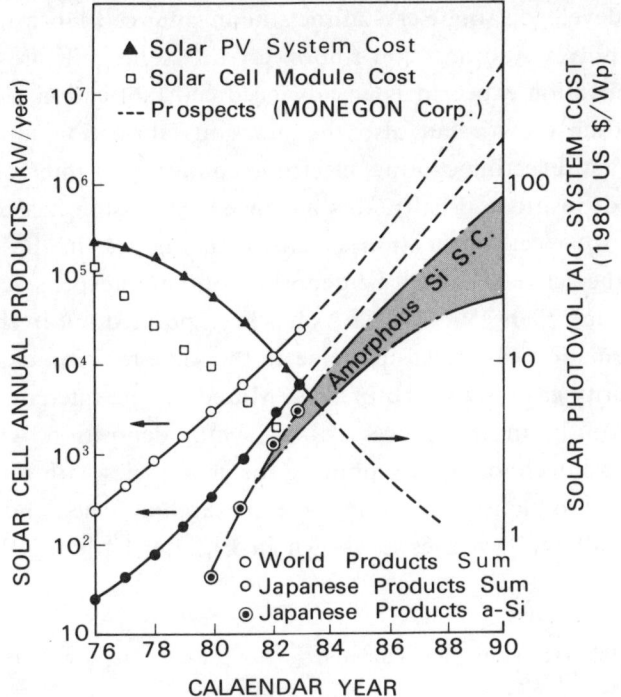

Fig. 1.2 Growth of world and domestic annual production of solar cell module and prospect of module cost.

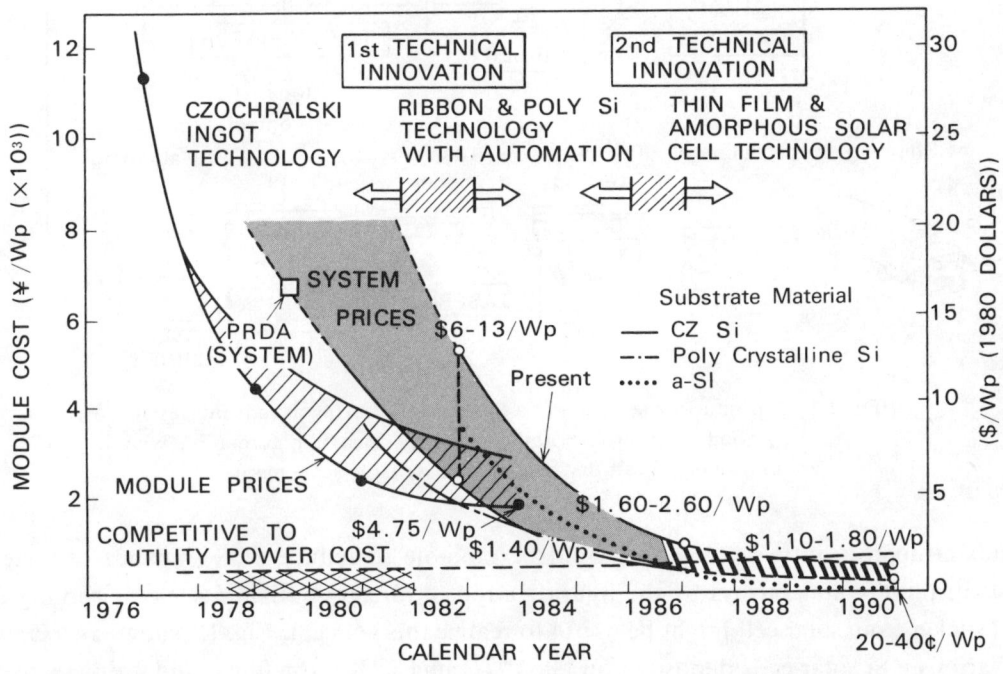

Fig. 1.3 Cost transitions of solar cell module and photovoltaic system with prospective technological innovations.

applicable to well developed single crystalline silicon solar cell fabrication process. In this category, low cost polycrystalline,[3] and ribbon crystal[4] solar cells are in progress.

Another innovation expected is an advanced cell fabrication technology involving not only the substrate growth but also the junction fabrication and other multi-layer processing such as antireflecting coating, electrode contact with interconnection by successive vapour phase depositions. Really, this advanced processing has been realized in the amorphous silicon solar cell fabrication as shown in Fig. 1.4. In the case of amorphous silicon solar cells, the active layer can be deposited on any inexpensive substrate with low temperature process less than 250~350°C. As has been pointed out in the early stage of the work,[5] junction formation can be easily made in the same reaction chamber by mixing of substitutional impurity gases into SiH_4 or SiF_4. Moreover, the interconnection of cells can be made simultaneously in the process of a-Si film deposition with a conventional integrated-circuit mask technology. Combining the advantages with this technology, automatization of a mass-production line could be easily accomplished and direct scale-up can be expected with a all-dry processes as shown in Fig. 1.4. Utilizing the concept of non-

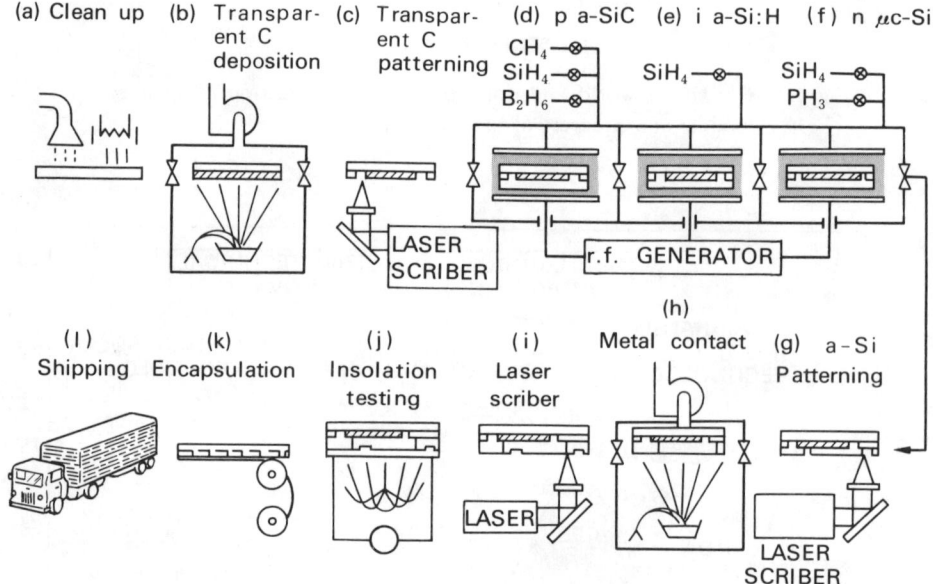

Fig. 1.4 A production sequence of a-Si solar cells having the advantages of ; 1) good mass produceability, 2) large area film formation with nonepitaxy, 3) all dry process, and 4) large scale merit.

epitaxial deposition technology, it could be possible to reduce BOS (balance of system) costs in photovoltaic arrays by the hybridization of already-built units. Solar roofing tile and sticker from solar cell might be useful to realize this concept. Fig. 1.5 shows an example of various a-Si solar cells deposited on glass,[6] ceramics,[7] Kapton films[8] and stainless steel.[9]

An increase of cell efficiency is also directly connected to the cost down project. In the amorphous silicon solar cell project, tremendous efforts to improve the cell efficiency

(a) a-Si solar cell roofing tile (presented by Sanyo Electric Co.)

(b) NEDO size module (presented by Fuji Electric Co.)

(c) Ceramic substrate a-Si solar cell (presented by Kyocera Co.)

(d) Polymer film substrate a-Si solar cell (produced by Teijin Co.)

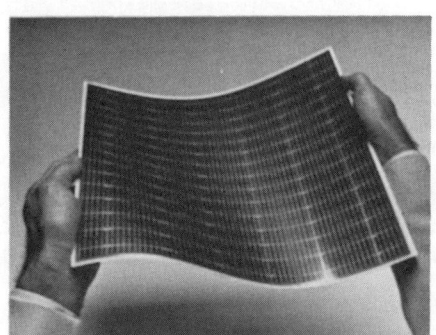

(e) Stainless steel substrate a-Si solar cell (produced by role-to-role mass production process presented by Sharp-ECD solar Inc.).

Fig. 1.5 a-Si solar cells deposited on various inexpensive materials.

INTRODUCTION

have also been made with various approaches such as the film quality improvement, new junction structures with a-SiC, c-Si, a-SiGe and also new electrode materials as will be discuseed in Chapter 5 in this work. For example, a-SiC/a-Si heterojunction solar cell, which firstly broken through the 8% efficiency barrier in 1980,[10] is a typical successful one, and now it becomes a routine technology for the high efficiency solar cells. In fact, the recent records more than 10% efficiency attained by RCA,[11] Sanyo,[12] Komatsu,[13] and TDK-SEL[14] etc. were all attained with this a-SiC/a-Si heterojunction structure. Fig. 1.6 shows the

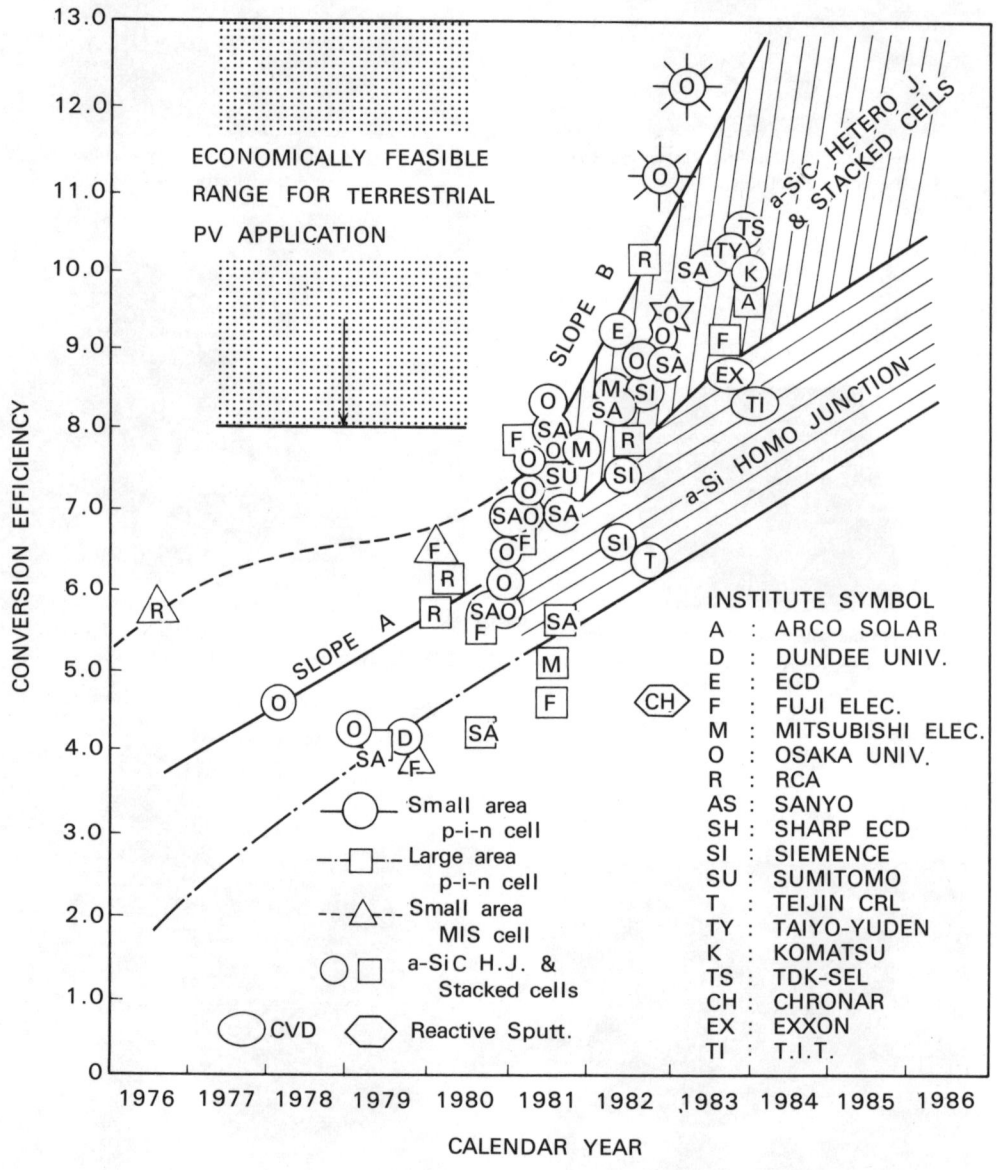

Fig. 1.6 Progress of a-Si solar cell efficiencies for various types of junction structures as of March 1984. It is seen a steep slope change with appearance of new amorphous materials around 1980.

transitions of cell efficiency for various types of a-Si solar cells since 1976. As can be seen from this figure, a step-like increase of the cell efficiencies is seen in the region -1981, while the slope A before 1981 corresponds to the improvement of the film quality and routine cell fabrication progresses. The key technologies that produced the step slope change from A to B at 1981, were development of heterojunction solar cells with a-SiC : H[15] and a-SiGe : H.[16]

Another possibility of further improvement in the amorphous solar cell efficiency is a heterostructure stacked junctions by utilization of the internal carrier exchange effect through the localized states at the interface of heterojunction. The details of this new type of stacked solar cell will be demonstrated in the Chapter 5.2. At present, more than 12.5% efficiency are obtained by the a-Si/poly-Si heterostructure stacked cell.[17] With this new technological innovation by the amorphous silicon technology, it would be expected that the module cost will be 30~50 $/Wp in 1990 as shown in Fig. 1.3. It should be notified here that present module cost estimated from the pocketable calculator a-Si solar cells is rather higher than that of CZ-Si solar cell, because of the big difference in the technological maturities as illustrated in Fig. 1.7. While the annual production scale in 1990 become 100

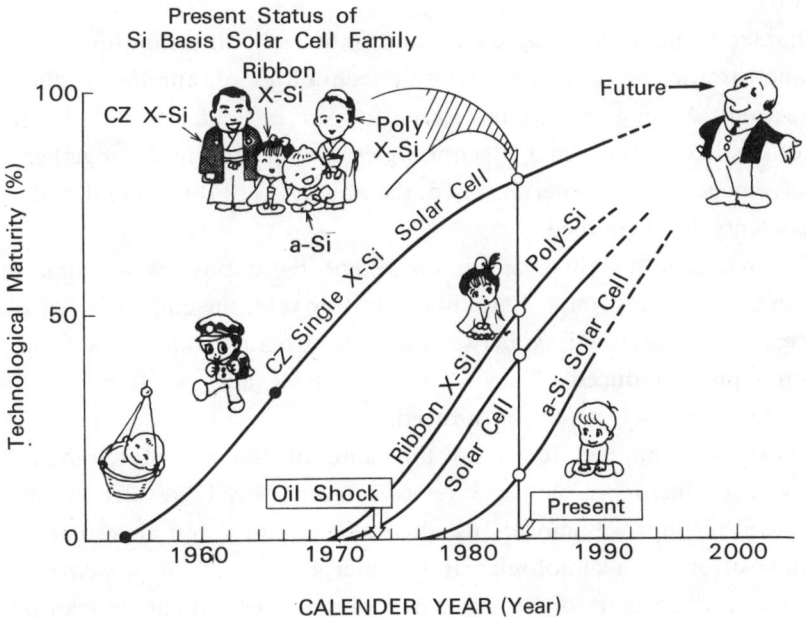

Fig. 1.7 Differences of the technological maturities among CZ-Si, poly-Si, ribbon-Si and a-Si solar cells.

MWatt/year as can be seen in Fig. 1.3 is that we may have the cross point in the module cost with the equivalent leveled utility power cost of presently available technology by the hydrodynamic power generation in Japan.

With these intensive efforts, amorphous silicon and its alloy now become champion materials for the low cost solar cell project. Moreover, a number of unique advantages in these materials such as a) structure sensitive with large photoconductivities, b) large area film formation, c) low temperature growth on any substrate material are well confirmed by recent basic researches, and recognized as a new electric materials for electronic and optoelectronic device applications such as, electrophotography, thin film transistors(TFT), three dimensional element like solid-state imaging pick up. In these view points, we should like to watch out the growth of the new material technologies with a warm mother's eye.

1.2 Organization of the Chapters

The most extensive efforts paid by the editorial board was the selection of the current topics on promising technology and their applications which have been conceived and developed through domestic efforts during the past one year. From a applied physics viewpoint, amorphous semiconductors can be classified into four categories: tetrahedrally bonded amorphous material, chalcogenide glass, oxide glass, and pnictide amorphous semiconductors, such as group V and group III-V compounds. However, no particular electronic and optoelectronic applications have so far been found for the last two. Therefore, emphasis has been placed on the present sratus of recent R & D efforts on preparation and basic property as well as their devices both of tetrahedrally bonded and chalcogenide semiconductors.

In Chapter 1, the recent progress of amorphous silicon technology is firstly pointed out with demonstrating the solar photovoltaic technology on annual production rate and cell cost reductions. Among various new technology in the field, a remarkable contributions of amorphous silicon and their technologies are introduced, together with some significant advantages of this material. Then, the objectives of this volume and the organization of its contents are explained.

In Chapter 2, a review is given on recent basic physics research on domestic amorphous semiconductor groups. Particularly in this year, the emphasis was placed on the optical process and its interactions between atomic structure, that is the topics on optical absorption and photo-induced effects in a-Si:H. ESR and NMR studies, Raman and ENDOR in a-Si:H are selectively introduced.

The next two chapters focus on the state of the art of tetrahedrally bonded amorphous semiconductors. Chapter 3 reviews recent R & D results in the fabrication technologies used to improve amorphous silicon film quality and also some new topics on high speed deposition rate technologies. In Chapter 4, activities on some new characterization techniques on the localized gap states and photogenerated carrier transport and their applications in device physics are introduced. The effects of impurity doping and surface-interface physics on both hydrogenated and fluorinated amorphous silicon and silicon carbide are also discussed. New process technologies of laser scribing and plasma etching are also introduced.

In Chapter 5, recent advances in a-Si solar cells and their technologies are introduced. Topics on single junction solar cell are reviewed. Then, theoretical limit of the multi-layer stacked cell and their experimental approaches in progress are demonstrated. Some new accomplishments on the mass production processes and on a-Si solar cell systems and their industrializations are also overviewed.

In Chapter 6 and 7, the current state of the arts in the field of other amorphous semiconductor devices in the domestic activities are demonstrated. Significant efforts are now underway on TFT, amorphous optical memory. New mode of devices such as the intelligent copy, CCD imaging devices utilizing low temperature growth of a-Si hybridized with c-Si are introduced, and recent data on related new microfabrication technologies are also discussed.

References

1) "Annual Summary Report on Optoelectronic Devices and Their Industrial Standard (JIS)", Japan Optoelctronic Industrial Association (1984) May, p. 434. and Y. Hamakawa: Sunshine Journal 4, No. 5 (1983) p. 1.
2) Y. Hamakawa: "Device Physics and Optimum Design of the Amorphous Silicon Photovoltaic Devices", Ch. 4.1 of "Amorphous Semiconductor Technology & Devices" (1982), OHMSHA & North-Holland.
3) C. H. Seager: J. Appl. Phys. 52 (1982) p. 2. and Wacker Chemitronics Technical Information (1980) p. 12.
4) T. F. Ciszek and G. H. Schwnkke: Proc. 15th IEEE Photovoltaic Specialists Conf. (1981) p. 247.
5) For example, Y. Hamakawa: J. Phys. (Paris) $C4$, suppl. 10 (1981) p. C4-1131.
6) Y. Kuwano, M. Onishi, and H. Shibuya: Jpn. J. Appl. Phys. 20 (1981) p. 157. and Y. Uchida and H. Haruki: The 3rd New Energy Industrial Symposium, NEF, Session 2A-3 (1983) Nov. p. 24.
7) K. Ishibitsu and Y. Nitta: Spring Annual Meeting of Jpn. Appl. Phys. (1983) Apr.
8) H. Okaniwa, M. Asano, K. Nakatani, M. Yano, and K. Suzuki: Proc. 3rd Photovoltaic Sci. & Eng. Conf. in Japan (1982).
9) For example, "Amorphous Semiconductor, Technology & Devices", Ch. 5, 6 (1983) ed. Y. Hamakawa, OHMSHA & North-Holland, p. 182.
10) Y. Tawada, H. Okamoto, and Y. Hamakawa: PVSEC-2 (1980) p. 52.
11) T. Catalano, A. Firestar, and B. Fanghaman: Proc. 16th IEEE Photovoltaic Specialists Conf., San Diego (1982).
12) M. Onishi, H. Nishiwaki, K. Enomoto, Y. Nakashita, S. Tsuda, T. Takahama, H. Tarui, M. Tanaka, H. Dojo, and Y. Kuwano: J. Non-Cryst. Solids, $59\&60$ (1983) p. 1107.
13) A. Tanabe, A. Tajika, H. Minakami, T. Miyako, S. Sano, and O. Kuboi: No.44 Annual JSAA Meeting (1983) 25a-L-8.
14) S. Yamzaki, K. Itoh, S. Watabe, A. Mase, k. Urata, K. Shibata and H. Shirohara: Proc. 17th IEEE Photovoltaic Specialists Conf., Florida (1984) 1-1B-1.
15) Y. Hamakawa and Y. Tawada: Int. J. Solar Energy 1, No. 1 (1982) p. 125.
16) G. Nakamura, K. Sato, H. Kondo, Y. Yukimoto, and K. Shirahata: Photovoltaic Solar Energy Conf., Stresa (1982) p. 616.
17) K. Okuda, H. Okamoto, and Y. Hamakawa: Jpn. J. Appl. Phys., 22(1983) p. L605.

CHAPTER 2

SOME TOPICS IN THE BASIC PHYSICS

2.1 Optical Absorption Properties in Amorphous Silicon Alloys

Fumiko YONEZAWA*

Abstract

We theoretically study the experimental results of Cody and his coworkers concerning the optical absorption edge of hydrogenated amorphous silicon.

We assert that impurity atoms in amorphous silicon cause changes in bonding energy as well as changes in distortion. A simple theoretical model which takes into account both of these changes is shown to reproduce the experimentally observed behaviours of the optical edge and of the exponential tail in the absorption coefficient. The optical edge, E_G, determines the absorption efficiency, while the width, E_0, of the exponential tail defines the electron-hole pair extraction efficiency. Our theory suggests that E_G and E_0 can be varied individually if those two effects of impurity atoms as described above can be controlled separately. This indicates that it would be possible to control the absorption efficiency and the electron-hole pair extraction efficiency individually. Accordingly, the dilemma discussed by Cody et al. could be avoided.

2.1.1. Introduction

Amorphous semiconductors contain various kinds of disorder. Therefore, it is necessary to characterize the different types of disorder in order to study the physical properties of amorphous semiconductors. For this purpose, we classify models of amorphous semiconductors into two categories.

To the first category belong ideal amorphous semiconductors defined by the fully-interconnected continuous random networks (CRN) in which the valency, similar to that found in a corresponding crystalline structure, is perfectly satisfied, bond lengths almost ideal, bond angles less ideal and dihedral angles rather disordered. An ideal amorphous semiconductor is advantageous in that it can extract the effects of topological disorder alone.

For instance, the effects on the band width of different types of disorder in an ideal amorphous semiconductor have previously been estimated as follows.[1]

The fluctuations in bond lengths give a tail at each band edge whose width is estimated to be about 0.02 eV while the fluctuations in bond angles yield a tail with a width of about 0.2 eV. A rather serious degree of disorder in dihedral angles is reflected to the

---

* Department of Physics, Keio University 3-14-1 Hiyoshi, Kohoku-ku, Yokohama 223.

disorder in ppπ-interactions as well as to the disorder in rings in the sense that both odd- and even-membered rings take place. The outcome of the disorder in ppπ-interactions is the erosion of the top of the valence band while the disordered rings are responsible for the appearance of the Lifshitz limit again at the top of the valence band. These two effects altogether give rise to the narrowing of the valence band by about 0.3 eV. When all these effects are taken into account, the optical energy gap for ideal amorphous silicon is estimated to be about 1.4 eV compared with 1.0 eV or so for crystalline silicon.

To the second category belong non-ideal (but more realistic) amorphous semiconductors which would serve as models for a-SiH as well as for a-SiF. In these systems, in addition to the kinds of disorder described above, there exist dangling bonds and terminators such as H and F.

The effects of these terminators have been experimentally examined by Cody and his coworkers.[2,3] The characteristic quantities which appear in their analysis are:
(1) The width of the exponential tail, E_0.
(2) The optical energy gap, E_G.

The optical absorption coefficient within the exponential tail is described by

$$I(E, T) = I_0 \exp\left[\frac{E - E_1}{E_0(T, x_H)}\right] \qquad (2.1.1)$$

in terms of $E_0(T, x_H)$ which is referred to as the width of the exponential tail. On the other hand, the optical energy gap, E_G, is determined by the expression

$$[I(E,T)E]^{1/2} = \text{const.} \times [E - E_G(T)] \qquad (2.1.2)$$

for the optical-absorption coefficients $I(E,T)$ outside the exponential region. Note that E_0 is related to the electron-hole pair extraction efficiency while E_G determines the optical absorption efficiency.

In Fig. 2.1.1 are shown the experimental results due to Cody et al. on the optical gap, E_G, illustrated as a function of E_0 for three samples of a-SiH$_{0.13}$. The solid and open circles are for measurements at constant hydrogen concentration x_H and variable T_m (the measurement temperatures). The solid and open triangles are for measurements at constant T_m (300 K) and variable x_H. The measurement temperatures are higher in the direction towards the right-bottom corner for measurements at constant hydrogen concentration, while the hydrogen concentration is lower in this direction for measurement at constant measurement temperatures.

Cody and his coworkers have asserted that the effects of an increasing number of hydrogen atoms are equivalent to those of the decrease in the thermal and structural disorder where the thermal disorder and the structural disorder are considered to be additive. Their interpretation of this situation is that there exists a fundamental tradeoff between the optical absorption efficiency and the electronhole pair extraction efficiency. Note that the optical absorption efficiency is higher the smaller the values of E_G, while the electron-hole pair extraction efficiency, E_0, is higher the smaller the values of E_0.

Accordingly, the assertion by Cody and his coworkers[2,3] indicates that it seems to be difficult to improve the photovoltaic efficiency by changing either the measurement tempera-

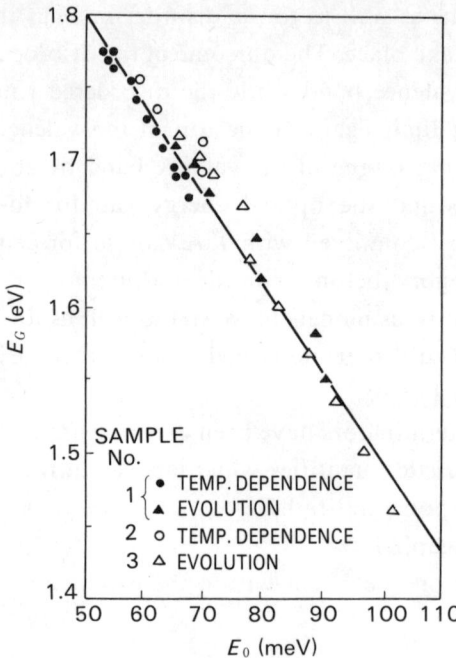

Fig. 2.1.1 Optical gap E_G (T,x_H) as a function of E_0 (T,x_H) for three samples of a-SiH$_{0.13}$. The solid and open circles are for measurements at constant x_H and variable T_m. The solid and open triangles are for measurements at constant T_m (300 K) and variable x_H (425 T_m 624 C).

ture, T_m, or the hydrogen concentration, x_H, since either the optical absorption efficiency or the electron-hole pair extraction efficiency is reduced when T_m or x_H is changed. Our observation, on the other hand, is that it would be possible to improve the photovoltaic efficiency provided we could control E_G and E_0 separately.[4]

Therefore, the purpose of the present paper is to find how we could do this by theoretically studying in more detail the origins determining E_G and E_0. In section 2.1.2, we interpret the dominant effects of the impurity atoms, propose a theoretical model which takes into account these effects, and evaluate the optical absorption coefficients on the basis of this model. In section 2.1.3, we discuss the indications of our results and propose the preferable properties for the impurity atoms to retain for improving the total efficiency of amorphous silicon alloys.

2.1.2 Model and Formulation

It is not difficult to infer that the increase in the number of terminators, such as H, causes two dominant effects. The first effect is that the inclusion of more hydrogen atoms is expected to yield an increase in average bonding energy since the bonding energy of a Si-H bond is larger than that of a Si-Si bond. The second effect is that, when the excess H-atoms happen to be in the neighbourhood of a highly distorted bond, the H-atoms are expected to break the distorted bond so that the distortions in the bond lengths, in the bond angles and, among other things, in the dihedral angles are relaxed and as a consequence the disorder associated with the bond under consideration is reduced.

Our purpose here, therefore, is to propose a model which takes into account both (1) the change in bonding energy and (2) the change in distortions due to H. Our model is a simple two-band system which is described by the following two Hamiltonians

$$H^c = \sum_n |nc> \varepsilon_n^c <nc| + \sum_{n \neq m}\sum |nc> V_{nm}^c <mc| \qquad (2.1.3)$$

$$H^v = \sum_n |nv> \varepsilon_n^v <nv| + \sum_{n \neq m}\sum |nv> V_{nm}^v <mv| \qquad (2.1.4)$$

each of which is the single s-band Hamiltonian. Here $|n\mu>$ describes the Wannier state associated with site n in the band μ ($\mu = c$ or v, c and v denoting the conduction and valence bands respectively). The diagonal element ε_n^μ is the site-energy of the band μ, and V_{nm}^μ expresses the transfer energy of the band μ. The diagonal terms ε_n^μ are taken to be random while the off-diagonal terms V_{nm}^μ are assumed to be regular.

A similar kind of model has previously been studied by Abe and Toyozawa[5, 6] for the calculation of the optical absorption coefficients. Our model is different from theirs in the following two points.

The first point is that;
we take $\gamma \equiv \varepsilon^c - \varepsilon^v$ as a function of x_H. This is based upon our observation that an increase in the number of hydrogen atoms is expected to give rise to an increase in the average bonding energy. The validity of our model is understood when we examine the behaviour of the top peak of the valence band and of the bottom region of the conduction band as shown in Fig. 2.1.2. A comparison of these two peaks with the schematic demonstration of our two-band model is shown in Fig. 2.1.3.

The second point is concerned with the distributions for ε^c and ε^v. When calculating the optical absorption coefficients $I(E)$ theoretically, the quantity we have to evaluate is

$$I(E) \propto \iint dE_1 dE_2 \delta(E_1 - E_2 - E) \ll Tr[|nv><nc|\delta(E_1 - H^c)|nc> \\ <nv|\delta(E_2 - H^v)] \gg \qquad (2.1.5)$$

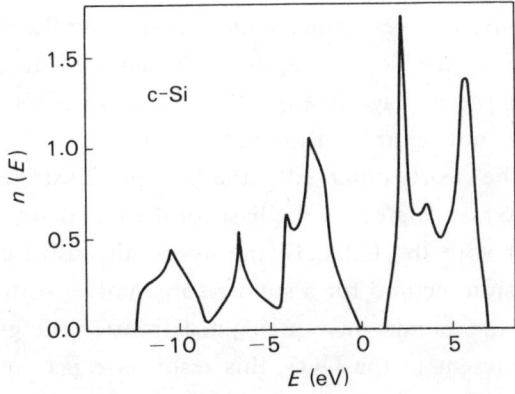

Fig. 2.1.2 The energy band of crystal silicon calculated by the tight-binding approximation. [after F. Yonezawa[14]]

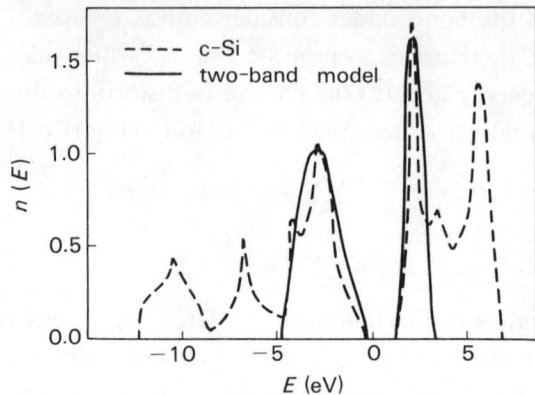

Fig. 2.1.3 A schematic demonstration of our two-band model (solid curve) compared to the crystal Si band.

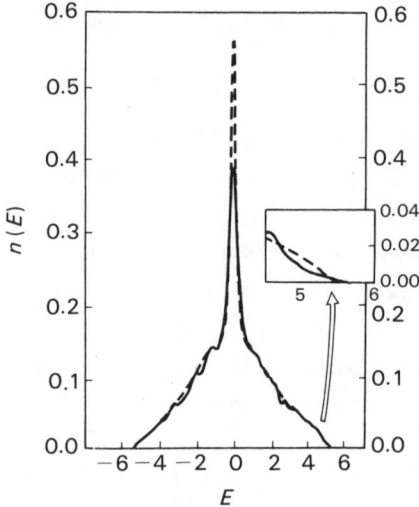

Fig. 2.1.4 Density of states for a disordered simple cubic lattice; solid curve, the result of the recursion method and broken curve, the result of the CPA.

in which the double angular brackets indicate the average over the random variables ε^c and ε^v. In previous work due to Abe-Toyozawa, they assumed the Gaussian distributions for ε^c and ε^v, and carried out the average in Eq. (2.1.5) within the CPA. It is widely known, however, that the CPA is not reliable in the tail region.[7~10]

The situation can be clearly observed in the example illustrated in Fig. 2.1.4 in which the density of states (DOS) calculated on the basis of the recursion method[11,12] is compared with the DOS evaluated with the CPA. In the figure, the solid curve denotes the DOS enumerated by the recursion method for a simple-cubic lattice with the lattice sites of $(21)^3$ in which 20 percent of the bonds have weakened transfer energies, V_{ij}. Except for the detailed fine structures present in the DOS, this result is expected to reflect the essential features of the exact DOS. On the other hand, the broken curve depicts the DOS calculated by the CPA. The fine structures present in the DOS by the recursion method are naturally due to the finite size of the system studied. From the size dependence of the shapes and of the fine structures in the DOS, it follows that these fine structures naturally tend to approach a smooth curve when the size of the system is increased. The important point in relation to the present problem is the behaviour of the DOS in the tail region. The DOS calculated by the recursion method has the tails extending up to 6 and down to -6 as it should while the DOS by the CPA is truncated at ± 5.3 or so. This result clearly shows that the CPA-DOS is not suitable to be employed in our problem. In other words, we must be very careful with the CPA, especially when we are concerned with those physical properties for which the states in the tail regions are considered to play an important role, which is exactly the case for the optical absorption coefficients we are interested in. With this situation in mind, we assume the Lorentz distributions for ε^c and ε^v since it has been shown by Lloyd that the average of the one-particle Green's function can be evaluated exactly for the Lorentz distributions.[13] It is also reasonable to assume that the joint probability distribution $\mathcal{P}(\{\varepsilon_\mu^n\})$ is given by statistically independent Lorentz distributions at each position, i. e.

$$\mathcal{P}(\{\varepsilon_n^\mu\}) = \prod_n P(\varepsilon_n^\mu) \qquad (\mu = c \text{ or } v) \tag{2.1.6}$$

with

$$P(\varepsilon_n^\mu) = \frac{1}{\pi} \frac{\sigma_\mu}{(\varepsilon_n^\mu - \overline{\varepsilon^\mu})^2 + \sigma_\mu^2} \qquad (\mu = c \text{ or } v) \tag{2.1.7}$$

The one-particle Green's function of Eq. (2.1.3), for instance, is defined as

$$\sum_{n'} \{(E - \varepsilon_n^c)\delta_{nn'} - V_{nn'}^c\} G_{n'm}^c(E) = \delta_{nm} \tag{2.1.8}$$

where E is the eigenvalue. The Green's function of Eq. (2.1.4) is also defined in a similar way. As stated above, the ensemble-averaged Green's function defined by

$$\ll G_{nm}^\mu(E) \gg = \int \cdots \int G_{nm}^\mu(E) \mathcal{P}(\{\varepsilon_n^\mu\}) \prod_s d\varepsilon_s^\mu \qquad (\mu = c \text{ or } v) \tag{2.1.9}$$

can be determined exactly when $\mathcal{P}(\{\varepsilon_n^\mu\})$ satisfies Eq. (2.1.7). The resulting expression becomes

$$\ll G_{nm}^\mu(E) \gg = G_{nm}^\mu(E - \overline{\varepsilon^\mu} + iS(E)\sigma_\mu) \qquad (\mu = c \text{ or } v) \tag{2.1.10}$$

where

$$S(z) = \text{sign}(Im\ z) \tag{2.1.11}$$

for z not real. Note that $G_{nm}^\mu(z)$ is the Green's function for the regular system in which the diagonal terms in Eqs. (2.1.3) and (2.1.4) are all set equal to zero.

The optical absorption coefficient given Eq. (2.1.5) can be rearranged as

$$I(E) = \iint dE_1 dE_2 \delta(E_1 - E_2 - E) S(E_1, E_2) \tag{2.1.12}$$

$$S(E_1, E_2) = \ll |p|^2 \sum <nc|\delta(E_1 - H^c)|mc><mv|\delta(E_2 - H^v)|nv> \gg \tag{2.1.13}$$

where p is the transition probability. By means of the Green's functions, $S(E_1, E_2)$ is expressed as

$$S(E_1, E_2) = \frac{1}{(2\pi i)^2}[K(E_1^+, E_2^+) - K(E_1^+, E_2^-) - K(E_1^-, E_2^+) + K(E_1^-, E_2^-)] \quad (2.1.14)$$

in which

$$K(z_1, z_2) = <nc| \ll G^c(z_1 - H^c) \gg |nc><nv| \ll G^v(z_2 - H^v) \gg |nv> \quad (2.1.15)$$

G^c and G^v are respectively the Green's function operators for Eqs. (2.1.3) and (2.1.4). These derivations indicate that the optical absorption coefficient $I(E)$ is evaluated exactly for the Lloyd model. As a consequence, we do not have to worry about reliability of the approximations. It is also possible to argue that the Lorentz distributions are physically plausible.

2.1.3 Results and Discussions

The optical absorption coefficients thus calculated are presented in Figs. 2.1.5 to 2.1.7 for various values of the standard deviations of the Lorentz distributions where the standard deviation σ for the conduction band is taken to be equal to that for the valence band without loss of generality.

The optical energy gap, E_G, is determined by Eq. (2.1.2) given for $I(E)$ outside the exponential tail, while the width, E_0, of the exponential tail is defined by Eq. (2.1.1).

The parameters not yet specified in our formulation are $\gamma = \overline{\varepsilon^c} - \overline{\varepsilon^v}$ and $\sigma_c = \sigma_v = \sigma$ which defines Eq. (2.1.6). The former is the function of (i) the difference of the bonding energies for a Si-H bond and for a Si-Si bond; and (ii) the hydrogen concentration x_H. The latter, on the other hand, is determined by the hydrogen concentration alone. Our simple but realistic model yields the optical gap, E_G, as a function of γ alone while the width, E_0, of the exponential tail is a function of σ exclusively.

When γ is appropriately expressed in terms of the bond energy difference as well as

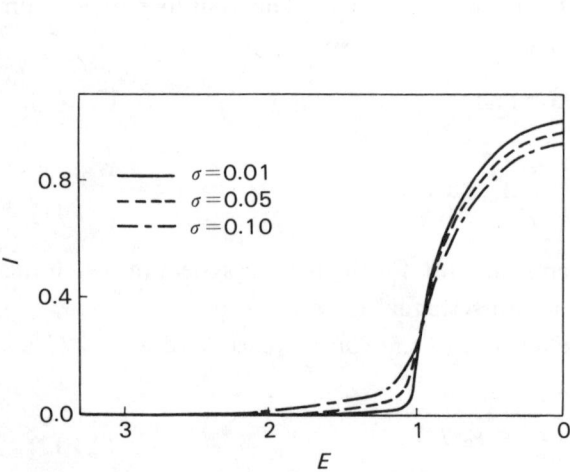

Fig. 2.1.5 Optical absorption coefficient I vs E.

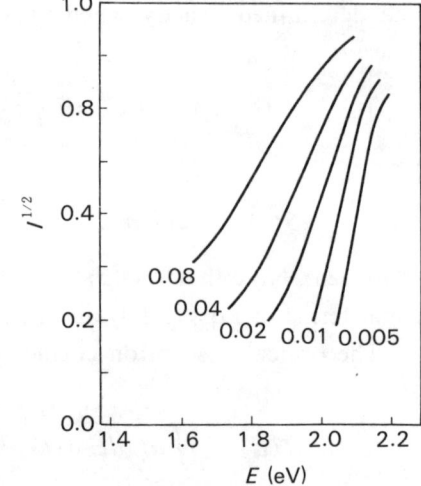

Fig. 2.1.6 $I^{1/2}$ vs E. The numbers denote the magnitude of the standard deviation.

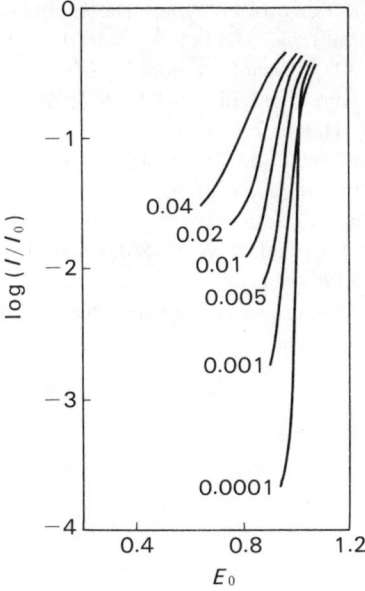
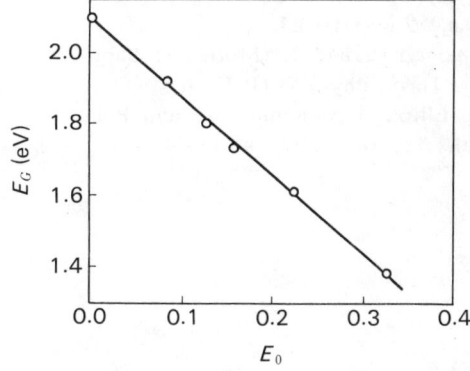

Fig. 2.1.7 Log (I/I_0) vs E.

Fig. 2.1.8 The optical energy gap, E_G, vs the width, E_0, of the exponential tail.

in terms of x_H, and σ in terms of x_H, then the relation between E_G and E_0 is obtained. The result is shown in Fig. 2.1.8. The fact that our theoretical model reproduces the behaviour found experimentally indicates that our interpretation can be regarded as one possibility for explaining these behaviours.

An important point, therefore, is that the optical energy gap, E_G, responsible for the optical absorption efficiency is determined by the average bonding energy denoted by γ as well as by the degree of disorder due to distortions denoted by σ while the width, E_0, of the exponential tail related with the electron-hole pair extraction efficiency is influenced by the degree of disorder, σ, alone. This indicates that the tradeoff between the optical absorption efficiency and the electron-hole pair extraction efficiency is not unavoidable. In other words, it would be possible to improve the total photovoltaic efficiency of a-Si alloys by using, as terminators, elements whose bonding energy with Si is lower than that of a Si-H bond. The candidates would be bromine or iodine whose bonding energy with Si is expected to be nearly equal to that of a Si-Si bond. The comparatively large ionic radius of Br or I may be a problem, but we cannot predict what will happen in amorphous structures and therefore we must say that this is still worth trying.

Detailed analyses and discussions on this subject will be found elsewhere.[14]

References

1) F. Yonezawa and M. H. Cohen: Theory of Electronic Properties of Amorphous Semiconductors, in Fundamental Physics of Amorphous Semiconductors, Springer Series in Solid-State Sciences 25, ed F. Yonezawa (Springer-Verlag, 1981) 119.
2) G. D. Cody, T. Tiedje, B. Abeles, B. Brooks, and Y. Goldstein: Phys. Rev. Lett., 16 (1981) 1480.
3) C.R. Wronski, B. Abeles, T. Tiedje, and

G. D. Cody : Solid State Commun., *44* (1982) 1982.
4) F. Yonezawa, Y. Ishida, S. Ogawa, and K. Tsujino : J. of Non-Crystalline Solids, *59-60* (1983) 69.
5) S. Abe : J. Phys. Soc. Japan, *49* (1980) 1179.
6) S. Abe and Y. Toyozawa : J. Phys. Soc. Japan, *50* (1981) 1717.
7) F. Yonezawa and K. Morigaki : Supp. Prog. Theor. Phys., *53* (1973) 1.
8) R. J. Elliott, J. A. Krumhansl, and P. L. Leath : Tev Mod. Phys., *46* (1974) 465.
9) F. Yonezawa : Prog. Theor. Phys., *40* (1968) 734.
10) J. M. Ziman : "Models of Disorder", Cambridge University Press (1979).
11) R. Haydock : in Solid State Physics, *35* (ed. Seitz and Turnbull)(1980) 215.
12) R. Haydock, V. Heine, and M. J. Kelly : J. Phys. C, Solis State Physics, *8* (1975) 2591.
13) P. Lloyd : J. Phys. C, Solid State Physics, *2* (1969) 1717.
14) F. Yonezawa : to be submitted to J. Phys. Soc. Japan.

2.2 Defects in Si-Based Films——ESR and NMR Studies——

Tatsuo SHIMIZU*

Abstract

Results are given of recent studies on defects in silicon-based amorphous films mainly by ESR and NMR. The origin of dangling bond production is discussed based on results of ESR, NMR and Raman studies. The role of incorporated H and the effect of alloying C and N with Si are described in relation to the local internal strain. ESR and IR studies on a-Si$_{1-x}$N$_x$: H ; NMR study on μc-Si : H ; ENDOR study on a-Si : H and ESR study on plasma-hydrogenation in CVD a-Si are also described. Understanding of the origin and the nature of defects in these films is indispensable for photovoltaic applications.

2.2.1 Introduction

Defects in silicon-based amorphous films are best characterized by electron spin resonance (ESR). Nuclear magnetic resonance (NMR) of H in these films reveals the incorporation scheme of H and the role of reducing the density of defects. In this paper, the results of recent investigations carried out in Japan on defects in silicon-based amorphous films and microcrystalline hydrogenated silicon (μc-Si : H) mainly by ESR and NMR are discussed.

Section 2.2.2 presents a summary of ESR and infrared (IR) absorption studies on a-Si$_{1-x}$N$_x$: H and a-Si$_{1-x}$N$_x$ and annealing effects on a-Si$_{1-x}$N$_x$: H. In Section 2.2.3, results of Raman, ESR and NMR studies show that internal local strain in overconstrained silicon-based amorphous films is the main origin in dangling bond production. The role of incorporated H and effect of alloying C and N with Si are studied in relation to the local strain. The structure of μc-Si : H is described in Section 2.2.4 based on NMR, ESR, IR, Raman and X-ray measurements. Section 2.2.5 is devoted to ENDOR study on a-Si : H, and the results of plasma-hydrogenation effects on defects in CVD a-Si is summarized in Section 2.2.6. In Section 2.2.7, the results are summarized.

2.2.2 ESR and IR Studies on a-Si$_{1-x}$N$_x$: H and a-Si$_{1-x}$N$_x$

Si-N films have recently attracted attention because of their importance in the field of silicon technology and the promise of their use as photovoltaic material. However, there has been no systematic investigation of these films from a microscopic viewpoint. This paper

---

* Department of Electronics, Faculty of Technology, Kanazawa University, Kanazawa 920.

presents results of ESR and infrared (IR) measurements by Shimizu et al., for hydrogenated and unhydrogenated amorphous silicon-nitrogen alloy films (a-Si$_{1-x}$N$_x$: H and a-Si$_{1-x}$N$_x$).[1,2] a-Si$_{1-x}$N$_x$ films were prepared by RF sputtering of Si target in Ar and N$_2$ gas mixtures. a-Si$_{1-x}$N$_x$: H films were prepared using two methods : RF sputtering of Si target in Ar + H$_2$ + N$_2$ gas mixture and glow discharge decomposition of SiH$_4$ and N$_2$. This experiment is a continuation of a similar experiment carried out by the group of the present author for a-Si$_{1-x}$Ge$_x$: H[3] and a-Si$_{1-x}$C$_x$: H.[4,5]

ESR spin density N_s, the linewidth and the g-value versus the N content x[1] can be seen in Figs. 2.2.1 and 2.2.2. It was found that N_s increases with x in a-Si$_{1-x}$N$_x$: H as in the case of a-Si$_{1-x}$C$_x$: H and a-Si$_{1-x}$C$_x$, whereas N_s decreases with x in a-Si$_{1-x}$N$_x$. The increase in the linewidth and the decrease in the g-value with an increase in x can be reproduced by calculation, assuming that the ESR signals originate from Si dangling bonds.[6] The increase in linewidth is caused by hyperfine interaction with neighboring ^{14}N nuclei, and the decrease in g-value is caused by the presence of N atoms around the Si dangling bond. Such agreement between the observed results and the calculated ones, and the absense of large hyperfine split ESR lines due to ^{14}N suggest that ESR originates mainly from Si dangling bonds and not from N dangling bonds. The absense of large density N dangling bonds with unpaired spin results from N atoms that are threefold coordinated and have lone pair electrons.

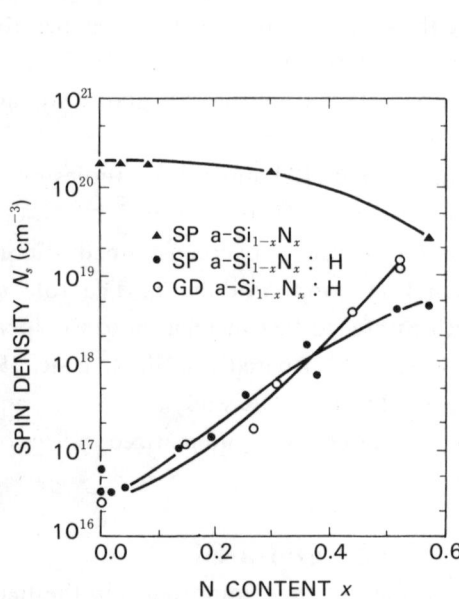

Fig. 2.2.1 The spin density versus the N content x for a-Si$_{1-x}$N$_x$ and a-Si$_{1-x}$N$_x$: H. Triangles, closed circles and open circles are for sputtered a-Si$_{1-x}$N$_x$, sputtered a-Si$_{1-x}$N$_x$: H and glow discharged a-Si$_{1-x}$N$_x$: H.

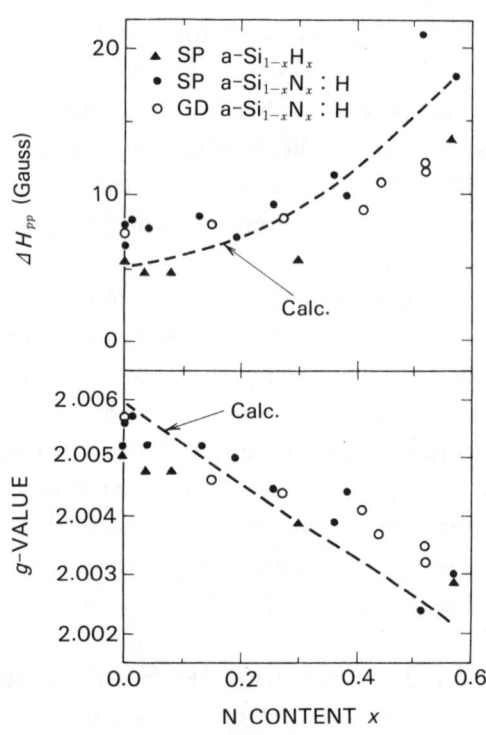

Fig. 2.2.2 The g-value and the linewidth ΔH_{pp} for the ESR signal versus the N content x. Dashed lines show the calculated ones. [after N. Ishii at al.[6]]

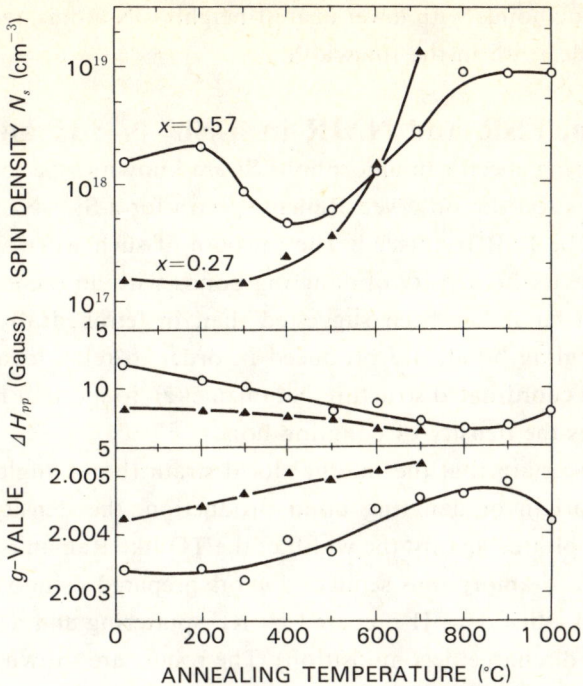

Fig. 2.2.3 The spin density N_s, g-value and linewidth ΔH_{pp} of the ESR signal for a-Si$_{0.73}$N$_{0.27}$: H (triangles) and a-Si$_{0.43}$N$_{0.57}$: H (circles) films as a function of annealing temperature.

N_s, g-value and the linewidth for ESR signals in a-Si$_{0.73}$N$_{0.27}$: H and a-Si$_{0.43}$N$_{0.57}$: H as a function of annealing temperature are shown in Fig. 2.2.3. N_s of a-Si$_{0.73}$N$_{0.27}$: H increases monotonically with an increase in annealing temperature accompanied by the effusion of H atoms. N_s of a-Si$_{0.43}$N$_{0.57}$: H is decreased by thermal relaxation and then increased by the effusion of H atoms. The increase in N_s is attributed to the formation of Si dangling bonds, because ESR signal is considered to originate from Si dangling bonds as previously mentioned. The increase in the density of Si dangling bonds for a-Si$_{0.43}$N$_{0.57}$: H is mainly caused by the decrease of H bonded to Si in the 400~600°C range and by the decrease of H bonded to N in the 600~800°C range. It is interesting that not only H atoms bonded to Si, but also those bonded to N reduce the density of Si dangling bonds. Similar behavior has previously been reported for a-Si$_{1-x}$Ge$_x$: H.[3]

The increment of N_s of a-Si$_{0.43}$N$_{0.57}$: H is smaller than that of N_s of a-Si$_{0.73}$N$_{0.27}$: H. This result is consistent with the result that the effusion of H atom bonded to Si atom is more difficult for a-Si$_{0.43}$N$_{0.57}$: H film than that for a-Si$_{0.73}$N$_{0.27}$: H film. The difference in the difficulty of the H effusion arises because the former film has Si-H bond with more nearest-neighbor N atoms than the latter film does. The g-value increases and the linewidth decreases with an increase in the annealing temperature for both films as shown in Fig. 2.2.3. It is known that in a-Si$_{1-x}$N$_x$: H films the g-value of Si dangling bond increases from 2.0025 to 2.0055 and the linewidth for Si dangling bond decreases from 20 to 8 G with a decrease in the number of nearest-neighbor N atoms. Hence, the effusion of H atom from Si with fewer nearest-neighbor N atoms by annealing brings about the increase in the number of Si dangling bonds with fewer nearest-neighbor N atoms, resulting in the increase in *g*-value and the decrease in the linewidth.

2.2.3 Raman, ESR and NMR in a-Si$_{1-x}$N$_x$: H and a-Si$_{1-x}$C$_x$: H

Raman scattering spectra in amorphous Si are known to be sensitive to local strain in the film.[1~3] In this section, observed Raman spectra for a-Si$_{1-x}$N$_x$: H and a-Si$_{1-x}$C$_x$: H are correlated with the ESR results.[10,11] The purpose of such work was to investigate the origin of the increase in the density of dangling bonds with an increase in N or C content x in these films. So far it has been suggested that, in tetrahedrally bonded amorphous semiconductors, dangling bonds are produced in order to relax local strain due to over-constrained fourfold coordinated structure. Shimizu et al. found that randomly dispersed H bonded to Si reduces the density of dangling bonds.[12,13]

In order to ascertain that the internal local strain (bond angle and/or bond length fluctuation) is the origin of dangling bond production, the density of dangling bonds derived from ESR is plotted against the width of the TO-like Raman band for a wide variety of tetrahedrally bonded amorphous semiconductors prepared using various methods (a-Si, a-Si : H, a-Si : F and a-Si$_{1-x}$N$_x$: H prepared by RF sputtering and a-Si : H and a-Si$_{1-x}$C$_x$: H prepared by glow discharge decomposition). The results are shown in Fig. 2.2.4.[10] So far it has been reported that (1) the increase in bond length distortion increases the width of the TO band and (2) the increase in bond angle and dihedral angle distortion increases the width of the TO band and the enhancement of the LO band. Since it is rather difficult to separate the TO band and the LO band, $\Delta\omega(TO)'$ is defined as corresponding to full width at half maximum of the TO band, and N_s is plotted against $\Delta\omega(TO)'$ shown in Fig. 2.2.4. Both the increase in the width of the TO band and the enhancement of the LO band make

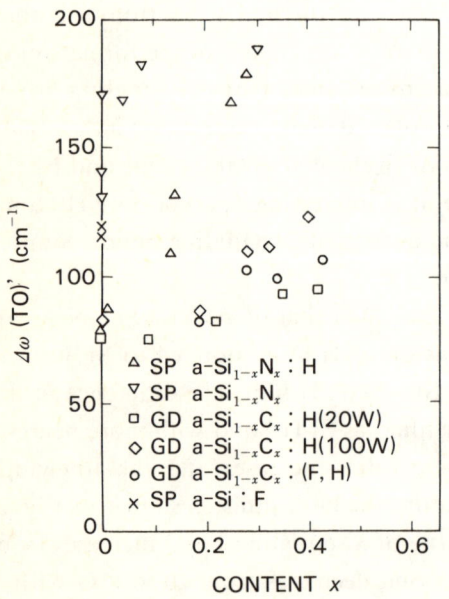

Fig. 2.2.4 $\Delta\omega(TO)'$ versus x. [after T. Shimizu[11]]

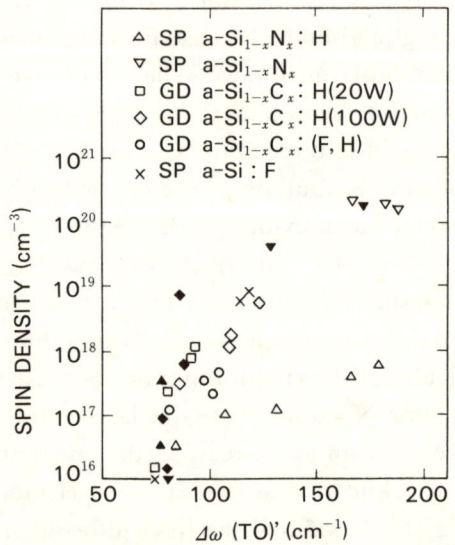

Fig. 2.2.5 The spin density versus $\Delta\omega(TO)'$ (closed symbols mean $x = 0$). [after T. Shimizu[11]]

$\Delta\omega(TO)'$ larger. Figure 2.2.5 shows how $\Delta\omega(TO)'$ increases with the N or C content x in a-$Si_{1-x}N_x$:H, a-$Si_{1-x}N_x$ and a-$Si_{1-x}C_x$:H. The results suggest that local strain increases with x in these films, resulting in the increase in N_s with x as shown in Fig. 2.2.6. From Fig. 2.2.5, it is found that $\Delta\omega(TO)'$ depends largely on a preparation method for films with $x = 0$.

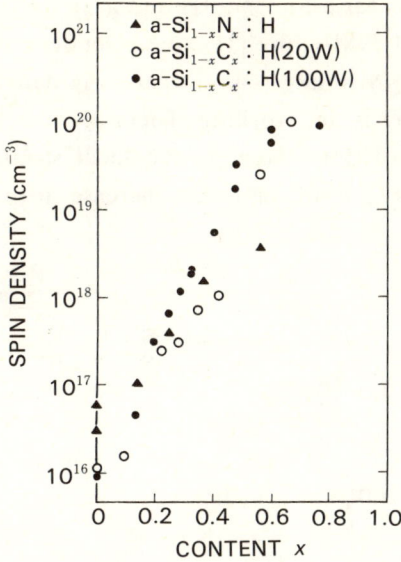

Fig. 2.2.6 The spin density versus x. [after T. Shimizu[11]]

The width of TO band $\Delta\omega(TO)$ obtained from doubling the value measured from the peak to the half-intensity point at the high frequency side of the TO band, which is free from the intensity of the LO band, also increases with x. Therefore, the increase in $\Delta\omega(TO)'$ can be attributed to the increase in $\Delta\omega(TO)$ as well as in the intensity of the LO band.

It is assumed that $\Delta\omega(TO)'$ is a measure of local strain in the amorphous network structure although in alloy systems Si-N and Si-C, the presence of N or C atoms near Si atoms may affect the Si-Si vibrational spectrum. More specifically, the fluctuation in bond angle and bond length for Si atoms in the network makes $\Delta\omega(TO)'$ larger. N_s of films with $x = 0$ is found to increase largely from $10^{16} cm^{-3}$ to $10^{20} cm^{-3}$ with an increase in $\Delta\omega(TO)'$ shown in Fig. 2.2.4. This finding indicates that the density of dangling bonds for a-Si increases with an increase in local strain as had been speculated. So far it has been reported that the increase in $\Delta\omega(TO)'$ or $\Delta\omega(TO)$ has a close correlation with the H content, substrate temperature, optical gap and dark conductivity. The present results show the increase in $\Delta\omega(TO)'$ has good correlation with an increase in the density of dangling bonds.

N_s of films with $x > 0$ is also found to increase with an increase in $\Delta\omega(TO)'$ caused by the incorporation of N or C atoms. The incorporation of N or C atoms with atomic radius and bonding configurations different from those of Si atoms is expected to cause the increase in local strain. The relation between $\Delta\omega(TO)'$ and N_s is similar for all films including a-Si$_{1-x}$C$_x$: H, except for a-Si$_{1-x}$N$_x$: H. This can be seen in Fig. 2.2.5. Therefore the assumption that the increase in $\Delta\omega(TO)'$ is brought about mainly by the local strain around Si atoms appears to be valid except for a-Si$_{1-x}$N$_x$: H. For a-Si$_{1-x}$N$_x$: H, it is possible that the presence of N which has electronegativity far larger than Si influences $\Delta\omega(TO)'$.

H NMR measurements for a-Si$_{1-x}$N$_x$: H and a-Si$_{1-x}$C$_x$: H show that the content of H contributing to the narrow NMR line $[H]_n$ has a tendency to decrease with x as shown in Fig. 2.2.7.[14] The narrow NMR line is known to originate from randomly dispersed H when the motional narrowing is not working. Therefore it is likely that the incorporation of N or C atoms not only directly increases the local strain around Si atoms but also increases it by decreasing $[H]_n$, resulting in the increase in the density of dangling bonds.

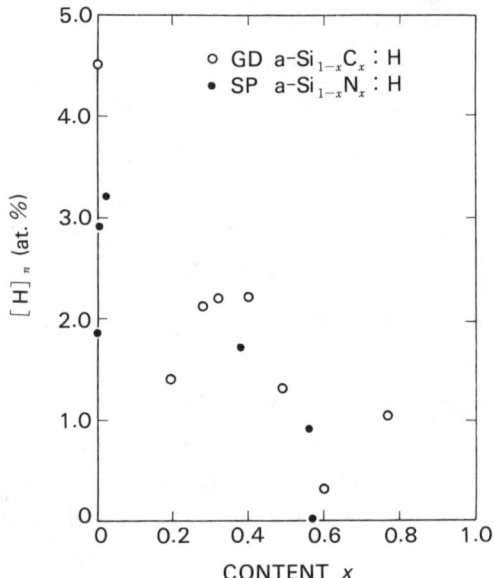

Fig. 2.2.7 The content of H contributing to the narrow NMR line $[H]_n$ versus x for a-Si$_{1-x}$C$_x$: H and a-Si$_{1-x}$N$_x$: H.

2.2.4 NMR Study on μc-Si : H

The H incorporation scheme in hydrogenated Si films containing both crystalline and amorphous phases (μc-Si : H) is not as well known as hydrogenated amorphous Si (a-Si : H). NMR has turned out to be a useful tool in investigating the H incorporation scheme in a-Si : H.[15] The H incorporation scheme is divided into two kinds, one in randomly dispersed form which contributes to a narrow NMR line and the other in gathered form which contributes to a broad NMR line.

In this section the results of NMR, IR, ESR, Raman scattering and X-ray diffraction measurements for μc-Si : H by Kumeda et al. are presented, and the structure of μc-Si : H is discussed from these results.[16,17] The purpose of such a study on μc-Si : H is to clarify the

H incorporation scheme and the structure of μc-Si : H. Imura et al. found that IR spectra due to Si-H vibration is different in μc-Si : H from those in a-Si : H and moreover non-bonded hydrogens are likely to be present in the film.[18,19] μc-Si : H films were prepared by conventional sputtering, magnetron sputtering and glow discharge decomposition. Sample preparation conditions are shown in Table 2.2.1. The average crystallite size and the volume fraction of the crystalline phase are estimated from the Raman spectrum and the X-ray diffraction pattern. The diameter of crystallites ranges from 7 to 20 nm and the volume fraction ranges from 36 to 43%.

Table 2.2.1 Sample preparation conditions.

Sample No.	MG109	MG111	MG38	MG41	SP114	GD45
Method of preparation	magnetron sputtering	magnetron sputtering	magnetron sputtering	magnetron sputtering	conventional sputtering	glow discharge decomposition
Gas (%)	H_2(100)	H_2(100)	H_2(96), Ar(4)	H_2(95), Ar(5)	H_2(60), Ar(40)	H_2(98.2), SiH_4(1.8)
RF power (W)	800	500	200	100	200	20
Substrate temperature (°C)	280	280	290	290	water cooled	350

Table 2.2.2 Two types of H incorporation scheme in μc-Si : H.

Sample	MG 38, MG 41, SP 114, GD 45	MG 109, MG 111
IR	similar to a-Si : H	sharp peaks
NMR	similar to a-Si : H (superposition of a broad Gaussian and a narrow Lorentzian)	one broad Lorentzian or superposition of two broad Gaussians
	motional narrowing due to $(SiH_2)_n$ or SiH_3	
H content	H content estimated from IR agrees with that from NMR	H content estimated from IR is larger than that from NMR (oscillator strength change, not non-bonded H)
ESR g-value	similar to a-Si : H 2.0051~2.0057	larger 2.0062~2.0064

As a result of this investigation, it was found that H incorporation scheme in μc-Si : H films can roughly be divided into two types as shown in Table 2.2.2. The first type has a behavior qualitatively similar to H incorporation in a-Si : H. As shown in Fig. 2.2.8, temperature dependence (4.2 ~ 300 K) of proton NMR shows that these hydrogens exhibit a prominent motional narrowing in contrast to those in a-Si : H. This is probably because most of them are in the form of $(SiH_2)_n$ or SiH_3. IR results are consistent with this interpretation. We can visualize a schematic structure for this type of μc-Si : H as shown in Fig. 2.2.9 (a). H-rich amorphous region is considered to enclose microcrystallites and the width of the enclosing shell can be estimated to be 0.38 ~ 0.73 nm from NMR results. Dangling bonds are present in the remaining amorphous region with less H. On the other

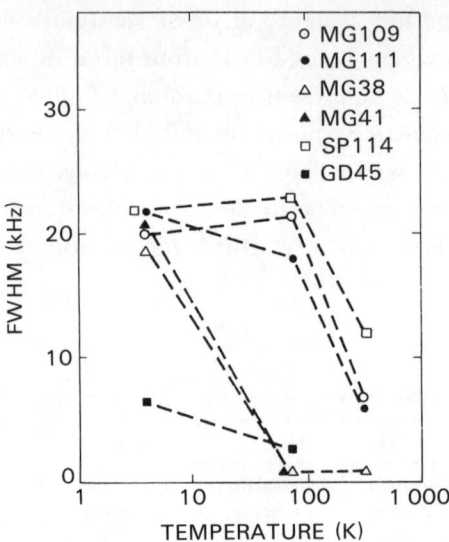

Fig. 2.2.8 Full width half maximum (FWHM) of the H NMR line in various μc-Si : H films versus measurement temperature.

hand, μc-Si : H films of the second type exhibit a very different behavior: (1) The IR absorption peaks sharply split as reported by Imura et al.,[18] (2) NMR lines are one broad Lorentzian or a superposition of two Gaussian lines and (3) g-values of dangling bond ESR are larger than usual 2.0055 for a-Si : H. Moreover the H content estimated from the IR measurement is larger than that estimated from NMR measurement. This discrepancy, however, is likely to result from the oscillator strength change, not from non-bonded hydrogens as suggested by Imura et al. This is because the behavior of the spin lattice relaxation time for proton NMR is inconsistent with the presence of a large amount of H_2 molecules and the ESR signal with a hyperfine splitting of about 500 G for atomic H is not observed. A schematic structure for this type is shown in Fig. 2.2.9 (b). No definite amorphous region is present, and dangling bonds are present in the H-covered surface region of crystallites because the g-value of the dangling bond ESR is different from the normal one.

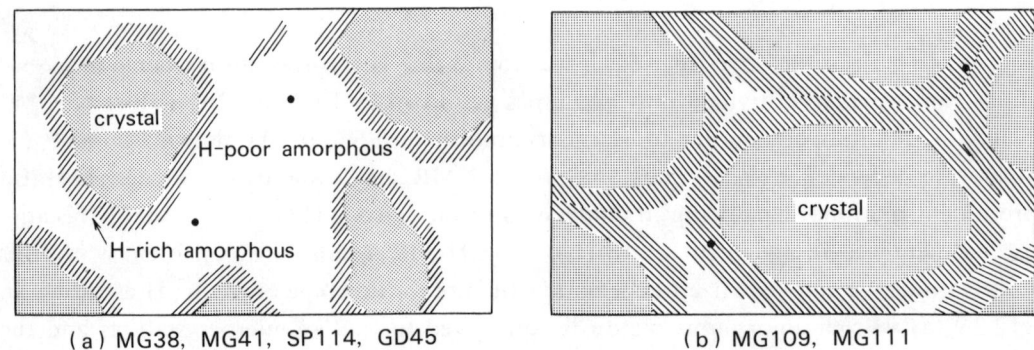

Fig. 2.2.9 Schematic models for μc-Si : H. Closed circles denote dangling bonds.

Hayashi et al. also studied the behavior of H in μc-Si : H deposited from glow discharge plasma of SiH_4/H_2 mixture by NMR. The measurements were made at temperatures between 100 ~ 300K.[20] The NMR spectrum is composed of a strong narrow line and a weak broad line (~25kHz width). The width of the narrow line is very small at about 0.5kHz, being much narrower than that of a-Si : H (~3kHz). Spin echo experiments confirmed that this narrow line is a motionally narrowed one even at 100K. The results are consistent with that of Kumeda et al.[16,17]

2.2.5 ENDOR Study on a-Si : H

A lot of information on the microscopic structure of a-Si : H has been provided by ESR and NMR. Incorporation scheme of H and the role of reducing the density of dangling bonds by NMR and ESR has been studied by Shimizu et al.[11~13] However, it is expected that more direct information on the relation between H incorporation scheme and dangling bonds can be obtained from the electron nuclear double resonance (ENDOR) technique.

Yamasaki et al. has carried out ENDOR measurements on undoped a-Si : H prepared by glow discharge decomposition, and studied how the ENDOR spectrum changes with isochronal annealing.[21] ESR and NMR measurements beside ENDOR measurements

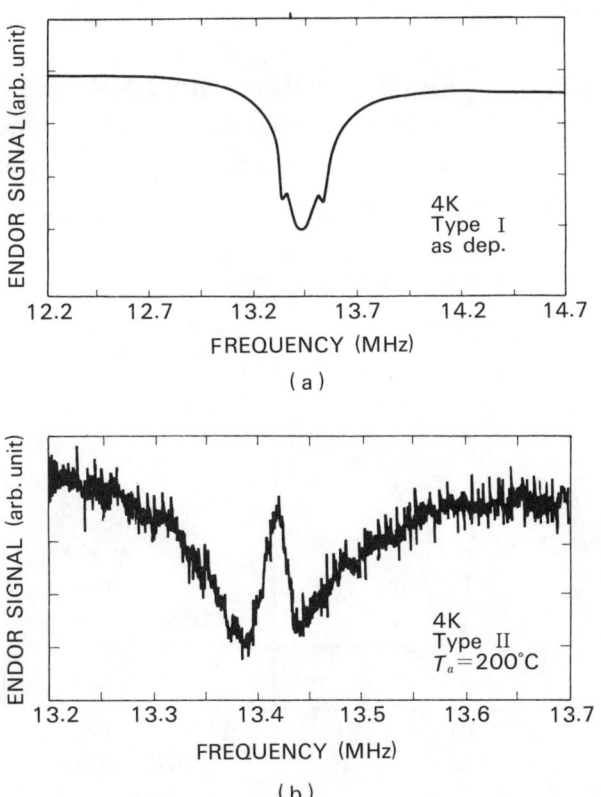

Fig. 2.2.10 ENDOR spectra obtained at the microwave power $P = 100$ mW of the a-Si : H film deposited at room temperature (a) and at $P = 1$ mW of the a-Si : H film annealed at $T_a = 200°C$ (b). [after S. Yamasaki et al.[21]]

were made of these films, ENDOR signals have been clearly detected at around the free proton frequency ν_0 and show a characteristic line shape of protons in amorphous substances. The proton ENDOR spectrum can be approximately divided into two typical types I and II as shown in Fig. 2.2.10 : Type I has a main peak at ν_0 with subpeaks on both shoulders at $\nu_0 \pm \alpha$ ($\alpha \sim 100$ kHz), and type II is essentially a doublet at $\nu_0 \pm \alpha'$ ($\alpha' \sim 30$ kHz), i. e., no peak at ν_0. For films annealed at temperatures between 200 and 400°C, only the type II spectrum has been observed with its signal maximum at $P \sim 1$ mW, and the lineshape is almost independent of P and RF field. Here P means the microwave power for saturating the ESR signal. At 500°C annealing, the ENDOR lineshape shows type I again.

The ENDOR spectra with a peak at ν_0 is known as a distant ENDOR resulting from the depolarization mechanism. Therefore, at least the type II ENDOR is not associated with the depolarization effect. The type II spectrum can be interpreted as resulting from electron-nuclear dipolar interaction between dangling bonds and the nuclear spin packet with a width of about 3 kHz. Compared with NMR results, the type II spectrum is suggested to be associated with the narrow component of the proton NMR spectrum. In contrast to the type II spectrum, the origin of the type I spectrum is not clear. By comparing the results of the NMR study, it was suggested that the type I spectrum results from the spin interaction between dangling bonds and H associated with the broad component of the proton NMR spectrum.

2.2.6 Plasma-Hydrogenation Effects on ESR in CVD a-Si

Besides ESR with $g = 2.0055$ originating from dangling bonds, two ESR signals with $g = 2.0043$ and $g = 2.013$ are observed in a-Si : H by P and B doping, respectively, or

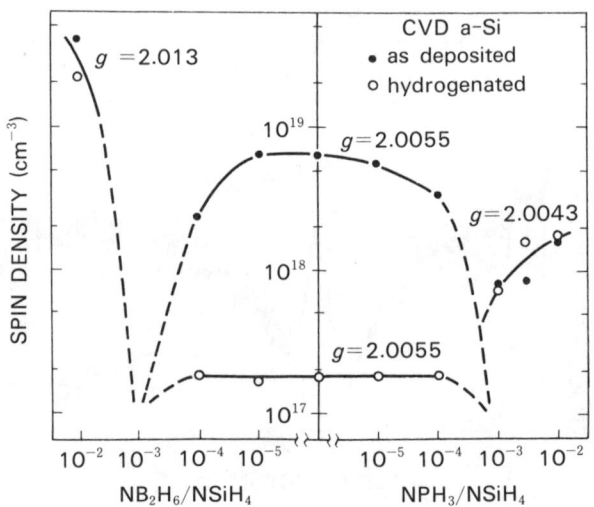

Fig.2.2.11 The spin density for the ESR signals with $g = 2.0055$, 2.0043 and 2.013 as a function of the doping gas ratios $NPH_3/NSiH_4$ and $NB_2H_6/NSiH_4$ for CVD a-Si films as deposited (closed circles) and annealed in a hydrogen-plasma at 300°C for 30 min (open circles). [after S.Hasegawa et al.[26)]

by band gap light illumination.[22~24] As the origin of these ESR centers, three models have been proposed ; (1) localized conduction and valence band tail states occupied, respectively, by an electron ($g = 2.0043$) and a hole ($g = 2.013$) which are induced by fluctuation of bond angle and length,[23,24] (2) negatively ($g = 2.0043$) and positively ($g = 2.013$) charged states of Si-Si weak bonds,[22] and (3) negatively ($g = 2.0043$) and positively ($g = 2.013$) charged twofold coordinated Si atoms.[25]

Hasegawa et al. observed how these three types of ESR signals ($g = 2.0055$, $g = 2.0043$ and $g = 2.013$) change with plasma-hydrogenation in chemically vapor deposited (CVD) a-Si doped with P or B.[26] It is found that the signal with $g = 2.0055$ decreases largely by hydrogenation whereas the signals with $g = 2.0043$ and $g = 2.013$ do not appreciably change by hydrogenation. These results can be seen in Fig. 2.2.11. Similar behavior has also been observed by Magarino et al. for CVD a-Si[27] and by Hasegawa et al. for glow discharge a-Si : H.[28] These results might give information concerning the origin of the ESR signals with $g = 2.0043$ and $g = 2.013$. However, at present, it is difficult to discriminate among the three models mentioned above from these results. The comparison between the observed g-values and electron correlation energies and the calculated ones by Ishii et al. suggests that model (2) is more likely than model (3).[29]

2.2.7 Conclusion

(1) ESR signals from a-Si$_{1-x}$N$_x$: H and a-Si$_{1-x}$N$_x$ mainly originate from Si dangling bonds, and the linewidth and the g-value increases and decreases, respectively, with an increase in x by the presence of N atoms near the Si dangling bond.

(2) The internal local strain in overconstrained silicon-based amorphous films is seen as the main origin of dangling bond production. The presence of randomly distributed onefold coordinated H atoms decreases and the incorporation of N or C atoms increases the local strain, resulting in a change of the density of dangling bonds.

(3) H incorporation scheme in μc-Si : H is shown to be different from that in a-Si : H using NMR. There is no indication of the presence of non-bonded hydrogens as suggested before.

(4) Two types of ENDOR signals are found in a-Si : H, possibly corresponding to the broad and narrow lines observed by H NMR.

(5) The intensities of ESR signals with $g = 2.0043$ and $g = 2.013$ do not appreciably change by using plasma-hydrogenation in doped CVD a-Si, whereas that with $g = 2.0055$ originating from dangling bonds largely decreases.

(6) Understanding of the origin and the nature of defects described in this article is indispensable for photovoltaic applications of Si-based films.

References

1) T. Shimizu, S. Oozora, A. Morimoto, M. Kumeda, and N. Ishii : Solar Energy Mater., *8* (1982) 311.
2) A. Morimoto, S. Oozora, M. Kumeda, and T. Shimizu : Phys. Status Solidi, (*b*) *119* (1983) 715.
3) A. Morimoto, T. Miura, M. Kumeda, and T. Shimizu : Jpn. J. Appl. Phys., *20* (1981) L 833.
4) A. Morimoto, T. Miura, M. Kumeda, and T. Shimizu : Jpn. J. Appl. Phys., *21* (1982) L 119.

5) A. Morimoto, T. Miura, M. Kumeda, and T. Shimizu: J. Appl. Phys., *53* (1982) 7299.
6) N. Ishii, S. Oozora, M. Kumeda, and T. Shimizu: Phys. Status Solidi, (*b*) *114* (1982) K111.
7) T. Ishidate, K. Inoue, K. Tsuji, and S. Minomura: Solid State Commun., *42* (1982) 197.
8) J. S. Lannin, L. J. Pilone, S. T. Kshirsager, R. Messier, and R. C. Ross: Phys. Rev., *B26* (1982) 3506.
9) R. Tsu, J. Gonzalez-Hernandez, J. Doehler, and S. R. Ovshinsky: Solid State Commun., *46* (1983) 79.
10) A. Morimoto, S. Oozora, M. Kumeda, and T. Shimizu: Solid State Commun., *47* (1983) 773.
11) T. Shimizu: J. Non-Cryst. Solids, *59 & 60* (1983) 117.
12) T. Shimizu, K. Nakazawa, M. Kumeda, and S. Ueda: Jpn. J. Appl. Phys., *21* (1982) L351.
13) T. Shimizu, K. Nakazawa, M. Kumeda, and S. Ueda: Physica, *117 B & 118 B* (1983) 926.
14) K. Nakazawa, S. Ueda, M. Kumeda, A. Morimoto, and T. Shimizu: Jpn. J. Appl. Phys., *21* (1982) L176.
15) J. A. Reimer: J. Physique, *42* (1981) C4-729.
16) M. Kumeda, Y. Yonezawa, K. Nakazawa, S. Ueda, and T. Shimizu: Jpn. J. Appl. Phys., *22* (1983) L194.
17) M. Kumeda, Y. Yonezawa, A. Morimoto, S. Ueda, and T. Shimizu: J. Non-Cryst. Solids, *59 & 60* (1983) 775.
18) T. Imura, K. Mogi, A. Hiraki, S. Nakashima, and A. Mitsuishi: Solid State Commun., *40* (1981) 161.
19) N. Fukada, T. Imura, A. Hiraki, T. Itahashi, T. Fukada, and M. Tanaka: Jpn. J. Appl. Phys., *21* (1982) L532.
20) S. Hayashi, S. Yamasaki, A. Matsuda, and K. Tanaka: J. Non-Cryst. Solids, *59 & 60* (1983) 779.
21) S. Yamasaki, S. Kuroda, and K. Tanaka: J. Non-Cryst. Solids, *59 & 60* (1983) 141.
22) S. Hasegawa, T. Kasajima, and T. Shimizu: Phil. Mag., *B 43* (1981) 149.
23) H. Dersch, J. Stuke, and J. Beichler: Phys. Status Solidi, (*b*) *105* (1981) 265.
24) R. A. Street and D. K. Biegelsen: Solid State Commun., *33* (1980) 1159.
25) D. Adler: J. Physique, *42* (1981) C4-3.
26) S. Hasegawa, D. Ando, Y. Kurata, and T. Shimizu: Jpn. J. Appl. Phys., *22* (1983) L815.
27) J. Magarino, D. Kaplan, and A. Friederich: Phil. Mag., *B 45* (1982) 285.
28) S. Hasegawa, S. Shimizu, and Y. Kurata: Phil. Mag., *B* (in press).
29) N. Ishii, M. Kumeda, and T. Shimizu: Phys. Status Solidi, (*b*) *116* (1983) 91.

2.3 Raman Study on Bonding Structure of a-Si and Related Materials

Toshikazu SHIMADA* and Yoshifumi KATAYAMA*

Abstract

Raman studies performed in Japan, on the bonding structure of amorphous and microcrystalline silicon and related materials are reviewed. Characterisic features of Raman bands for amorphous and microcrystalline silicon alloys are discussed referring to those for single crystalline silicon. Raman spectra for various members of the amorphous silicon family and their applicability as a process monitor are summarized.

2.3.1 Introduction

In order to clarify the bonding structure of networks in amorphous silicon (a-Si) and related materials, many intensive studies have been done and are in progress. This is because the bonding structure of an amorphous network is the basis of the electronic structure and has a great deal of variation depending on the preparation conditions, in contrast to that of crystalline silicon. From this point of view, the importance of studying bonding structure cannot be overemphasized in research on amorphous materials. In addition, the number of such studies is increasing with the expansion of the amorphous silicon family (that is, multi-component alloys a-SiC : H, a-SiGe : H, a-SiSn : H, a-Si : F : H, etc.) and microcrystalline materials.

The atomic vibrational spectrum is one of the most beautiful reflections of the bonding structure. Lattice vibration is a resonance phenomenon even in amorphous materials. Raman scattering, infrared absorption, inelastic neutron-, X-ray- and electron scattering methods are used to take the vibrational spectrum.

In this section, Raman studies performed in Japan, on the bonding structure of amorphous and microcrystalline silicon alloys are reviewed.

Single crystalline silicon shows a strong Raman peak at 520cm^{-1} corresponding to the "first-order"-allowed TO (Γ) phonon mode and weak peaks corresponding to the "second-order"-allowed (usually called "forbidden") phonon modes.[1] On the other hand, a Raman spectrum of amorphous silicon mainly consists of disorder-induced Raman modes which roughly correspond to the density distribution of vibrational states. This means that the density distribution of vibrational states in amorphous silicon is quite similar to that broadened in crystalline silicon. Also, the selection rule for wave vectors in Raman

---

* Centeal Research Laboratory, Hitachi, Ltd., Kokubunji, Tokyo 185.

scattering vanishes for amorphous sillicon. Therefore, character of Raman spectra dramatically changes from first- or second-order-allowed bands to disorder-induced bands with the change of the structure from a crystalline to an amorphous state. This change is a good measure of the structural change of the materials.

Great attention is being paid to microcrystalline silicon, μc-Si, as a new material with higher mobility and doping efficiency than a-Si : H, and a larger optical absorption constant than that in crystalline silicon. The simplest structural model of μc-Si is that of a inhomogeneous material consisting of amorphous and crystallites.

A volume fraction for amorphous or crystalline regions is estimated by the ratio of Raman intensities for amorphous to those of crystallite. It is not clear at this stage however, whether surface/interface or inhomogeneity-induced effects on a Raman spectrum are negligible. Careful consideration will be necessary for estimation of the volume fraction.

Another important point of view from the application side is that Raman scattering can be used as a process monitor to rapidly check the structure of materials (such as amorphus or μc ones) and the bonding structure of hydrogen (Si-H, Si-H$_2$, etc.) without using a special substrate such as a c-Si wafer for IR transmission measurement.

In Section 2.3.2, the origin, shape and intensity of the Raman spectra and the corresponding structure of the silicon networks are reviewed.

In Section 2.3.3, work on compositional and structural analysis of the a-Si family are summarized.

In Section 2.3.4, examples of availability as a process monitor are shown.

2.3.2 Raman Spectrum and Its Origin

The Raman spectrum for crystalline silicon contains only one Raman active lattice vibrational mode at 520cm^{-1} which has three-fold degenerate Γ_{25}' (zone center optic mode with a wave vector of phonon $k \cong 0$) symmetry. Weak Raman bands corresponding to higher order Raman processes also appear in the spectrum. In the case of second-order scattering, the sum of the wave vectors of two phonons must be approximately zero. On the other hand, in amorphous silicon all phonon modes become first-order Raman active because of the lack of translational symmetry. The origins of the Raman bands for μc-Si of which microstructure is believed to be a mixture of amorphous and crystallites and/or an intermediate state in between the two are interesting.

Figure 2.3.1 shows examples of the Raman spectra of single crystalline (A), microcrystalline (B) and (C), and amorphous sillicon (D).[2] Assignments for Raman spectra of single crystalline and amorphous states are shown at bottom.[3] The characteristic features of Raman bands are summarized as follows.

The 520cm^{-1} band corresponds to the first-order-allowed band in the crystalline state and to the highest phonon frequency in the Si-Si network. The band for single crystalline silicon shows a very narrow and symmetric shape with full width at half maximum (FWHM) of about 2.6cm^{-1} as shown in Fig. 2.3.2. In the spectrum for μc-Si, the width greatly increases, the shape of the band becomes asymmetric (Fig. 2.3.3) and the peak position shifts with strain in material.[6]

The 480cm^{-1} band which arises from the amorphous part, originates from the

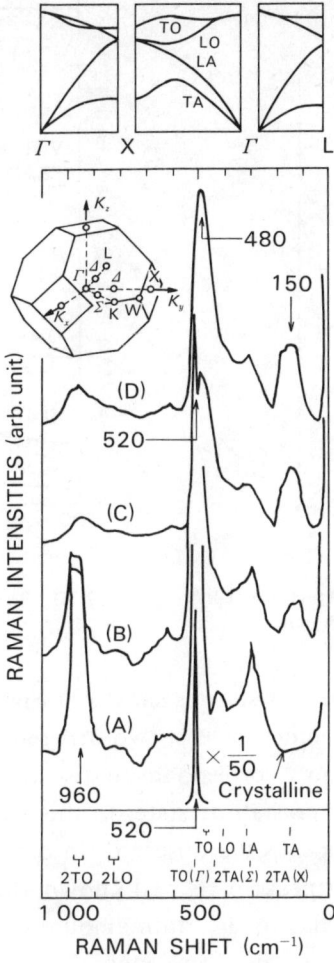

Fig. 2.3.1 Raman spectra of crystalline Si (A), μc-Si (B) and (C), and amorphous Si (D). Assignments for crystalline and amorphous states are shown at bottom. [after T. Shimada et al.[2]]

disorder-induced TO phonon modes and a well developed reflection of amorphous structure.

Kumeda et al.[5] and Hamasaki et al.[6] estimated the crystallite volume fraction in μc-Si by using the ratio of the integrated Raman intensities of the 520 and 480cm^{-1} bands. They found that conductivity is closely correlated with the volume fraction as shown in Fig. 2.3.3.

Saitoh et al. also used both Raman bands to check for the existence of crystallite in films prepared by a photo-CVD method.[7]

Ishidate et al.[8] discussed the hydrostatic pressure effect on Raman intensity; and also discussed the FWHM of the bands for amorphous silicon films prepared by glow-discharge and reactive sputtering methods in terms of changes in bond stretching and bond bending forces.

A Raman band appearing in the vicinity of the 300cm^{-1} region is a complex band which is assigned to the 2TA(X) overtone band for the crystalline state or to the disorder-

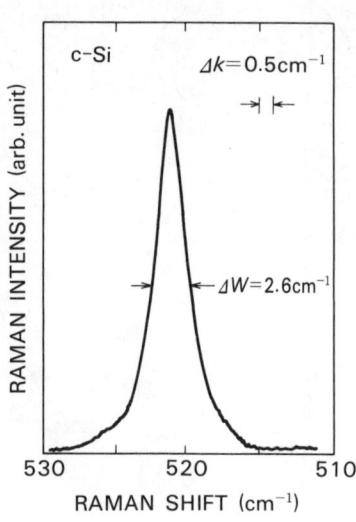

Fig. 2.3.2 High resolution Raman spectrum corresponding to the TO(Γ) phonons.

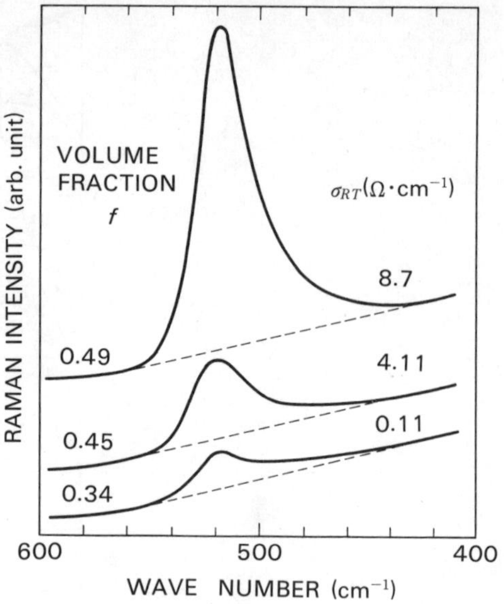

Fig. 2.3.3 Raman spectra of B-doped μc-Si:H. [after T. Hamasaki et al.[6]]

induced LA band for the amorphous state as shown in Fig. 2.3.1. This means that the peak at around 300 cm^{-1} in μc-Si includes at least two different characters.

The peak at around 150 cm^{-1} corresponds to disorder-induced TA phonons, and that is the first-order forbidden in a crystalline state. In this frequency region, neither second-order nor higher-order Raman peaks are observed. In other words, the 150 cm^{-1} peak is a "pure" disorder-induced Raman peak. As is well known, the 480 cm^{-1} band merges into the tail of 520 cm^{-1} band for μc-Si having a certain amount of crystalline. When the spectrum is separated into the two components, some ambiguity might be introduced due to the strong intensity of the 520 cm^{-1} band. It is concluded that the 150 cm^{-1} band is a better measure of randomness in the specimen than the 480 cm^{-1} band.[2]

Raman bands, which were observed in the Raman shift region of greater than 520 cm^{-1}, should be assigned as multiphonon Raman bands, except for Si-H local vibrational bands appearing at around 2000 cm^{-1}.[3] The Raman band at around 960 cm^{-1} (as seen in spectrum (A) for single crystalline silicon) is assigned as 2TO which is a second-order-allowed band. In the spectrum for amorphous silicon (D), a Raman band at the same wave number region of 960 cm^{-1} is observed. Raman bands at around 960 cm^{-1} in spectra (A) and (D) seem to have the same origin. This evidence indicates that the frequency distribution of lattice vibration in the amorphous state is roughly the same as that in the crystalline state. A broadening of the 960 cm^{-1} band however, is seen in the spectrum (D). Namely, a considerable broadening of vibrational frequency distribution is caused and van-Hove singularity features become weaker.

The optical absorption coefficient α, at an incident laser photon energy of around 2 eV, strongly depends upon the structure of the silicon network. At that energy of single crystalline silicon, α is in the order of 10^4 cm^{-1} and typical α at the same energy as that of a-Si:H is about 10^5 cm^{-1}. Within the framework of polarizability theory with appropriate assumption, the Stokes scattering cross section varies by the change in the complex dielectric constant ε.[3] Crystalline and amorphous silicon are about the same in having real parts of ε's that are ten times or more greater than imaginary parts of ε's at around 2 eV. Therefor, the Stokes scattering cross section is mainly determined by real parts of ε. The order of magnitude is almost constant when the microscopic structure changes from crystalline to amorphous states. The imaginary parts of ε affect on Raman intensity only through the change of radiating volume of Raman light. Incident laser light intensity exponentially decreases with an absorption coefficent α in the materials, and Raman light is also attenuated when passing through the materials. The observed Raman intensity is derived from the radiating volume which is proportional to $1/\alpha$. The Raman intensities from various silicon materials are normalized by multiplying them by α. The situation is essentially the same as both first-order and second-order Raman processes.[1] These facts lead to the conclusion that the second-order allowed Raman band at 960 cm^{-1} is a good measure to normalize Raman intensities for silicon materials having various microscopic structures.

In order to obtain a relative Raman cross section $R(k_1)$, observed Raman intensities $I(k_1)$ were normalized using absorption coefficients as both incident and scattered light is attenuated differently in specimens with different absorption coefficients ($\alpha(k_0)$ or $\alpha(k_1)$), and is expressed as:

$$R(k_1) = \text{const.} \times I(k_1) \times [\alpha(k_0) + \alpha(k_1)] = \text{const.} \times N(k_1)$$

If this normalization is plausible, the R (960 cm^{-1}) should remain unchanged in the amorphous, μc and crystalline states.

Figure 2.3.4 shows the normalized Raman intensity $N(k_1)$'s of the 960, 520, 480, and 150 cm^{-1} bands against an absorption coefficient α at an incident laser light wavelength of 5308 A, which represents the wavelength of the interband transition region. The α strongly correlates with the bonding structure. The α of 0.8×10^4 cm^{-1} is for single crystalline Si,

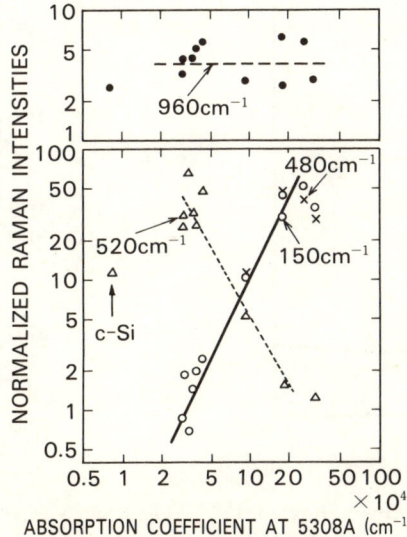

Fig. 2.3.4 Plots of normalized Raman intensities vs. absorption coefficient at 5308 A. [after T. Shimada et al.[2]]

and that of about $20 \times 10^4 \mathrm{cm}^{-1}$ is for amorphous Si. As seen at the topmost plots in Fig. 2.3.4, $N(960 \mathrm{cm}^{-1})$ is roughly constant as was expected. This fact implies that the normalization procedure for a Raman cross section is applicable to other Raman bands as shown in the lower part of Fig. 2.3.4. As seen in the figure, $N(150 \mathrm{cm}^{-1})$ decreases parallel with $N(480 \mathrm{cm}^{-1})$ as α decreases. The measure of randomness f is defined as: $f = N(150 \mathrm{cm}^{-1})/\{N(150 \mathrm{cm}^{-1}) + N(520 \mathrm{cm}^{-1})\}$, instead of $N(480 \mathrm{cm}^{-1})/\{N(480 \mathrm{cm}^{-1}) + N(520 \mathrm{cm}^{-1})\}$. The f is a more sensitive measure of randomness than $N(480 \mathrm{cm}^{-1})$. However, further study is necessary to confirm that f is identical to the amorphous volume fraction in the μc-Si.

2.3.3 Raman Spectra for the a-Si Family

Amorphous $Si_{1-x}C_x$: H and amorphous $Si_{1-x}Ge_x$: H are important materials for photo-electric devices, especially for solar cells and photo-receptors.

Raman experiments on a-SiC alloys have been conducted by Inoue et al.,[10] Shimizu et al.,[11,12] and Watanabe et al.[13] The spectra contain two obvious broad bands in the 300~600 and 1300~1600 cm^{-1} regions corresponding to Si-Si and C-C bonds. Inoue et al.[10] studied the relation between the intensities of these bands and the composition X. The intensity ratio $I_{\mathrm{Si-Si}}/I_{\mathrm{C-C}}$ changes in proportion to $(1-X)^3/X^3$. Three tentative explanations have been proposed; first, that the Raman efficiencies depend upon X; second, is the effect of incorporated hydrogen atoms; and third, is a change in the bonding structure of the C-C bonds.

Shimizu et al.[11,12] discussed the relation between dangling bond density N_s and local strain by comparing the results of ESR with those of the Raman experiment which is described in another section. They concluded that the incorporation of carbon atoms

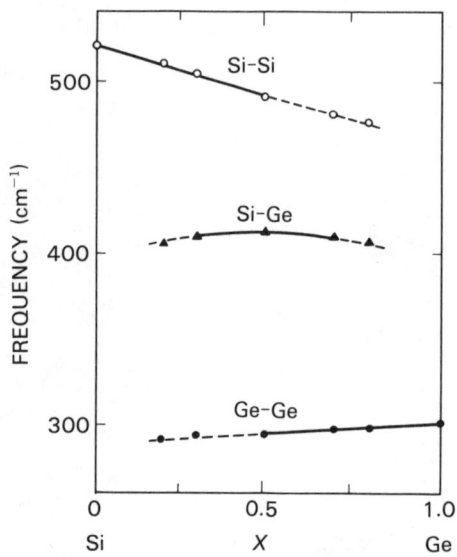

Fig. 2.3.5 Variations of c-$Si_{1-x}Ge_x$ alloy TO phonon frequencies with Ge content X. [after S. Minomura et al.[14]]

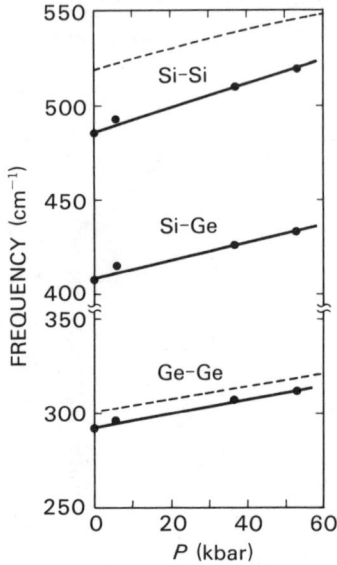

Fig. 2.3.6 Variations of c-Si (dashed line) and c-$Si_{0.5}Ge_{0.5}$ TO phonon frequencies with hydrostatic pressure. [after S. Minomura et al.[14]]

induces strain and increases the width of the TO-like peak at around 480cm^{-1}. The strain may come from the electronegativity difference between Si and C atoms. This evidence is emphasized in a-Si$_{1-x}$N$_x$: H alloys because nitrogen has an electronegativity far larger than Si and C.

Minomura et al.[14] systematically studied the Raman spectra of c-Si$_{1-x}$Ge$_x$ and a-Si$_{1-x}$Ge$_x$: H against the alloy composition X and hydrostatic pressure. Figure 2.3.5 shows the X dependences of three main peaks relating to TO phonons of Si-Si, Si-Ge and Ge-Ge bonds in c-Si$_{1-x}$Ge$_x$. The same result was reported by Morimoto et al.[15] The lower frequency shift of the Si-Si peak, the higher frequency shift of the Ge-Ge peak and the peaking of the Si-Si frequency are interpreted in terms of charge transfer from Ge to Si and change in bond length. Hydrostatic pressure effects on these peaks, as shown in Fig. 2.3.6, support this interpretation. And they also concluded that the remarkable asymmetry of these peaks occurring by alloying, results from a relaxation of the k-conservation selection rule.

Morimoto et al.[15] examined the crystallization processes of a-Si$_{1-x}$C$_x$, a-Si$_{1-x}$N$_x$ and a-Si$_{1-x}$Ge$_x$ alloys by thermal annealing using Raman scattering. Figure 2.3.7 shows a example of the change of the Raman spectra for a-Si$_{1-x}$N$_x$ alloys by annealing. Great decreases are observed in TO mode linewidth and in TA mode relative intensity by annealing. They revealed that segregation occurs in Si-C and Si-N alloys and with an increase in crystallization temperature. In contrast to this, no segregation is observed in Si-Ge and it has a lower crystallization temperature.

Yamamoto et al.[16] measured the Raman spectra of the a-Si : F alloys. They interpreted the result with the aid of the calculations using the valency force field model and confirmed the existence of SiF$_4$ molecules embedded in the voids in a-Si : F films.

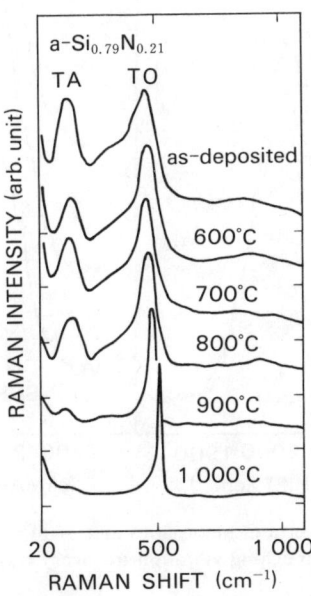

Fig. 2.3.7 Annealing temperature dependence of the Si$_{0.79}$N$_{0.21}$ Raman spectrum. [after A. Morimoto et al.[15]]

2.3.4 Raman Measurements as a Process Monitor to Check the Bonding Structure

Finally, examples of the availability of Raman measurements as a process monitor are shown. In order to check bonding structures of members of the amorphous silicon family, infrared absorption spectra are commonly used and give useful bonding structure information. For this measurement, a transparent substrate such as single crystalline silicon is used. On the other hand, Raman spectra can be measured in backscattering geometry and one can choose the photon energy of incident laser light at a high absorption coefficient α in the order of $10^4 \sim 10^5 \text{cm}^{-1}$. Under these conditions, any substrate can be used because there is no Raman light from the substrates. This means structural characterization by Raman experiment has several advantages as a process monitor. The first is that one can check the bonding structure (Si-H or Si-H_2 ; amorphous or microcrystalline) of the films on the device in the course of the fabrication process. The second is that one can check the substrate effect on the film structure. Figure 2.3.8 shows substrate dependence of Si-H bands as an example of the spectra for a process monitor. For the a-Si : H, no spectral change is seen in changing the substrate material. An infrared absorption spectra of the film on the c-Si substrate is also shown as a reference. However, a considerable difference is observed for μc-Si : H.

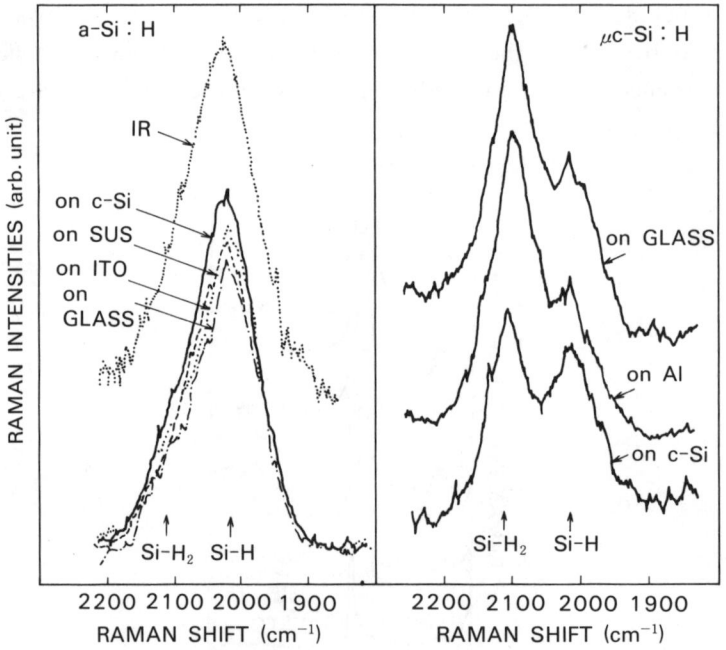

Fig. 2.3.8 Raman spectra of amorphous and μc-Si for various substrates in the Si-H stretching vibration frequency region.

2.3.5 Summary

Some topical work on the bonding structure of amorphous silicon and related materials from a viewpoint of Raman spectroscopy, recently performed in Japan, is briefly described. Further in depth study is necessary. Throughout these studies however, it becomes increasingly apparent that knowledge of structural properties is important.

References

1) P. A. Temple and C. E. Hathaway: Phys. Rev., *B 7* (1973) 3685.
2) T. Shimada, Y. Katayama, K. Nakagawa, H. Matsubara, M. Migitaka, and E. Maruyama: J. Non-Cryst. Solids, *59 & 60* (1983) 783.
3) D. Bermejo and M. Cardona: J. Non-Cryst. Solids, *32* (1979) 405.
4) J. B. Renucci, R. N. Tyte, and M. Cardona: Phys. Rev., *B 11* (1975) 3885.
5) M. Kumeda, Y. Yonezawa, A. Morimoto, S. Ueda, and T. Shimizu: J. Non-Cryst. Solids, *59 & 60* (1983) 775.
6) T. Hamasaki, M. Ueda, Y. Osaka, and M. Hirose: J. Non-Cryst. Solids, *59 & 60* (1983) 811.
7) T. Saitoh, T. Shimada, M. Migitaka, and Y. Tarui: J. Non-Cryst. Solids, *59 & 60* (1983) 715.
8) T. Ishidate, K. Inoue, K. Tsuji, and S. Minomura: Solid State Commun. *42* (1982) 197.
9) Y. Tawada, K. Tsuge, M. Kondo, H. Okamoto, and Y. Hamakawa: J. Appl. Phys., *53* (1982) 5273.
10) Y. Inoue, S. Nakashima, A. Mitsuishi, S. Tabata, and S. Tsuboi: Solid State Commun., *48* (1983) 1071.
11) T. Shimizu: J. Non-Cryst. Solids, *59 & 60* (1983) 117.
12) A. Morimoto, S. Oozora, M. Kumeda, and T. Shimizu: Solid State Commun. in print.
13) K. Watanabe, Y. Hishikawa, H. Tarui, T. Takahama, N. Nakamura, S. Tsuda, H. Nishiwaki, M. Onishi, and Y. Kuwano: Abst. 44th Fall Meeting of Jpn. Soc. of Appl. Phys., (1982) 27p-k-4 [in Japanease].
14) S. Minomura, K. Tsuji, M. Wakagi, T. Ishidate, K. Inoue, and M. Shibuya: J. Non-Cryst. Solids, *59 & 60* (1983) 541.
15) A. Morimoto, M. Kumeda, and T. Shimizu: J. Non-Cryst. Solids, *59 & 60* (1983) 537.
16) K. Yamamoto, T. Nakanishi, H. Kasahara, and K. Abe: J. Non-Cryst. Solids, *59 & 60* (1983) 213.

2.4 Photoinduced Effect in Hydrogenated Amorphous Silicon

Kazuo MORIGAKI*

Abstract

Photoinduced effect in hydrogenated amorphous silicon is reviewed, particularly on luminescence, ODMR, ESR and photoinduced absorption. Evidence of defect creation by light exposure is shown and the mechanism of such defect creation is discussed.

2.4.1 Introduction

Since Staebler and Wronski[1] discovered that band gap illumination causes a drastic drop in subsequent dark conductivity, whereas thermal annealing at 160°C recovers the dark conductivity to the initial value, the photoinduced effect on various electronic properties of hydrogenated amorphous silicon (a-Si:H) has been extensively investigated. This photoinduced conductivity change has been considered in terms of defect creation by light exposure[2] and also light-induced properties of the surface and interface.[3] However, a definite conclusion has not been obtained on the mechanism underlying the photoinduced conductivity change although both effects play important roles in it. In this chapter we mainly review the photoinduced effect on luminescence, ODMR, ESR and photoinduced absorption, most of which have been investigated in our laboratory. These photoinduced phenomena have been interpreted in terms of the defect (dangling bond) creation by light exposure. We also discuss the evidence and mechanism for such defect creation.

2.4.2 Luminescence Fatigue and Creation of Defects by Light Exposure

The fatigue effect on the luminescence of a-Si:H, i.e. prolonged band gap illumination at low temperatures causes its luminescence intensity to decrease compared with before the illumination, was reported for the first time by Morigaki et al.[4] This fatigue of the luminescence was completely recovered by thermal annealing of the samples at 90~170°C. Thus, this is a reversible and reproducible photoinduced effect. Subsequently, Pankove and Berkeyheiser[5] discovered that the low energy part of the luminescence of a-Si:H is enhanced after light exposure and that the fatigue and enhancement effects are emphasized by increasing the illumination temperature up to 400K. This fatigue effect on the luminescence was interpreted in terms of the creation of those dangling bond centres which act as

---

* Institute for Solid State Physics, University of Tokyo, Roppongi, Tokyo 106.

nonradiative recombination centres that suppress the luminescence. In the following, we present evidence for such defect creation.

The evidence that light exposure creates dangling bond centres first came from the ESR and time-resolved luminescence experiments.[6] The ESR experiments showed that the dangling bond centres with $g = 2.005$ were created with a density of $1 \times 10^{17} \text{cm}^{-3}$ after prolonged exposure to an infrared-cut off 500 W mercury light at 77 K and remained, in part, stable even after the sample was annealed at room temperature. The enhancement of the decay rate[7] of the luminescence by light exposure was also consistent with the model of the creation of the dangling bond centres. The ESR measurements were also reported by Dersch et al.[8] in which the illumination by a focused 100 W tungsten halogen light for 15 hours at room tempearature (actual sample temperature was $90 \sim 100°C$) created dangling bonds of density $N_s = 1.8 \times 10^{16} \text{cm}^{-3}$, a factor of two larger than the initial value ($N_s = 9.1 \times 10^{15}$ cm^{-3}). The ODMR measurements[9,10] also provided direct evidence for the creation of dangling bond centres by light exposure at low temperatures. Figure 2.4.1 shows a drastic change in the ODMR spectrum after the sample was illuminated at 2 K with 360 mW of unfocused argon ion laser light at 514.5 nm for 30 min. The ODMR signals of dangling bond centres (D_2 centres) and trapped hole centres (A centres) were increased and decreased after light exposure, respectively. The decrease in the A centre signal intensity was caused by the enhancement of the nonradiative recombination channel that occurred as a result of the creation of dangling bond centres by light exposure. The annealing effect on the ODMR spectrum, as shown in Fig. 2.4.1, is also consistent with that on the luminescence fatigue.[4] It is also interesting to note that the relative change in the luminescence intensity, $(\Delta I/I)_{\text{ESR}}$,

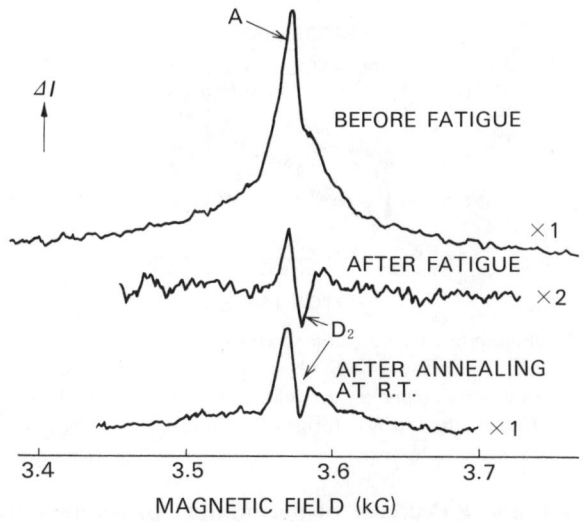

Fig. 2.4.1 ODMR spectra taken at 2 K before and after fatigue (illumination) and after annealing the sample at room temperature in a-Si : H sample No. 519 prepared at 300°C, monitoring the intensity of emitted light with 1.16 eV. The sample was excited by unfocused argon ion laser light of 25 mW at 514.5 nm for the ODMR measurement. [After K. Morigaki et al.[9]]

at the A centre resonance was increased after light exposure in the low energy part of the luminescence spectrum, where the luminescence intensity itself increased after light exposure from the initial value, as had already been observed by Pankove and Berkeyheiser.[5] Morigaki et al.[9] attributed this enhancement of the A centre ODMR signals to the creation of radiative electron centres responsible for the low energy luminescence. This is also consistent with the suggestion by Pankove and Berkeyheiser,[5] although they have not identified which types of centres, electrons or holes, are created by light exposure. This will be discussed later in more detail.

The creation of dangling bond centres by light exposure has also been suggested from the measurements of photoinduced absorption,[11] transient photoconductivity,[12] optical absorption,[13] thermally stimulated current,[14] ICTS[15] etc. Figure 2.4.2 shows the photo-induced absorption (PA) spectra[16] taken at 1.8 K for an a-Si:H sample No. 540 ($N_s = 2.9 \times 10^{18}$ cm^{-3}) before and after band gap illumination by 14 W/cm^2 of krypton ion laser light at 530.9 nm for 30 min. The PA spectrum has been interpreted in terms of the excitation of trapped holes at the A centres into the valence band and that of additional electrons at doubly occupied dangling bond centres (D$^-$) into the conduction band, whose threshold energies were estimated from the PA spectra to be 0.35 eV and 0.75 eV for this sample, respectively. The hole contribution to PA is significantly weakened as a result of the shortening of the lifetime of trapped holes owing to the enhancement of their recombination with electrons via dangling bond centres. On the other hand, the electron contribution to PA is determined by the competition between an increase of the number of dangling bond

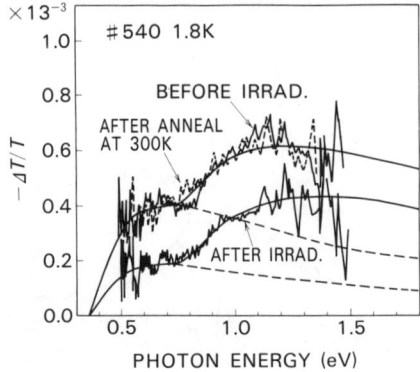

Fig. 2.4.2 Photoinduced absorption spectra taken at 1.8 K before and after light irradiation for a-Si:H sample No. 540 prepared at 75°C. The solid curves represent the calculated ones (See Ref. 11)). [After I. Hirabayashi and K. Morigaki[16]]

centres by light exposure and a decrease of the number of trapped electrons owing to the enhanced recombination with holes via dangling bond centres. The result shown in Fig. 2.4.2 seems reasonable for the electron excitation from the D$^-$ centres into the conduction band.

In the following, we discuss luminescence fatigue as a function of the dangling bond density and the hydrogen content[17] that is shown in Figs. 2.4.3 and 2.4.4. As shown in Fig.

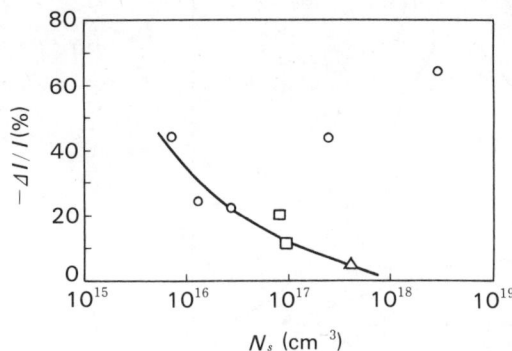

Fig. 2.4.3 Relative change in the luminescence peak intensity, I, between before and after illumination, $-\Delta I/I$, at 2 K plotted as a function of the dangling bond spin density, N_s for various samples; triangles: rf-bias sputtered a-Si, circles: glow-discharge a-Si:H and squares: glow-discharge a-Si:H, F. The sample was illuminated at 2 K with 360 mW of unfocused argon ion laser light at 514.5 nm for 30 min. [After Y. Sano et al.[17]]

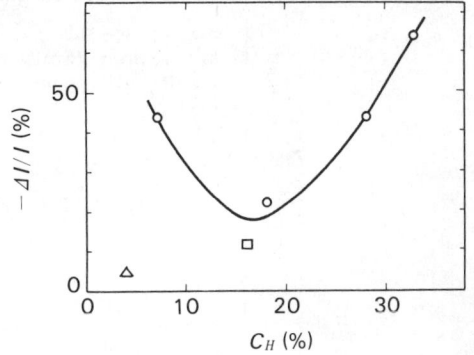

Fig. 2.4.4 Relative change in the luminescence peak intensity, I, between before and after illumination, $-\Delta I/I$, at 2 K plotted as a function of the hydrogen content, C_H (in units of at. %) for various samples. The same symbols as those in Fig. 2.4.3 are used for the sample classification. The illumination condition was the same as that for Fig. 2.4.3. [After Y. Sano et al.[17]]

2.4.3, the degree of the luminescence fatigue, $-(\Delta I/I)$, decreases as the dangling bond density increases except for two samples No. 541 and 540 containing a large amount of hydrogen, i.e. a hydrogen content of 28 at. % and 33 at. %, respectively. This is quite reasonable, because the density of dangling bonds created by light exposure is estimated to be in the order of magnitude of 10^{17} cm^{-3}, so that if the sample contains a higher density of dangling bonds than this, the illumination effect on the luminescence intensity would become insignificant. However, the luminescence fatigue depends on not only the dangling bond density, but also the hydrogen content. In spite of the large density of dangling bonds for samples No. 541 and 540, these samples exhibited a large luminescence fatigue, as shown in Figs. 2.4.3 and 2.4.4. For these samples, No. 540 and 541, it was also confirmed from ESR measurements[18] that a number of dangling bond centres with $N_s = 6.7 \times 10^{19}$ cm^{-3} and $N_s = 3.2 \times 10^{18}$ cm^{-3}, respectively, were created by light exposure at 77 K compared with the initial values, i.e. $N_s = 3.4 \times 10^{18}$ cm^{-3} and 2.8×10^{17} cm^{-3}, respectively. From this view of luminescence fatigue vs. hydrogen content, it is interesting that our recent measurements on the luminescence fatigue in a-Si:H samples prepared from glow-discharge decomposition of disilane at 16°C show a large luminescence fatigue in the high energy part of the luminescence spectrum compared with the low energy part as shown in Fig. 2.4.5.[19] In reminding that the high energy part of the luminescence spectrum arises from the region containing a large amount of hydrogen, this inhomogeneous behaviour of the fatigue in the luminescence spectrum seems to be related to the presence of hydrogen that is distributed inhomogeneously in the sample. The ODMR result[19] shown in Fig. 2.4.6 suggests that the luminescence fatigue for this sample is also due to the creation of dangling bond centres by light exposure. It is also interesting to note that the relative change in the luminescence

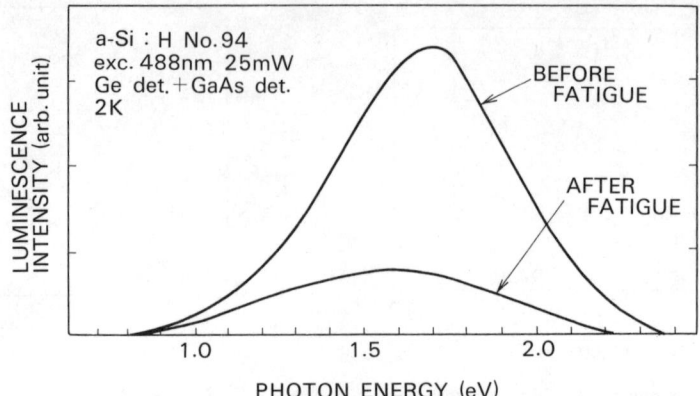

Fig. 2.4.5 Luminescence spectra taken at 2 K before and after illumination for a-Si : H sample prepared by glow-discharge decomposition of disilane at 16°C. The sample was illuminated at 2 K with 360 mW of unfocused argon ion laser light at 488 nm for 30 min. The luminescence spectra were measured under excitation by unfocused argon ion laser light of 25 mW at 488 nm. [After M. Yoshida et al.[27]]

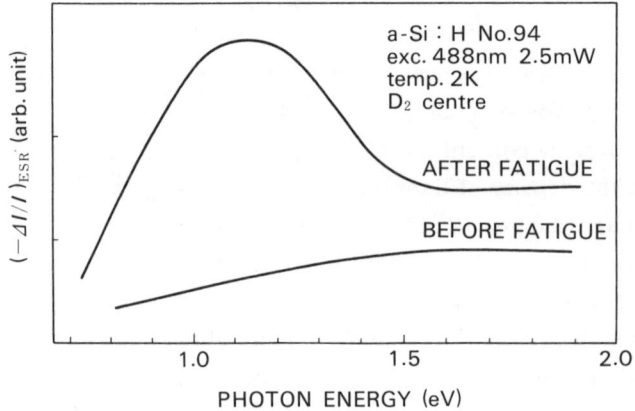

Fig. 2.4.6 Spectral dependences of the ODMR signal of dangling bond centres at 2 K before and after illumination for the same sample used for Fig. 2.4.5. The illumination condition was the same as that for Fig. 2.4.5. [After M. Yoshida et al.[19]]

intensity at the dangling bond centre resonance, $(-\Delta I/I)_{ESR}$, becomes much greater after illumination than before illumination at 0.8~1.5 eV, compared with $(-\Delta I/I)_{ESR}$ taken above 1.5 eV. This suggests that the creation of dangling bond centres by light exposure works more effectively for the luminescnece fatigue in the high energy part above 1.5 eV than that in the remaining part of the luminescence spectrum.

The above results on luminescence fatigue vs. hydrogen content suggest that the creation of dangling bond centres by light exposure is related to the flexibility of the amorphous network, as will be discussed in the following section.

2.4.3 Mechanism for Defect Creation by Light Exposure

In the following, we discuss the mechanism underlying defect creation by light exposure. As was discussed in the preceding sections, the dangling bond centres are obviously created by light exposure. Hirabayashi et al.[6] suggested that a possible site for a dangling bond created by light exposure may be a Si-Si bond neighboring a Si-H bond for the following reason. Since the Si-H bond strength is greater than the Si-Si bond's and the electronegativity of hydrogen is greater than that of silicon, the electrons at a central Si atom are more localized towards the Si-H bond, so that the Si-Si bonds adjacent to the Si-H bond could be more easily broken than normal Si-Si bonds under illumination. Furthermore, Morigaki et al.[4] pointed out that when nonradiative recombination occurs at the dangling bond centre created by light exposure, the released energy associated with this recombination serves to cause local structural changes around the dangling bond which stabilize atomic configuration of the dangling bonds. From the above consideration, we suggest a possible model for the dangling bond centre created by light exposure as shown in Fig. 2.4.7 (a), where an unpaired electron lies in the nonbonding orbital of the threefold coordinated Si centre having a Si-H bond. However, taking into account the presence of the positively charged threefold coordinated Si centre as shown in Fig. 2.4.7 (a), the hydrogen atom might switch a position shown in Fig. 2.4.7 (a) to that in Fig. 2.4.7 (b) to stabilize these T_3^0 and T_3^+ centres. An advantage of such T_3^0 centres is that the stretching vibration of a Si-H bond has higher energy than that of the bulk phonon mode, so that nonradiative

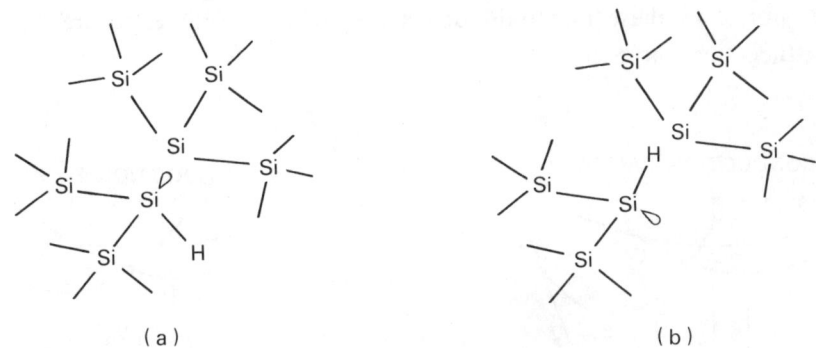

Fig. 2.4.7 Schematic diagram of the threefold coordinated Si atoms, T_3^0 and T_3^+. Figures (a) and (b) represent different coordinations of a hydrogen atom with a Si atom.

recombination occurs more effectively than for the normal T_3^0 centre, as was first suggested by Mott.[20] This can be shown in the following way. The nonradiative recombination process can be expressed in terms of the configurational coordinate model, as shown in Fig. 2.4.8 (a). Then, the nonradiative recombination rate, P_{NR}, via a dangling bond centre can be given by the following equation,[21] using the Englman and Jortner formula[22] in the strong electron-phonon regime :

$$P_{NR} = W_0 \left(\frac{\hbar\Omega_0}{kT^*}\right)^{1/2} \exp\left(-\frac{2R}{R_e} - \frac{E_a}{kT}\right) \tag{2.4.1}$$

$$T^* = \left(\frac{\hbar\Omega_0}{2k}\right) \coth\left(\frac{\hbar\Omega_0}{2kT}\right) \tag{2.4.2}$$

where W_0, Ω_0, R, R_e and E_a designate a preexponential factor, the average phonon frequency, the distance between a trapped electron and a dangling bond centre, the radius of the wave function of trapped electrons and the activation energy defined in Fig. 2.4.8 (a). Here we assume that the radius of the wave function of trapped electrons is greater than that of the dangling bond centre and that a hole recombines quickly with a doubly occupied dangling bond centre after a trapped electron is transferred to a dangling bond site through either the activated transition or the tunneling transition. At low temperatures, T^* in Eq. (2.4.2) is approximated by $(\hbar\Omega/2k)$, so that a larger $\hbar\Omega_0$ gives a larger P_{NR} in Eq. (2.4.1), as just mentioned above.

As was mentioned in Section 2.4.2, the sample containing a large amount of hydrogen exhibits a large luminescence fatigue that is caused by the creation of a number of dangling bond centres by light exposure. This may be related to the above consideration on a possible site of dangling bond centres and also the flexibility of the amorphous network in which a hydrogen atom is onefold-coordinated with a Si atom and thus the average coordination number is reduced from four, i. e. the coordination number of the overconstrained network according to Phillips.[23] This consideration is consistent with the case for chalcogenide glasses in which the luminescence fatigue easily occurs by light exposure.[24] Chalcogenide glasses have a flexible network composed of twofold or threefold coordinated atoms. In these materials, defect creation by light exposure has also been observed by Biegelsen and Street.[25]

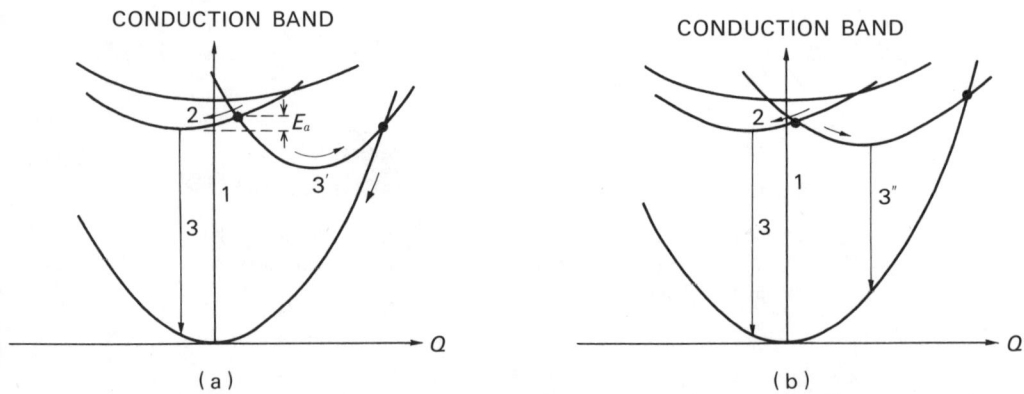

Fig. 2.4.8 (a) Schematic diagram of the configurational coordinate model for radiative and nonradiative recombination processes. (a): The symbols of 1, 2, 3 and 3′ designate the following processes; optical excitation, thermalization, radiative recombination and nonradiative recombination, respectively. (b): For this figure, a deep centre acts as a radiative centre whose radiative transition is indicated by 3″.

As regards the radiative electron centre created by light exposure, we have no idea about the model of its atomic structure. However, we think that it would be worthy to consider that the T_3^+ centre illustrated in Fig. 2.4.7 (b) acts as a radiative electron centre. Although we have previously suggested that a single T_3^+ centre in the normal amorphous Si network acts as a nonradiative centre,[26] the T_3^+ centre of the present concern has a different environment from such a single T_3^+ centre, so that it would still have a possibility to act as a radiative centre, depending on the detail of the adiabatic energy curves in the configurational coordinate model, as shown in Fig. 2.4.8 (b). From the experimental point of view, the radiative centres giving rise to the low energy luminescence as well as nonradiative centres (dangling bond centres) appear to be created at the same time by light exposure at low temperatures.

2.4.4 Conclusion

From the experimental evidence we have discussed in the preceding sections, it can obviously be concluded that the defects (dangling bond centres) are created by prolonged illumination with the photons whose energy exceeds the bandgap energy. However, further investigation is required to explain the mechanism of such defect creation by light exposure.

Acknowledgements

The author wishes to thank Ms. Mihoko Yoshida, Dr. Izumi Hirabayashi and Professor Shoji Nitta for their collaboration on the experiments for a-Si : H samples prepared from glow-discharge decomposition of disilane and also for stimulating discussions on the topics in this chapter. He is also grateful to Professor Hitoshi Sumi for helpful discussions on the nonradiative recombination process.

References

1) D. L. Staebler and C. R. Wronski : Appl. Phys. Lett., *31* (1977) 292.
2) D. L. Staebler and C. R. Wronski : J. Appl. Phys., *51* (1980) 3262.
3) I. Solomon, T. Dietl, and D. Kaplan : J. Physique,*39* (1978) 1241.
4) K. Morigaki, I. Hirabayashi, M. Nakayama, S. Nitta, and K. Shimakawa : Solid State Commun.,*33* (1980) 851.
5) J. I. Pankove and J. E. Berkeyheiser : Appl. Phys. Lett., *37* (1980) 705.
6) I. Hirabayashi, K. Morigaki, and S. Nitta : Jpn. J. Appl. Phys., *19* (1980) L357.
7) I. Hirabayashi, K. Morigaki, and S. Nitta : J. Phys. Soc. Japan,*50* (1981) 2961.
8) H. Dersch, J. Stuke, and J. Beichler : Appl. Phys. Lett., *38* (1981) 456.
9) K. Morigaki, Y. Sano, and I. Hirabayashi : J. Phys. Soc. Japan, *51* (1982) 147.
10) K. Morigaki : in Hydrogenated Amorphous Silicon, ed. J. I. Pankove, Semiconductors and Semimetals *Vol. 21 Part C*, Academic Press, New York, 1984.
11) I. Hirabayashi and K. Morigaki : J. Non-Cryst. Solids, 59 & 60 (1983) 433.
12) R. A. Street : Appl. Phys. Lett., *42* (1983) 507.
13) N. M. Amer, A. Skumanich, and W. B. Jackson : Physica, *117 B & 118 B* (1983) 897.
14) M. Yamaguchi : J. Non-Cryst. Solids, 59 & 60 (1983) 425.
15) H. Okushi, A. Asano, M. Miyakawa, S. Yamasaki, and K. Tanaka : J. Non-Cryst. Solids, 59 & 60 (1983) 393.
16) I. Hirabayashi and K. Morigaki : to be published.
17) Y. Sano, K. Morigaki, and I. Hirabayashi :

Solid State Commun., *43* (1982) 439.
18) I. Hirabayashi, K. Morigaki, and M. Yoshida : Solar Energy Mat., *8* (1982) 153.
19) M. Yoshida, K. Morigaki, and S. Nitta : Solid State Commun., *51* (1984) 1.
20) N. F. Mott : Phil. Mag., *36* (1977) 413.
21) K. Morigaki, Y. Sano, and I. Hirabayashi : Solid State Commun., *39* (1981) 947.
22) R. Englman and J. Jortner : Mol. Phys., *18* (1970) 145.
23) J. C. Phillips : J. Non-Cryst. Solids, *34* (1979) 153.
24) F. Mollot, J. Cernogora, and C. Benoit à la Guillaume : Phys. Stat. Sol., (*a*) *21* (1974) 281.
25) D. K. Biegelsen and R. A. Street : Phys. Rev. Lett., *44* (1980) 803.
26) K. Morigaki : Jpn. J. Appl.Phys., *22* (1983) 375.
27) M. Yoshida, K. Morigaki, I. Hirabayashi, H. Ohta, A. Amamou, and S. Nitta : Proc. Intern. Topical Conf. Optical Effects in Amorphous Semiconductors, Snowbird, Utah, American Institute of Physics, 1984, in press.

CHAPTER 3

NEW ASPECTS OF PREPARATIONS AND GROWTH KINETICS

3.1 Plasma Spectroscopies and Film Deposition

Kazunobu TANAKA*, Nobuhiro HATA* and Akihisa MATSUDA*

Abstract

Progress in the area of plasma spectroscopies is reviewed in relation to the deposition mechanism of a-Si : H film from various kinds of plasmas. A variety of new plasma diagnostic tools are described such as coherent anti-Stokes Raman spectroscopy (CARS), a triode-chamber technique for measuring the lifetime of a precursor, and a pulsed rf-discharge technique for studying chemical processes involved in a-Si : H deposition.

It is demonstrated that these new diagnostic tools play an important role in understanding the deposition kinetics of a-Si : H.

3.1.1 Introduction

As is well known, hydrogenated amorphous silicon (a-Si : H) deposited by silane glow discharge shows a variety of structural, chemical and electronic properties depending strongly on the deposition conditions.[1] That is the reason why there has been considerable variation among the results from various research groups in the early stage of this field. For the past several years, many efforts have been made to understand silane glow discharge in order to deduce microscopic key parameters of the species reaching and leaving the surface from the external parameters of the system.[2] Actually, many plasma diagnostic techniques have been reported and applied to SiH_4 glow discharge; optical emission spectroscopy (OES),[2~6] ion mass spectrometry,[4~6] probe I-V characteristics measurement,[7] infrared emission and absorption spectroscopy,[8,9] laser induced fluorescence (LIF),[10~12] and coherent anti-Stokes Raman spectroscopy (CARS).[13~15]

The mechanisms resulting in amorphous silicon deposition from silane glow discharges are classified into (1) primary processes involving electron impact on the silane molecule and its resulting dissociation, (2) secondary processes in terms of diffusion and reactions of species within the plasma and (3) final processes on the growing surface of the film such as sticking, desorption, diffusion and reactions of precursors.

Recently, Schmitt has reported cross sections for electron impact on SiH_4 as functions of electron energy (see Fig. 3.1.1.) and has determined the branching ratio of SiH_4 into various species induced by 70-eV electron impact. The result is illustrated in Fig. 3.1.2.[11] Although it appears that ionization and neutral dissociation are of comparable importance, the amount of ionic species is much less than that of neutral radicals in a realistic deposition

---

* Electrotechnical Laboratory, 1-1-4 Umezono, Sakura-mura, Niihari-gun, Ibaraki 305.

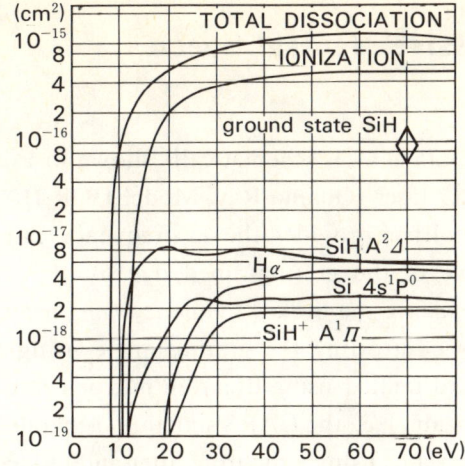

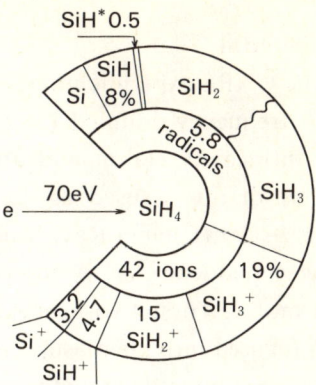

Fig. 3.1.1 Cross sections for electron impact on silane. [after J. P. M. Schmitt[11]]

Fig. 3.1.2 Probability manifold for the silane dissociative channels due to 70-eV electron impact. [after J. P. M. Schmitt[11]]

condition because the average kinetic electron energy lies in the range between 3 and 10 eV (see Fig. 3.1.1.). The most striking indication is that SiH_3 and/or SiH_2 might dominate in the silane plasma and could be candidates for the main precursors of a-Si : H deposition. Concurrently, according to LIF measurement,[11] ground state SiH is a minor component among the radicals created in the primary processes.

SiH_2 has been directly detected by Hata et al. for the first time using the CARS (coherent anti-Stokes Raman spectroscopy) technique.[14] Quite recently also, SiH_2 has been spectroscopically assigned by Inoue et al. using LIF.[12] However, SiH_3 has not yet been observed directly so far, probably due to the assignment problem, although it is considered to be a most probable precursor for a-Si : H deposition since SiH_3, differing from SiH_2, does not react with silane.[11,16]

Scott et al. have initially speculated, assuming the identical precursor (SiH_2) for Si_2H_6 as well as SiH_4 plasma, that a higher deposition rate should be ascribed to a higher rate of SiH_2 formation from Si_2H_6.[17] On the other hand, Matsuda et al. have measured the lifetimes of dominant precursors for both deposition gas plasmas and have demonstrated in connection with the OES data that a film precursor for Si_2H_6 plasma is clearly different from that for SiH_4 plasma.[18] Furthermore, recently, Jasinski et al. have shown, using Frequency Modulation Absorption Spectroscopy, that the concentration of SiH_2 in disilane discharges is less than that in silane discharges.[19]

In this report, several advancements achieved in this field for the past year in Japan are described. Emphasis is placed on new diagnostic tools or methods such as CARS for detecting non-emissive radicals, the triode chamber system for measuring the lifetime (reaction time) of precursors, and the pulsed rf-discharge technique associated with chemical processes.

3.1.2 CARS-Detection of Non-Emissive Neutral Species

Experimental

The CARS experimental setup reported by Hata et al. is schematically shown in Fig. 3.1.3.[14] A frequency doubled Q-switched Nd:YAG laser (Quanta-Ray, Model DCR-IIA) with an intracavity etalon and an extracavity amplifier provides the ω_1-frequency laser beam. Part of this is used as the pumping source of the grating-tuned dye oscillator/amplifier system (Quanta-Ray, Model PDL-1) producing the Stokes frequency (ω_2) beam. The Stokes frequency ω_2 is computer-scanned by controlling the grating angle using a stepping motor. These two beams are collinearly collimated using dichroic mirror optics, and then focused into the plasma. The anti-Stokes beam, i. e., the CARS signal ω_3 generated at the focus region is separated from the ω_1 and ω_2 beams using a dichroic filter and guided into a double monochromator (Jovin Yvon, Model U 1000).. A silicon photodiode-array detector (EG & G PAR, Model 1420) equipped with a multichannel intensifier plate has been operated under "GO ON EXTERNAL" mode and triggered to start scanning and data reading-out by the Q-switch pulse from the Nd:YAG laser. The data is sent to a minicomputer (DEC, MINC-11) from the controller through a specially designed DMA (direct memory access) interface circuit.

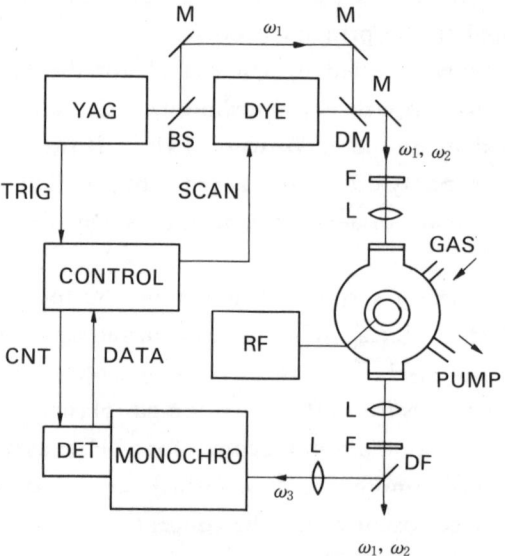

Fig. 3.1.3 Block diagram of the CARS setup in the present study. [after N. Hata et al.[14]]

A capacitively-coupled rf (13.56 MHz) glow-discharge plasma chamber was used to produce the silane plasma. The diameters of the cathode and anode were 90 mm and 110 mm, respectively, being separated by 40 mm. The cylindrical chamber wall is 250 mm from the center of the electrodes. Except where noted, semiconductor-grade 100% silane was fed at a flow rate of 10 SCCM into the chamber, with a gas pressure of 200 mTorr. Silane gas diluted with hydrogen or helium was also studied under the constant silane partial pressure.

the RF power supplied and the anode heater temperature (T_s) were controlled in the range of 0~70 W and 25~250°C, respectively, in the present CARS experiment.

SiH₄ spectra

Silane is a spherical top molecule, and its fundamental frequencies are shown in Ref. 20). Only a $\Delta J = 0$ transition being Raman-allowed for the totally symmetric ν_1 mode, and a strong Raman line is reported for the ν_1 Q-band of silane. That strong Raman line is further enhanced in the CARS signal because the CARS signal intensity is quadratically proportional to the Raman scattering cross section.[21]

The observed ν_1 Q-band spectra are shown in Fig. 3.1.4 for anode heater

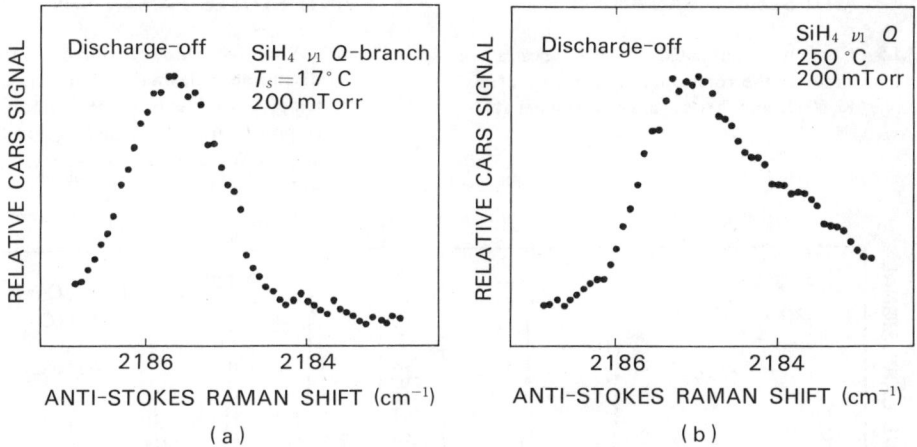

Fig. 3.1.4 ν_1 Q-band CARS spectra of 200-mTorr SiH₄ at the anode temperature of (a) 17°C and (b) 250°C. [after N. Hata et al.[15]]

temperatures, (a) $T_s = 17°C$ and (b) 250°C, under the discharge-off condition (zero rf power).[15] As shown in the figure, the spectrum becomes broader and its peak position moves towards the lower Raman shift for higher T_s. This apparent peak shift of the convoluted spectrum of a series of ν_1 Q-branch rotational levels is simply due to a change in the probability distribution.

Figure 3.1.5 shows the computer simulated ν_1 Q-band spectra of silane for rotational temperatures of 300 K, 500 K and 700 K.[15] For this calculation, the transition probability for each rotational level is considered to be equal.[20] From the results of Figs. 3.1.4 and 3.1.5, it is clear that the rotational temperature of SiH₄ under the present gas flow and pressure conditions increases with the anode heater temperature.

Figure 3.1.6 shows the ν_1 Q-band spectrum of silane under the discharge-on condition, namely, the spectrum of SiH₄ in rf glow-discharge plasma for $T_s = 250°C$ and rf power = 50 W.[15] In comparison to the data of Fig. 3.1.4 (b) for the discharge-off condition at the identical T_s, the spectrum shows a broader peak with a lower Raman shift, suggesting direct or indirect rotational excitation by electrons in the plasma.

Figure 3.1.7 (a) and (b), shows the ν_3 P- and O-band spectra, which were observed only when the silane pressure was increased to above 1 Torr. Since these lines are weaker

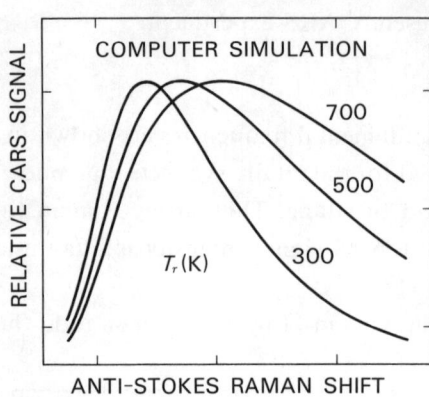

Fig. 3.1.5 Computer simulated CARS spectra of silane at the rotational temperature of 300 K, 500 K and 700 K. [after N. Hata et al.[15]]

Fig. 3.1.6 CARS spectrum of silane in the glow-discharge plasma, where the anode temperature, gas pressure, gas flow rate and rf power are 250°C, 200 mTorr, 10 SCCM and 50 W, respectively. [after N. Hata et al.[15]]

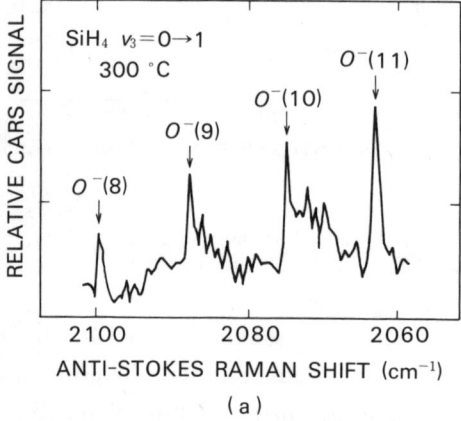
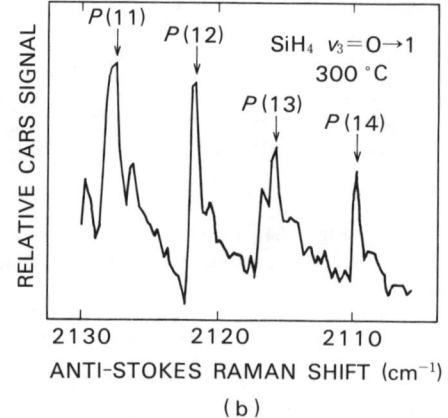

Fig. 3.1.7 ν_3 (a) O- and (b) P-band CARS spectra of SiH$_4$. These are observed only at higher silane pressure (> 1 Torr) than in the usual glow-discharge plasmas. [after N. Hata et al.[15]]

than ν_1 Q-band, they are less suited for plasma diagnosis.[15]

H$_2$ spectra

CARS measurement of hydrogen fundamental Q-band at 4170 cm^{-1} in a hydrogen discharge was demonstrated by Péalat et al. in 1981.[22] Similar spectra have been obtained in our experiment as shown in Fig. 3.1.8. Each rotational level appears separately in the spectum, so that it is easy to determine the distribution among the rotational levels. Hot-band signals from vibrationally excited hydrogen molecules have also been reported in Ref. 22). Hydrogen CARS spectra would thus be applied to measure rotational and vibrational temperatures of hydrogen molecules in the hydrogen diluted, and probably 100-% silane, glow-discharge plasmas.

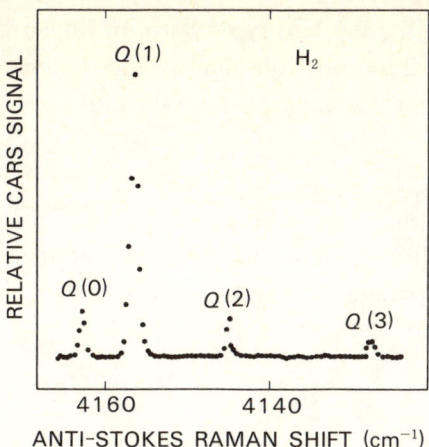

Fig. 3.1.8 CARS spectrum of hydrogen. [after N. Hata et al.[15]]

SiH$_2$ spectra

Figure 3.1.9 shows the CARS spectrum measured in the range of 2018~2038 cm^{-1}.

According to Milligan and Jacox investigating the fundamental frequencies of SiH$_2$,[23] the observed *ir* peaks at 2022 cm^{-1} and 2032 cm^{-1} have been assigned to the relatively strong ν_3 and the weak ν_1 stretching line of the fragment, respectively, although there exists some ambiguity in assignment of the ν_1 line.

In contrast to the *ir* measurement, the ν_1 line is expected to appear stronger in the CARS signal associated with the Raman cross section, and actually the highest peak is observed near 2032 cm^{-1}, as shown in Fig. 3.1.9. It should be noted that the interference between the CARS signal from SiH$_2$ in the plasma and the glass windows might distort the spectral shape, depending on the degree of interferency as a function of the coherency of the laser beams and the precision of the optics components.

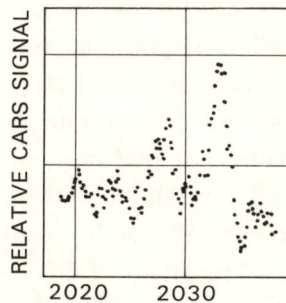

Fig. 3.1.9 SiH$_2$ ν_1 Q-band CARS spectra in the high power (1 W/cm^2) rf glow-discharge plasma of 2.4-Torr SiH$_4$-H$_2$ mixture (SiH$_4$/H$_2$ = 1/9). [after N. Hata et al.[14]]

A series of spectral peaks observed in a lower wavenumber range might be hot-band signals originating from vibrationally excited states of the identical species. According to the experiment done by Knights et al.,[8] the vibrational temperature of SiH amounts to 2000

K. Therefore, considering a higher density of states in higher vibrational levels of the triatomic fragment SiH_2, it is quite possible that a large number of SiH_2 molecules stay in vibrationally excited states, which might produce the multi-structured spectrum as shown in the figure.

Discussions

The volume in which 75% of the CARS signal are generated is given by the beam waist, ϕ, and the confocal parameter length, l, multiplied by a factor of six,[24]

$$\phi = \frac{4\lambda f}{\pi d}$$

and

$$l = \frac{\pi \phi^2}{2\lambda} \qquad (3.1.1)$$

where λ, f and d are the wavelength of the laser light, the focal length of the focusing lens and the beam diameter at the lens, respectively. Substituting the experimental values of $\lambda = 532$ nm, $f = 300$ mm and $d = 13$ mm into Eq. (3.1.1), $\phi = 16\,\mu$m and $6l = 4.3$ mm are obtained. These values are relatively small compared to the electrode distance in the present experiment.

The CARS signal is quadratically proportional to the magnitude $|\chi^{(3)}|$ of the third order electric susceptibility,[21] which is in turn linear to the number of the molecules in resonance. Therefore, the density can be deduced from the CARS signal intensity when it is calibrated to the signal from the reference sample with a known density.[13] Hata et al. have reported the data on the number density of SiH_4 involved in the DC glow-discharge plasma.[13]

In the silane CARS spectra shown in Fig. 3.1.4, the rotational structure is not resolved because the dye laser band width $0.3\,\text{cm}^{-1}$ is greater than the transition wavenumber difference between each level. However, the rotational distribution is clearly reflected in the spectral shape. In this situation, a computer simulation of the CARS signal has been done for determining the spatial distribution of SiH_4 number density and its rotational temperature.[24] This has made clear that the rotational tamperature is constant throughout the surveyed volume, while the silane number density varies drastically as shown in Fig. 3.1.10. It takes the minimum value at the position near the cathode where the optical emission looks most luminous. The origin of this density distribution is supposed to be volume expansion due to translational temperature increase,[25] dissociation, and the electronic or vibrational excitation of silane.

It is difficult to determine the number density of SiH_2 precisely, because the exact data of the SiH_2 Raman scattering cross section is not available. Furthermore its population might extend over a large number of rotationally and vibrationally excited states. In spite of this limitation, assuming that spectral response and Raman scattering cross section are identical between the $\nu_1 Q$-band of SiH_2 ($2032\,\text{cm}^{-1}$) and of SiH_4 ($2187\,\text{cm}^{-1}$) CARS signals, the number density of SiH_2 (n_{SiH_2}) in its ground state is estimated as $1.7 \times 10^{14}\,\text{cm}^{-3}$ from the

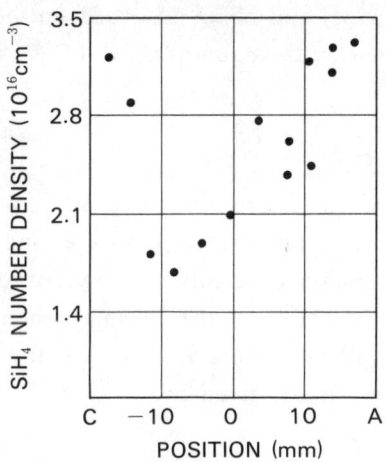

Fig. 3.1.10 Silane number density distribution between the cathode (C) and anode (A). [after N. Hata et al.[15]]

measured CARS data. This was done by simply comparing the signal intensity with that of SiH$_4$ of known pressure taking account of its quadratic dependence on the number density, while $n_{SiH_4} = 5 \times 10^{15}$ cm^{-3} in non-discharge gas flow of the same pressure as in the SiH$_2$ study.

The SiH$_4$ dissociation rate was determined from the decay rate of the CARS signal intensity (ν_1 Q-band of SiH$_4$) measured outside the plasma volume just after turning on the rf power supplied to the gas mixture in the closed chamber. The result is shown in Fig. 3.1.11.[14] The dissociation rate, γ, of SiH$_4$ in the present plasma against the rf power density

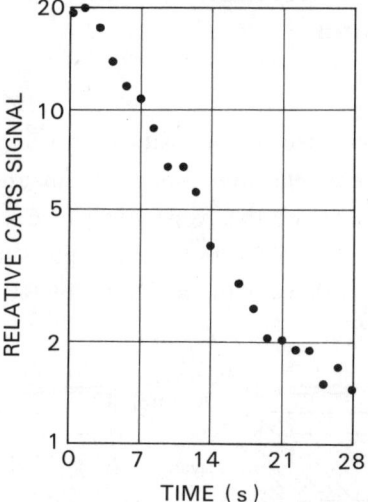

Fig. 3.1.11 SiH$_4$ ν_1 Q-band relative CARS signal intensity plotted against the time after the rf power turning on. [after N. Hata et al.[14]]

of 1 W/cm^2 was calculated to be 6.2 s^{-1} from the gradient of the time-decay curve of the SiH$_4$ CARS signal shown in the figure, taking into account the ratio of plasma to chamber volume. If one assumes that one-third of the dissociated SiH$_4$ molecules become SiH$_2$

radicals diffusing to the electrodes and being adsorbed on the surface without desorption, the diffusion constant, D, given by the equation,

$$D = \frac{1}{3}\left(\frac{8kT}{\pi m}\right)^{1/2}\frac{1}{4\sqrt{2}\pi r^2 n} \tag{3.1.2}$$

amounts to as much as $10^2 \text{cm}^2/\text{s}$, where $r = 1.5 \times 10^{-8}$ cm is assumed as the collision radius, $T = 500\,\text{K}$ as the gas translational temperature, and k, m, n are the Boltzmann constant, the mass and the number density of the colliding molecules, respectively. Diffusion time of the SiH_2 radical from the plasma center to the electrode surface is estimated as $(L/2)^2/D = 4 \times 10^{-2}$ s, where $L = 4$ cm is the electrode distance. Thus the number density, n_{SiH_2}, in the plasma is given by,

$$n_{SiH_2} = \frac{n_{SiH_4}\gamma}{3}\frac{(L/2)^2}{D} = 4.1 \times 10^{14}\,\text{cm}^{-3} \tag{3.1.3}$$

Within the framework of the present estimation, the value given in Eq. (3.1.3) shows a fairly good agreement with that estimated from the CARS signal intensity. However, obviously, the validity of the assumptions made in the present work should be checked through a more detailed study in the future.

It should be noted that SiH_2 radicals have been observed only in the glow-discharge plasma of a SiH_4-H_2 mixture (see Fig. 3.1.9), but not in the pure SiH_4 (100%) plasma. It suggests that SiH_2 has a very short reaction time (lifetime) in the 100-% SiH_4 plasma, while SiH_2 can survive for a longer time in H_2-diluted SiH_4 plasma.

3.1.3 Triode Chamber Method—Transport of Precursors in SiH_4 and Si_2H_6 Plasma

Experimental

Matsuda et al. have performed the deposition of a-Si : H films from Si_2H_6 as well as SiH_4 glow discharge using a conventional diode system and a specially-designed triode system, as shown in Fig. 3.1.12 (a) and (b), respectively.[18] Each gas material was introduced at a flow rate of 1, 5, 15 and 30 SCCM under a gas pressure of 20 or 50 mTorr into the evacuated chambers. rf power density in the range of $0.01 \sim 1\,\text{W/cm}^2$ was used fo

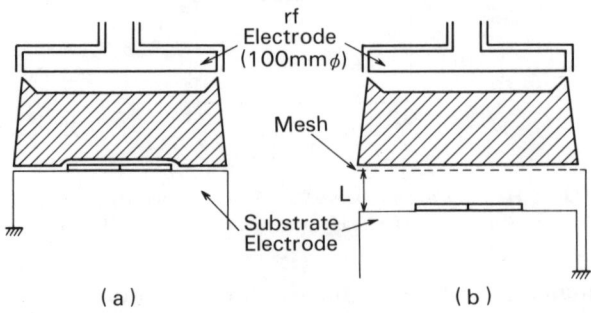

Fig. 3.1.12 Deposition systems used in the present work ; (a) diode-type, (b) triode-type reactor with variable L. [after A. Matsuda et al.[18]]

deposition of a-Si : H onto a glass substrate maintained at 300°C. Optical emission spectroscopy (OES) was employed as a diagnostic tool, the details of which were described earlier.[4] The triode system was used for measuring the lifetime of the precursor of each deposition gas, by changing the separation L between the mesh and the substrate electrode.

OES data

As is shown in Fig. 3.1.13, the deposition rate (r) of a-Si : H from SiH_4 plasma is

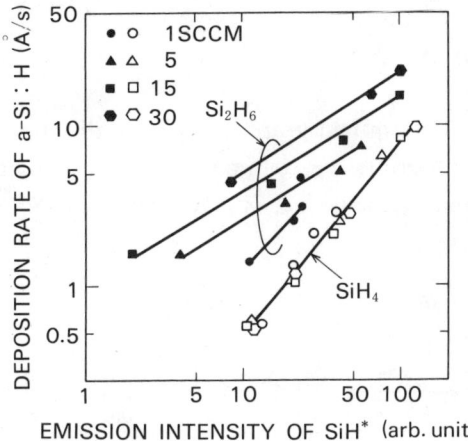

Fig. 3.1.13 Deposition rate vs. SiH* emission intensity. [after A. Matsuda et al.[6]]

proportional to the intensity of SiH* emission (4127 A) from the plasma in the entire range of the present plasma parameters, while r for Si_2H_6 plasma shows no such simple behavior against SiH*.[6] This fact suggests that a dominant radical in SiH_4, responsible for the film deposition, is strongly correlated with SiH* emissive radicals in contrast to that in Si_2H_6 plasma.

From the quantitative measurement of the luminosity of a 4127-A light from SiH_4 plasma (0.01 W/cm², 5 SCCM and 50 mTorr), the density of excited SiH radicals (i. e., SiH*) has been estimated as $3 \times 10^6/cm^3$ taking into account the radiative transition probability.[26] This value is three or four orders of magnitude lower than the density of species required for explaining the observed deposition rate, indicating that SiH* is not a direct precursor for a-Si:H deposition, but other species correlated with SiH* are responsible for the deposition from SiH_4 plasma.[18]

Lifetime (reaction time) of precursors

In order to determine the lifetime (reaction time) of each dominant radical in each gas, the deposition rate was measured as a function of the distance, L, between mesh and substrate in the triode system shown in Fig. 3.1.12 (b). The results are shown in Fig. 3.1.14. A deposition is determined by the amount of the radicals reaching the substrate after travelling the distance, L, without reactive collisions with other neutral species. Therefore, the decay curve of the figure is directly correlated with the lifetime of the radical responsible for a-Si:H deposition under various kinds of collisions between species.

The deposition rate, r, at L is described as

$$r(L) = r_0 A^n \tag{3.1.4}$$

where $r_0 = r(0)$, A the survival probability of the radical per one collision process, and n the average number of collisions of the radical for diffusing the distance L. Since $n = (L/\lambda)^2$ and $\lambda \propto P^{-1}$ where λ and P are the mean free path of the radical and the gas pressure, respectively, Eq. (3.1.4) can be rewritten as

$$\log \frac{r}{r_0} \propto (PL)^2 \log A \tag{3.1.5}$$

Therefore, the value of A is determined graphically from the logarithmic plots of (r/r_0) vs. $(PL)^2$, and a corresponding lifetime, τ, can be calculated using the relation $\tau = \lambda/v_{th} \ln(A^{-1})$, since the mean thermal velocity, v_{th}, is given by $v_{th} = (8kT/\pi m)^{1/2}$.

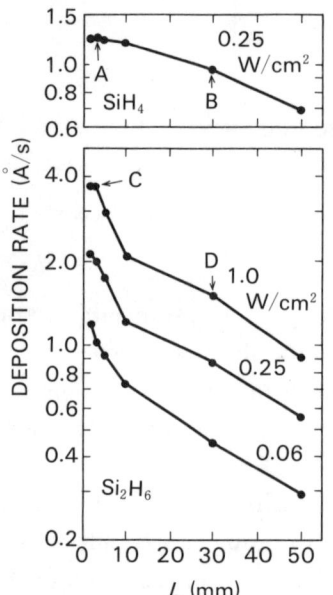
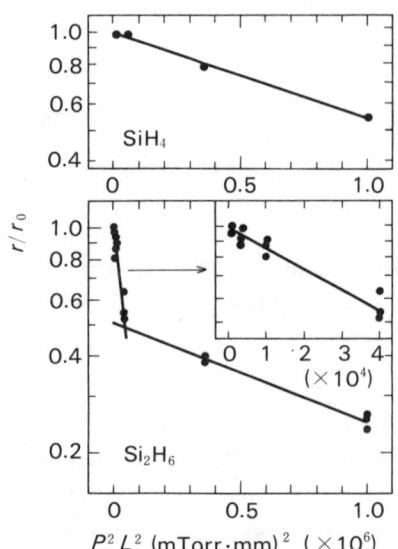

Fig. 3.1.14 Deposition rate as a function of L. [after A. Matsuda et al.[18]]

Fig. 3.1.15 Normalized deposition rate (r/r_0) plotted against $(PL)^2$. [after A. Matsuda et al.[18]]

Figure 3.1.15 shows (r/r_0) vs. $(PL)^2$ characteristics replotted from Fig. 3.1.14. As shown in the figure, the dominant radical from SiH_4 plasma shows a single straight line, i. e., dominated by a single precursor with $\tau = 1.6 \times 10^{-3}$ s, while Si_2H_6 plasma includes two different kinds of precursors with $\tau = 1.5 \times 10^{-4}$ s and 1.3×10^{-3} s, respectively. It clearly indicates that a dominant precursor with a shorter τ in Si_2H_6 plasma is quite different from that in SiH_4 plasma, while the other precursor in Si_2H_6 plasma, detected in a larger $(PL)^2$ range, might be the same as that in SiH_4 plasma judging from the similarity of the magnitudes between their lifetimes. Actually, depending on those two different precursors, the deposited films have shown dissimilar H structures. Figure 3.1.16 shows the H-evolution

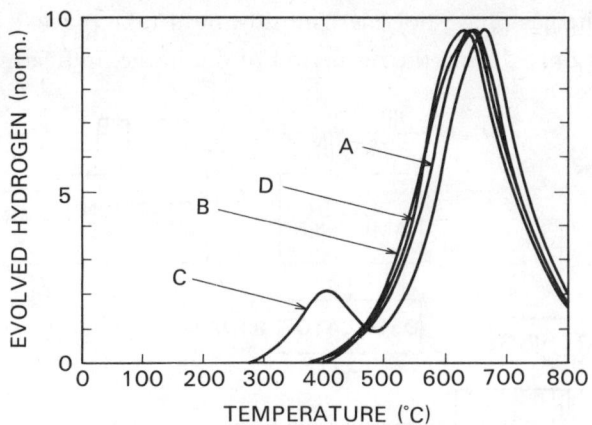

Fig. 3.1.16 H-evolution spectra of a-Si:H deposited under the conditions A, B, C and D indicated in Fig. 3.1.14. [after A. Matsuda et al.[18]]

spectra for four different films deposited under the conditions denoted by A, B, C and D in Fig. 3.1.14, where two peaks are observed only in the spectrum of film C associated with a shorter-lifetime precursor. Consequently, it is reasonably considered that the dominant radical with a shorter lifetime in Si_2H_6 plasma gives the origin of the high deposition rate of a-Si:H. It clearly opposes Scott's assumption,[17] as mentioned in 3.1.1, that precursors in Si_2H_6 as well as SiH_4 plasma are identical.

3.1.4 Pulsed rf-Discharge Technique

Hamasaki et al. have studied the growth rate and bonded hydrogen content of a-Si:H prepared by a pulsed rf discharge technique as functions of repetition frequency.[27] On the basis of the results, they have discussed whether the predominant chemical process in a-Si:H growth is homogeneous (gas-phase reaction) or heterogeneous (surface reaction).

Amorphous hydrogenated silicon was deposited using a conventional parallel-electrode type reactor shown in Fig. 3.1.17, where substrates were placed on the grounded electrode. The substrate temperature, total pressure, concentration of silane gas and total flow rate were, respectively, kept at 300°C, 0.18 Torr, $SiH_4/H_2 = 11\%$ and 67 SCCM. Power to sustain the plasma was supplied from an rf generator which was modulated by a function generator with a rising time faster than 300 ns. The peak to peak height of the rf voltage was maintained at a constant level by monitoring an oscilloscope trace, and the pulse width, t_{on}, and the pulse interval, $t_{on} + t_{off}$, were changed over a wide range.[27]

Plots of the growth rate vs repetition frequency of pulsed discharges of silane and ethane are shown in Fig. 3.1.18 (a), where the total time of the on-period and duty cycle were kept constant for each frequency. The growth rate of a-Si:H is almost independent of repetition frequency up to 100 kHz. If the lifetime of the reactive species in the silane plasma is as long as 1 ms and chemical reactions proceed even during the off-period, the growth rate should exhibit an abrupt increase at a frequency around 1 kHz as in the case of plasma polymerization of ethane (dashed curve in Fig. 3.1.18 (a)). The effect of duty cycle on the growth rate is shown in Fig. 3.1.18 (b) for both the silane and ethane plasma. For the SiH_4 plasma, the growth rate of the film is proportional to the duty cycle, while it is not for the methane plasma where the gas-phase polymerization should take place.[28] This means that a-Si : H film is deposited only during the on-period of discharge, indicating that gas-phase

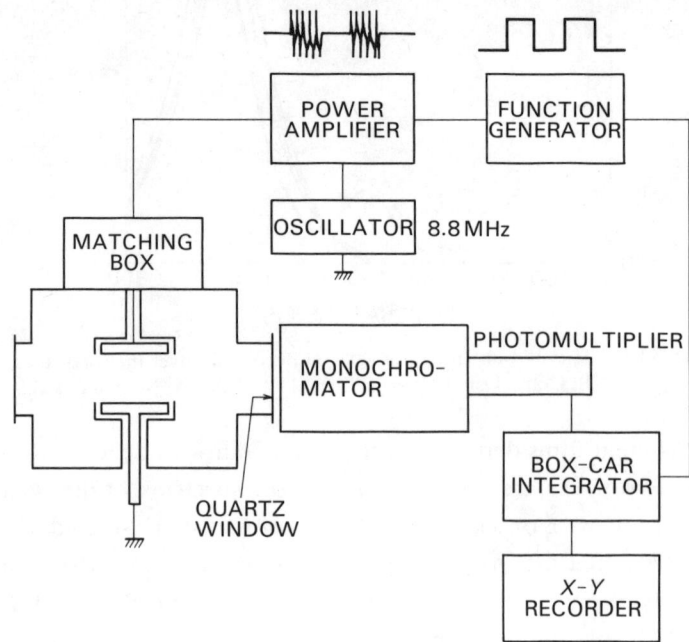

Fig. 3.1.17 Schematic diagram of an experimental apparatus. [after T. Hamasaki et al.[27]]

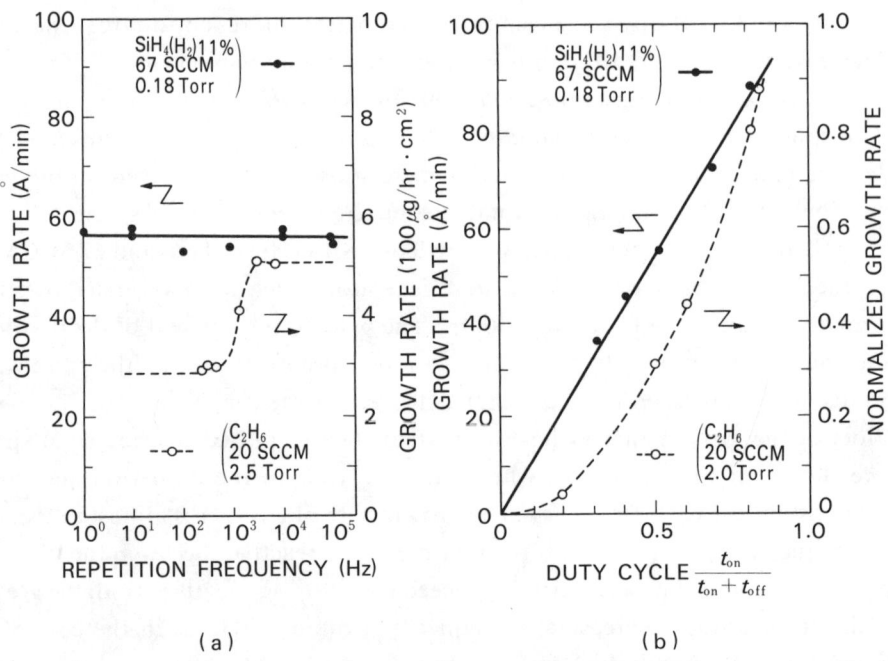

Fig. 3.1.18 (a) The growth rate vs. repetition frequency. Duty cycle for a-Si : H deposition is 50%. (b) Effect of duty cycle on the growth rate. [after T. Hamasaki et al.[27] and J. W. Vinzant et al.[28]]

polymerization is not the predominant process in a-Si : H deposition.[27]

Further confirmation of the absence of gas phase polymerization is obtained from the SiH stretching absorption spectra of a-Si : H which is almost independent of the repetition frequency.[27] On the other hand, for the case of the ethylene plasma, Yasuda and Hsu have found that the growth rate and ESR signals of free spins in polymer films are considerably changed by a pulsed discharge probably due to polymerization reactions in the gas phase during the off-period of the pulsed discharge.[29]

3.1.5 Conclusion

(1) Coherent anti-Stokes Raman spectroscopy (CARS) has been applied to investigate SiH_4 plasma for the first time, and concentrations of neutral species such as H_2, SiH_4 and SiH_2 have been detected. From the ν_1 Q-band spectra of silane, the spatial distribution of SiH_4 density and its rotational temperature has been determined. It has been demonstrated that SiH_2 has a shorter lifetime in pure SiH_4 plasma and a longer lifetime in SiH_4 diluted H_2.

(2) Lifetimes (reaction times) of precursors in Si_2H_6 as well as SiH_4 glow-discharge plasma have been determined using a triode reactor with a variable electrode-substrate distance. It has been indicated that the dominant radical as a film precursor from Si_2H_6 plasma is different from that from SiH_4 plasma.

(3) A pulsed rf discharge experiment has demonstrated that the growth of a-Si : H proceeds through heterogeneous reactions among chemical species on the growing film surface.

References

1) K. Tanaka, K. Nakagawa, A. Matsuda, M. Matsumura, H. Yamamoto, S. Yamasaki, H. Okushi, and S. Iizima : Jpn. J. Appl. Phys., Suppl., *20-1* (1981) 267.
2) A. Matsuda, K. Nakagawa, K. Tanaka, M. Matsumura, S. Yamasaki, H. Okushi, and S. Iizima : J. Non-Cryst. Solids, *35 & 36* (1980) 183.
3) R. G. Griffith, F. J. Kampas, P. E. Vanier, and M. D. Hirsch : J. Non-Cryst. Solids, *35 & 36* (1980) 391.
4) A. Matsuda and K. Tanaka : Thin Solid Films, *92* (1982) 171.
5) M. Hirose, T. Hamasaki, Y. Mishima, H. Kurata, and Y. Osaka : in R. A. Street, D. K. Biegelsen and J. C. Knights (eds.), Tetrahedrally Bonded Amorphous Semiconductors, American Instiutute of Physics, New York (1981) 10.
6) A. Matsuda, T. Kaga, H. Tanaka, L. Malhotra, and K. Tanaka : Jpn. J. Appl. Phys., *22* (1983) L 115.
7) J. Perrin, J. P. M. Schmitt, G. de Rosny, J. Huc, and A. Lloret : Chem. Phys., *73* (1982) 383.
8) J. C. Knights, J. P. M. Schmitt, J. Perrin, and G. Guelachvilli : J. Chem. Phys., *76* (1982) 3414.
9) T. Hamasaki, M. Hirose, and Y. Osaka : Proceedings of the Sixth Symposium on Ion Sources and Ion-Assisted Techhology (Tokyo, Japan, June 7-9, 1982) 263.
10) H. Lee, J. P. deNeufville, and S. R. Ovshinsky : J. Non-Cryst. Solids, *59-60* (1983) 671.
11) J. P. M. Schmitt : J. Non-Cryst. Solids, *59-60* (1983) 649.
12) G. Inoue and M. Suzuki : Chem. Phys. Lett., *105* (1984) 641.
13) N. Hata, A. Matsuda, K. Tanaka, K. Kajiyama, N. Moro, and K. Sajiki : Jpn. J. Appl. Phys., *22* (1983) L 1.
14) N. Hata, A. Matsuda, and K. Tanaka : J. Non-Cryst. Solids, *59-60* (1983) 667.

15) N. Hata, A. Matsuda, and K. Tanaka: in Proc. Int'l Ion Engineering Congress — ISIAT'83 & IPAT'83, ed. by T. Takagi (Kyoto, Japan, 1983) 1457.
16) B. A. Scott, J. A. Reimer, and L. A. Longeway: J. Appl. Phys., *54* (1983) 6853.
17) B. A. Scott, M. H. Brodsky, D. C. Green, P. B. Kirby, R. M. Plecenik, and E. E. Simonyi: Appl. Phys. Lett., *37* (1980) 727.
18) A. Matsuda, T. Kaga, H. Tanaka, and K. Tanaka: J. Non-Cryst. Solids, *59-60* (1983) 687.
19) J. M. Jasinski, E. A. Whittaker, G. C. Bjorklund, R. W. Dreyfus, R. E. Estes, and R. E. Walkup: Appl. Phys. Lett., in press.
20) G. Herzberg: Molecular Spectra and Molecular Structure II, Infrared and Raman Spectroscopy of Polyatomic Molecules (D. Van Nostrand Co. Inc., Princeton, USA, 1945).
21) W. Kiefer: in Non-Linear Raman Spectroscopy and Its Chemical Applications, NATO Advanced Study Institute Series, ed. by W. Kiefer and D. A. Long (D. Reidel Publishing Co., Boston, USA, 1982).
22) M. Péalat, J. P. E. Taran, J. Taillet, M. Bacal, and A. M. Bruneteau: J. Appl. Phys., *52* (1981) 2687.
23) D. E. Milligan and M. E. Jacox: J. Chem. Phys., *52* (1970) 2594.
24) J. W. Nibler: in Non-Linear Raman Spectroscopy and Its Chemical Applications, NATO Advanced Study Institute Series, ed. by W. Kiefer and D. A. Long (D. Reidel Publishing Co., Boston, USA, 1982).
25) W. G. Breiland and M. J. Kushner: Appl. Phys. Lett., *42* (1983) 395.
26) W. H. Smith: J. Chem. Phys., *51* (1969) 520.
27) T. Hamasaki, M. Ueda, M. Hirose, and Y. Osaka: J. Non-Cryst. Solids, *59-60* (1983) 679.
28) J. W. Vinzant, M. Shen, and A. T. Bell: Plasma polymerization, eds. by M. Shen and A. T. Bell (Amer. Chemical Soc., 1979) 79-85.
29) H. Yasuda and T. Hsu: J. Polymer Science *15* (1977) 81.

3.2 a-Si:H Deposition from Higher Silanes

Masataka HIROSE*

Abstract

Deposition kinetics and physical properties of hydrogenated amorphous silicon (a-Si:H) produced by the glow discharge, thermal CVD and photochemical vapor deposition (photo-CVD) of disilane (Si_2H_6) are reviewed. Optical emission spectroscopy of glow discharge reveals that dominant radicals as film precursors in SiH_4 and Si_2H_6 plasmas are different, and a significant amount of hydrogen atoms which are not bonded with silicon are incorporated in the films deposited from Si_2H_6 plasma. However, at a deposition rate of 11 Å/s a 5.47% conversion efficiency is obtained in n-i-p type solar cells. Films produced by thermal CVD of Si_2H_6 in the temperature range 450~600°C are studied as a function of the doping ratio of phosphine or diborane. The growth rate is decreased by phosphorus doping and increased by boron doping. It is found that the electrical properties of disilane CVD films are improved compared with those of monosilane CVD films which contain very little hydrogen. As a new alternative approach to the preparation of CVD films at temperatures below 300°C, the photochemical vapor deposition technique is developed. The photoconductivity as high as $3.7 \times 10^{-4} \Omega^{-1} \cdot cm^{-1}$ (AM1, 100 mW/cm^2) is obtained in the direct photo-CVD a-Si:H, and the film exhibits no light induced degradation. Also, mercury photosensitized CVD produces a-Si:H and the p-i-n solar cell achieves a 4.39% conversion efficiency.

3.2.1 Introduction

Hydrogenated amorphous silicon (a-Si:H) produced by the glow discharge of monosilane has been extensively studied as a new electronic material for fabricating novel devices such as solar cells and TFTs.[1,2] The SiH_4 plasma deposition rate is rather limited, typically of the order of 1~5 Å/s. This is a drawback in the mass production of a-Si:H. In this respect, the plasma decomposition of Si_2H_6 has attracted attention as a new technique for achieving high deposition rates of a-Si:H films.[3~5] Compared to the film growth from a SiH_4 plasma, the deposition rate is enhanced by a factor of 20, which corresponds to a rate of 20~60 Å/s. The chemical vapor decomposition of Si_2H_6 at low temperatures is another candidate for producing a-Si:H thin films because this technique achieves low autodoping, high reproducibility and absence of ion damage. In the monosilane CVD, low-cost substrates such as glass or stainless steel cannot be used because of a high deposition

---

* Department of Electrical Engineering, Hiroshima University, Higashihiroshima 724.

temperature of more than 600°C. The resulting films contain no significant bonded hydrogen which reduces the silicon dangling bonds. Recently, Gau et al.,[6] Mishima et al.,[7] and Dalal[8] have prepared hydrogenated CVD amorphous silicon at temperatures below 500°C using higher silanes. Dark conductivity and AM-1 photoconductivity of the films deposited were 10^{-9} and $5 \times 10^{-5} \Omega^{-1} \cdot cm^{-1}$, respectively. The film exhibited no significant Staebler-Wronski effect.[8] Multiple gap-stacked pin cells utilizing this material for the i layer achieved a 4.5% photovoltaic efficiency.[8] Furthermore, a new attempt to prepare a-Si:H films by employing a Si_2H_6 photochemical vapor decomposition technique was successful.[9,10] In the next section, recent progress in a-Si:H deposition from glow discharge, thermal CVD and photo-CVD of Si_2H_6 is described.

3.2.2 Glow-Discharge Deposition

Deposition of a-Si:H films from Si_2H_6 glow discharge is performed in conventional plasma CVD apparatus under a wide range of plasma conditions.[5,11,12] The glow discharge has shown diagnostically through optical emission spectroscopy (OES) a correlation between the deposition rate of a-Si:H films and the emission intensity from SiH radical (4 127 Å). The deposition rate from SiH_4 plasma is proportional to the SiH intensity in the entire range of plasma parameters investigated, while rather different behaviour is observed in Si_2H_6 glow discharge, as illustrated in Fig.3.2.1. At higher flow rates, the deposition rate can be expressed as the following relation:

$$\text{Deposition Rate} \propto [\text{SiH}]^{1/2} \tag{3.2.1}$$

This result implies that deposition kinetics from the Si_2H_6 plasma is different from that of the SiH_4 discharge. To determine total hydrogen content, the resulting a-Si:H films were

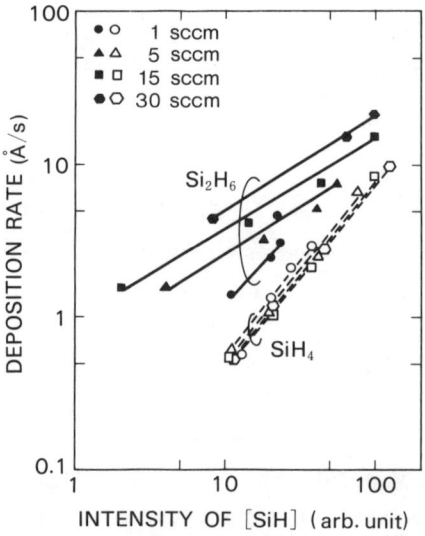

Fig. 3.2.1 Deposition rate of a-Si:H films plotted against the line intensity of [SiH] emissive radicals in glow discharge plasma during deposition. rf power and gas flow rate are changed from 1 to 80 W and 1 to 30 sccm, respectively, keeping gas pressure constant (50 mTorr). [After A. Matsuda et al.[5]]

heated in a vacuum from room temperature to 800°C at a rate of 20°C/min. A quadrapole mass filter placed between the heating furnace and the vacuum system analyzed the H_2 gas thermally evolved. Results showed H_2 evolution starts at lower temperatures in Si_2H_6 films compared to those obtained from SiH_4 glow discharge. Total hydrogen content in the films was calculated from the area of the respective thermal evolution chart. Variation of the total H content as a function of rf power used for film deposition is shown in Fig. 3.2.2. The content of bonded H in the films determined by ir absorption is also plotted. The figure indicates that while almost all incorporated H is bonded to Si in SiH_4 films, a good amount of H not bonded to Si is also incorporated in films prepared from Si_2H_6. Such tendencies are enhanced in the higher rf power range. A reason for the incorporation of a great deal of H (except for SiH bonds) in these films may be due to heterogeneous surface reactions between monomeric and dimeric species. Another reason may possibly be ascribed to the difference in oscillator strength of Si-H vibrational modes between SiH_4- and Si_2H_6-based films, as seen in Fig. 3.2.2. Regardless of these results, electrical and optical measurements performed to evaluate film quality offered encouraging results. Dark conductivity and photoconductivity under AM1, $100 mW/cm^2$ simulation, varied from 10^{-10} to $10^{-9} (\Omega \cdot cm)^{-1}$ and 10^{-5} to $10^{-4} (\Omega \cdot cm)^{-1}$, respectively.[12] These electrical properties do not depend on deposition rates up to 18.7 Å/s. Furthermore, a-Si:H films from disilane can be doped with either n or p type by incorporating appropriate amounts of phosphorus or boron, respectively. Conductivities on the order of $4 \times 10^{-2} (\Omega \cdot cm)^{-1}$ and $2 \times 10^{-3} (\Omega \cdot cm)^{-1}$ were obtained for n- and p-doped films, respectively. The optical band gap deduced from $\sqrt{\alpha h\nu} - h\nu$ characteristics, where α and $h\nu$ are the absorption coefficient and photon energy, respectively, was in the 1.75~1.8 eV range. This suggests that the resulting films are somewhat similar to monosilane-produced a-Si:H films. The difference is that disilane-produced a-Si:H films have a wider optical band gap. From these results, it is clear that a significantly high growth rate can be achieved using Si_2H_6 glow discharge. Properties of Si_2H_6 films are basically compatible with those of SiH_4 films.

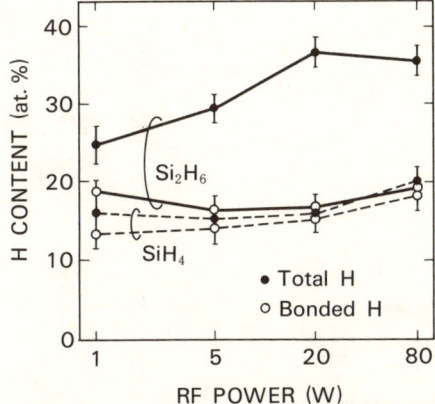

Fig. 3.2.2 Total hydrogen content of a-Si:H films measured by thermal evolution experiments and bonded hydrogen content measured by ir absorption technique plotted against rf power for the depositions. [After A. Matsuda et al.[5]]

3.2.3 a-Si : H Produced by Pyrolysis of Disilane

A horizontal CVD reactor was used for the deposition of pyrolytic a-Si : H. It was composed of a cold-wall quartz tube, in which a silicon susceptor was placed and heated by external halogen lamps. The substrate temperature was measured with a thermocouple which had been calibrated before CVD. Disilane gas diluted with 99% He was decomposed at a substrate temperature of 500°C. Either PH_3 or B_2H_6 gas diluted with H_2 was admixed at a flow rate of 45 cc/min with Si_2H_6 whose flow rate was 450 cc/min. Nitrogen gas (500 cc/min) was used as a carrier gas for atmospheric pressure CVD, in which all preparations were carried out. In preparation of a-Si_xN_{1-x} : H, ammonia was added to disilane whose flow rate was kept constant.

Undoped films

Disilane gas was thermally decomposed at a temperature between 450~600°C. Growth rate vs. substrate temperature is plotted in Fig. 3.2.3. Gau et al.[6] reported an appreciably high growth rate compared with that of the present study; partly because their gas contained a series of higher silanes. Substrate temperature dependence of the hydrogen content calculated from the absorption coefficient due to the SiH stretching mode, is shown in Fig. 3.2.4. It is interesting to note that an appreciable shift of wavenumber for the SiH stretching mode from 2000 cm^{-1} to 1970 cm^{-1} was observed and no SiH_2 stretching absorption at 2090 cm^{-1} appeared.

The optical absorption spectrum for a disilane CVD film produced at a substrate temperature of 500°C is shown in Fig. 3.2.5. The absorption tail below the optical band gap E_{opt} is still large. This indicates that the localized states in the gap are less compensated with

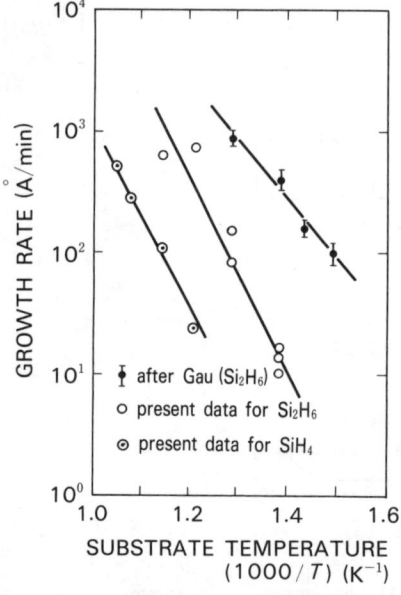
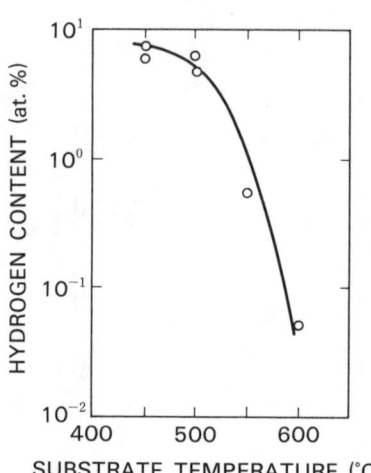

Fig. 3.2.3 Growth rate of CVD a-Si produced from monosilane and disilane against the substrate temperature.

Fig. 3.2.4 Hydrogen content plotted as a function of substrate temperature.

hydrogen. The upper inset shows substrate temperature dependence of the optical band gap determined from $(\alpha h\nu)^{1/2}$ versus $h\nu$ plot. Increase in the optical band gap with a decrease

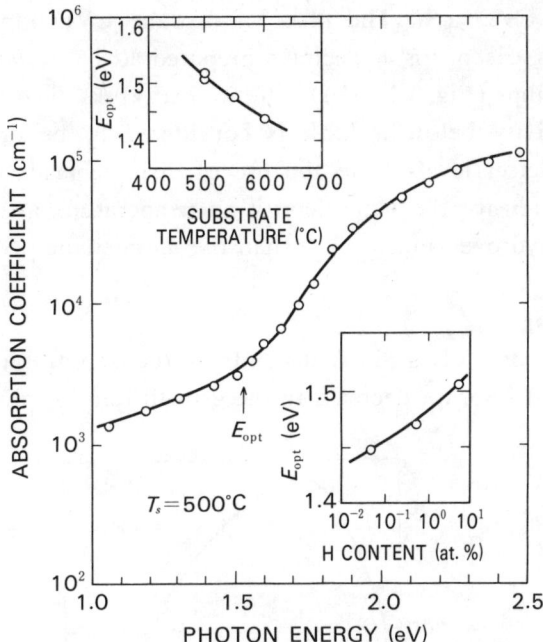

Fig. 3.2.5 Spectral dependence of optical absorption coefficient for a CVD a-Si film prepared at a substrate temperature of $T_s = 500°C$.

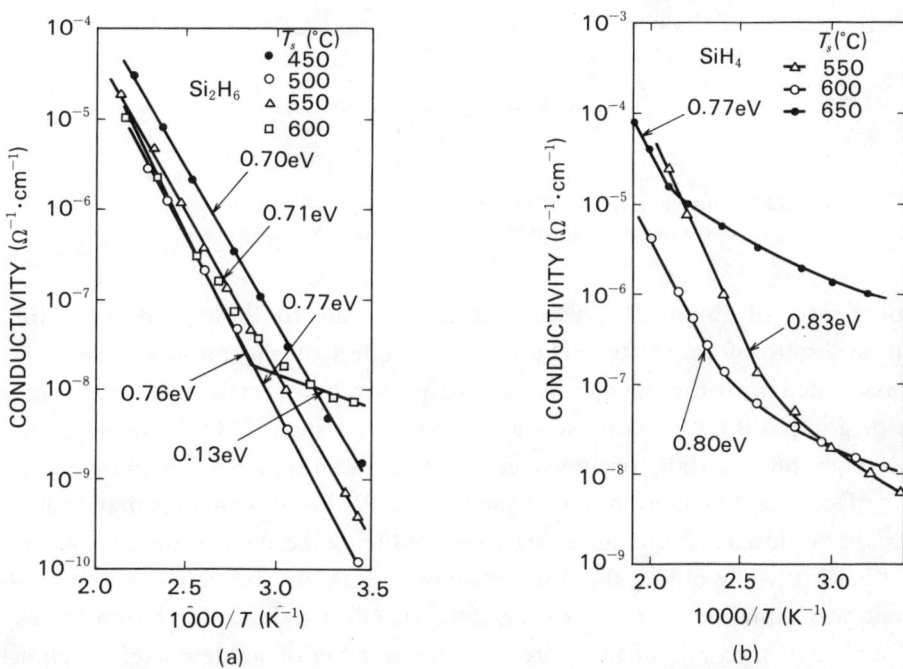

Fig. 3.2.6 Reciprocal temperature plots of conductivity for disilane CVD a-Si (a) and for monosilane a-Si (b)

in the substrate temperature can be explained in terms of hydrogen concentration dependence of the optical band gap as shown in the lower inset. A plot of conductivity versus reciprocal temperature is shown in Fig. 3.2.6 (a) and (b), both refer to disilane CVD a-Si and monosilane CVD a-Si. The plots of disilane CVD films (Fig. 3.2.6 (a)) are definitely straight lines, except for a specimen prepared at $T_s = 600°C$. Comparatively the plots for monosilane films (Fig. 3.2.6 (b)) exhibit clear kinks even for the film with $T_s = 550°C$. Here, conductivity below a kink is considered to be dominated by hopping conduction through defect levels.[13] So the use of disilane leads to a reduction in the hopping conduction even for the same deposition temperature, as the disilane CVD film incorporates bonded hydrogen much more than the monosilane CVD.

Substitutional doping

The growth rate of CVD a-Si : H depends on the doping ratio of PH_3 and B_2H_6 as shown in Fig. 3.2.7.[14] An abrupt decrease in the growth rate for phosphine doping occurs

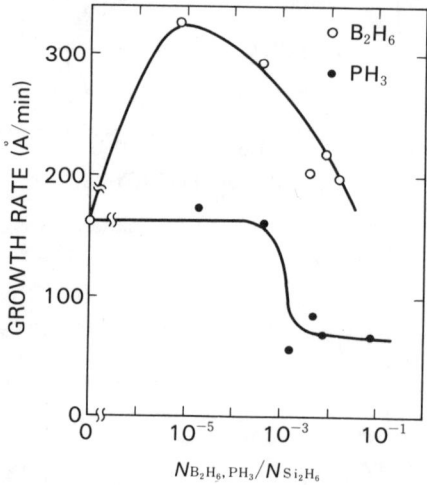

Fig. 3.2.7   Growth rate of CVD a-Si : H plotted against phosphine or diborane molar fraction to disilane [After Y. Ashida et al.[14]]

at a doping ratio of about 10^{-3}, while the growth rate for boron doping exhibits the maximum at a ratio of about 10^{-5}. The inhibition effect of phosphine on the growth rate may be associated with the phosphine catalyst-poisoning effect in the same manner that hydrides of group VB elements poison a metal catalyst such as Pt.[15] Namely, phosphine molecules might preferentially occupy chemically active sites such as dangling bonds on the growing surface. The dissociation rate of the surface PH bond is smaller than that of SiH,[16] and hence the surface reactions are deactivated reducing the deposition rate. As for boron doping, it is worthwhile noting that the influence of B_2H_6 in silicon epitaxial growth using monosilane was explained in terms of the catalytic effect by holes or boron atoms on the growing surface of silicon,[17] or the increase in the number of stable nuclei which act as an SiH_4 adsorption site.[18] A similar phenomenon is also observed in conventional CVD of monosilane, because the growth rate increases with the boron doping ratio.[19] However, such simple enhancement of the growth rate by boron doping is not the case for Si_2H_6 CVD since the growth rate decreases in the high doping range. An alternative model would be as follows: The decomposition rate of Si_2H_6 may be accelerated by the presence of B_2H_6 in the gas phase and therefore the concentrations of the decomposed species are increased with an increase of the doping ratio, resulting in enhancement of the growth rate. However, when the film precursor concentration becomes too high, gas phase nucleation of silicon particles takes place and the effective flux density of the precursor towards the substrate surface is reduced.

The hydrogen content C_H and optical band gap E_{opt} determined from a $(\alpha h\nu)^{1/2}$ vs. $h\nu$ plot as a function of doping ratio is show in Fig. 3.2.8. For PH_3 doping, both C_H and

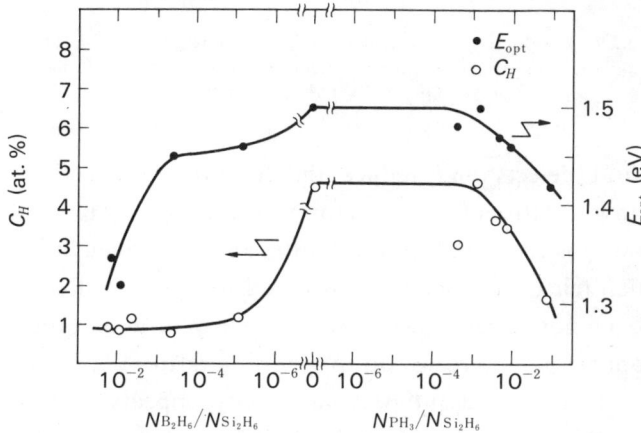

Fig. 3.2.8 Hydrogen content C_H and the optical band gap E_{opt} as a function of doping ratio. [After Y. Ashida et al.[14]]

E_{opt} are decreased while increasing doping ratio, and the relationship $E_{opt} \propto C_H$ is seen as that previously reported for discharge-produced a-Si:H.[20] On the other hand, there exists no such correlation between E_{opt} and C_H when boron is doped. Although C_H remains almost constant at boron doping ratios above 10^{-5}, E_{opt} rapidly decreases from 1.44 to 1.33 eV at ratios exceeding 10^{-3}. A possible explanation of this is to assume that the silicon-boron alloying effect reduces the optical band gap, as suggested by a previous work.[21] Another explanation is to connect the decrease in E_{opt} with an increase of defect states or gap states, since the gas phase nucleation of silicon particles induced by boron doping disturbs the formation of the silicon random network on the growing surface and deteriorates the electronic properties of the resulting films. A pronounced increase in the gap states results in remarkable enhancement of optical absorption at photon energies below the band gap and hence E_{opt} measured should be lowered. The latter explanation is more likely because effective boron-doping efficiency in disilane CVD films is inferior to that in monosilane CVD, as discussed below.

The dark conductivity at room temperature σ_{RT} and its activation energy E_a are plotted as a function of doping ratio in Fig. 3.2.9. Doping curves for SiH_4 CVD a-Si are also given as a reference. The σ_{RT} vs. N_{PH_3}/N_{SiH_4} plot has a minimum because phosphorus atoms

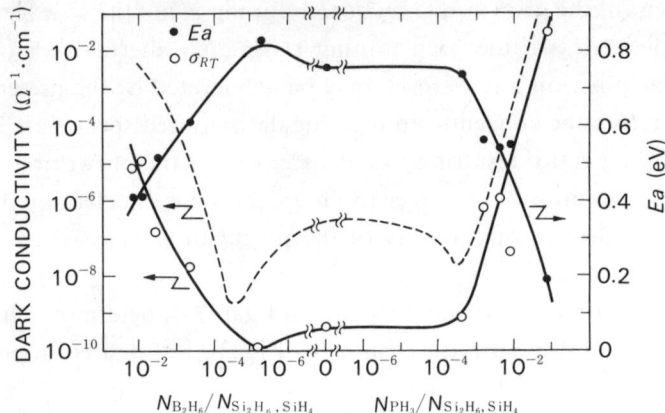

Fig. 3.2.9 Conductivity σ_{RT} and its activation energy E_a as a function of doping ratio. Solid line refers to Si_2H_6 CVD and dashed line to SiH_4 CVD. [After Y. Ashida et al.[14]]

compensate structural defects and reduce the hopping conduction component which dominates σ_{RT} below a ratio of 10^{-3}.[22] However, σ_{RT} for Si_2H_6 CVD films exhibits no hopping conduction, possibly because the film contains bonded hydrogen amounting to about 5 at.%,[7] which relax the silicon network and reduce the silicon dangling bonds. As a consequence, σ_{RT} monotonically increases with phosphorus doping.

In boron doping, σ_{RT} decreases up to a doping ratio of 7×10^{-6} because the Fermi level shifts towards midgap and p-type conductivity appears at ratios above 7×10^{-6}. Apparently low efficiency boron doping in Si_2H_6 CVD compared with that in SiH_4 CVD is basically attributable to diborane-induced gas phase nucleation of silicon particles in Si_2H_6 CVD. In fact, under no gas phase nucleation conditions achieved by lowering T_s to 300°C, σ_{RT} as high as $5 \times 10^{-3} \Omega^{-1} \cdot cm^{-1}$ is obtained at a boron-doping ratio of 1.6×10^{-2}.

3.2.4 Photochemical Deposition

Mercury-photosensitized deposition

In plasma deposition, careful optimization of discharge conditions is needed, because the positive column in the reactor contains high energy particles which induce radiation damage in a growing layer or an underlying film. A new attempt to deposit SiO_2 or Si_3N_4 at low temperatures utilizing the mercury-photosensitized decomposition of SiH_4 with O_2 or with N_2H_4 to avoid damage has been reported.[23,24] This technique has also been applied to obtain amorphous hydrogenated silicon.[25~27] The decomposition mechanism of silane is inferred to involve collisional energy transfer from UV excited Hg* ($^3P_1$) state to reacting species. Little is known about more detailed deposition mechanisms. In the mercury photo-CVD system, an ozone-free low-pressure mercury lamp was used as a light source which emitted a strong 2537 Å resonance line ($\sim$6 mW/cm² at 3-cm distance). The light intensity of another resonance line of 1849 Å is heavily reduced by absorption in the lamp wall (fused silica). The reactant gas used was 10% disilane (Si_2H_6) diluted with 90% helium. Disilane has a higher probability of photochemical decomposition than that of monosilane.[27] The light source was kept just above the 17 cm diameter, 5 mm thick quartz window. The total flow rate is typically 200 sccm. Reactant Si_2H_6/He gases were premixed with a small amount of mercury vapor in a temperature-controlled mercury-vaporizer and introduced into the reactor. The deposition rate depended on the temperature of the mercury vaporizer and the gas pressure, as seen in Fig. 3.2.10. Substrate temperature is not

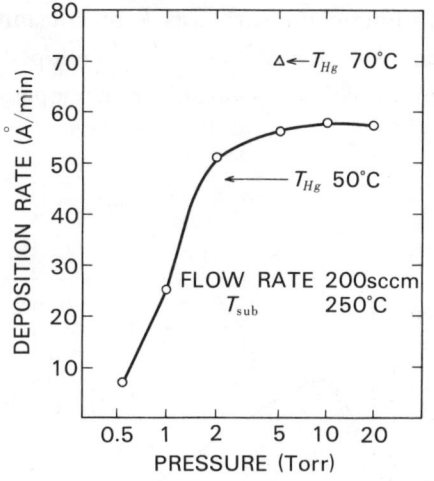 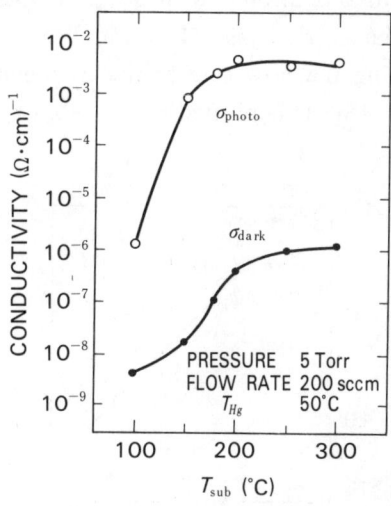

Fig. 3.2.10 Deposition rate of undoped photo-CVD films. [After T. Inoue et al.[26]]

Fig. 3.2.11 Photo- and dark-conductivities as a function of substrate temperature for undoped photo-CVD films. The photoconductivity was measured under the AM1, 100 mW/cm² insolation. [After T. Inoue et al.[26]]

an important factor in determining deposition rate. The deposition rate is mainly determined by the gas phase reaction, and limited by the interaction of excited mercury atoms and reactant gas in the low pressure regions and by the amount of excited mercury atoms in the high pressure regions. Variation of photo and dark conductivities with a substrate temperature for ~0.4 μm thick films is shown in Fig. 3.2.11. Dark conductivity is relatively high compared with that for GD films. Activation energy (0.69~0.89 eV) determined from temperature dependence of dark conductivity indicates that the Fermi level is near the center of the gap.

Direct photochemical vapor deposition

Presently, it is known that mercury incorporation into deposited thin-films appears to be negligible. Nevertheless, it is necessary to develop an appropriate way for achieving direct photochemical decomposition without employing mercury-photosensitization, because residual Hg contaminants in the reaction system should be eliminated. Sarkozy[23] claimed that the rates of direct photochemical decomposition of SiH_4 and Si_2H_6 are extremely low, because very little optical absorption by gas molecules occurs even in wavelength regions shorter than 1850 Å for SiH_4 and 2000 Å for Si_2H_6. In fact, no systematic work has so far been reported on silicon thin-film formation by direct photochemical reactions of SiH_4 or Si_2H_6. However, recently it was found that disilane gas can be decomposed at an appropriate reaction rate by direct UV light excitation from a Hg resonance lamp ($\lambda = 2537$ Å, 1849 Å), without employing mercury-photosensitized reaction.[10] In the direct photo-CVD, a horizontal quartz reactor with a substrate heated with external halogen lamps was employed, and a low pressure mercury-lamp (110 watts) as a UV radiation source (254 nm and very weak 185 nm) was placed just above the quartz tube. Disilane gas diluted with 99% or 95% He was admitted to the reactor together with a nitrogen carrier gas. The total flow rate at atmospheric pressure was kept constant by adjusting the flow rate of the nitrogen carrier gas. The deposition rate of undoped and doped a-Si : H is plotted in Fig. 3.2.12 as a function of the reciprocal substrate temperature.

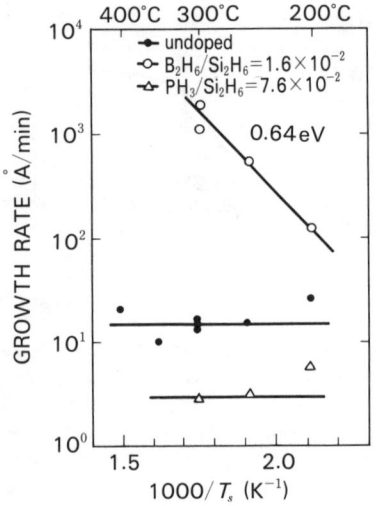
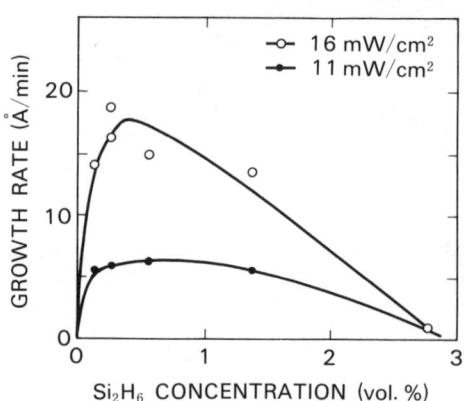

Fig. 3.2.12 Growth rate of Si_2H_6 photo-CVD films plotted against substrate temperature.

Fig. 3.2.13 Growth rate of undoped films against Si_2H_6 concentrations.

The direct photo-CVD used to prepare undoped and PH_3 doped films provides a constant growth rate below 350°C. In contrast to this, the deposition rate in B_2H_6 doping is remarkably increased and thermally activated with an energy of 0.64 eV. This activation energy is unchanged even in the thermal CVD although the growth rate is reduced to about one third the photo-CVD case. The growth rate of photo-CVD a-Si : H is significantly decreased by phosphorus doping and increased by boron doping. Very similar phenomena was observed in thermal Si_2H_6 CVD.[14] This is interpreted in terms of the catalyst poisoning effect by phosphine and of the catalytic effect of holes or boron atoms on the growing surface of silicon. The deposition of undoped films exhibits a characteristic photochemical process as shown in Fig. 3.2.13. Concentration of the photochemically decomposed species in the gas phase increases with an increase of Si_2H_6 concentration, resulting in enhancement of the growth rate. However, when the film-precursor concentration becomes too high, gas phase reactions to form inert molecules, such as Si_3H_8 etc. take place and hence effective flux density of precursors towards the substrate surface is lowered. For boron doped films, the photo-CVD process is likely to be determined by the thermal process although there exists enhancement of the reaction rate through UV excitation. Vibrational spectrum for a direct photo-CVD film is compared with that of thermal CVD silicon, as illustrated in Fig. 3.2.

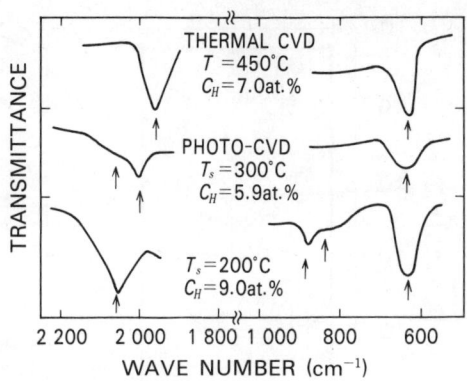

Fig. 3.2.14 Infrared absorption spectra of photo-CVD films and a thermal CVD film.

14. The photo-CVD film prepared at $T_s = 200°C$ exhibits the SiH_2 stretching mode at 2 060 cm^{-1} together with the corresponding bending modes at 840 cm^{-1} and 880 cm^{-1} and the wagging mode at 630 cm^{-1}. The vibrational spectrum of the specimen grown at 300°C is essencially identical to that of the thermal CVD film except for a little difference in the peak positions of stretching absorption. It is interesting to note that hydrogen content of the direct photo-CVD film formed at $T_s = 300°C$ is small compared to that of thermal CVD silicon obtained at $T_s = 450°C$. This apparent discrepancy of hydrogen content might be explained as follows; Pollock et al.[27] showed that an important primary step of the mercury-photosensitized decomposition of disilane is the cleavage of a single Si-H bond as:

$$Hg^*(^3P_1) + Si_2H_6 \rightarrow Hg(^1S_0) + H + Si_2H_5 \qquad (3.2.2)$$

As a final step, the SiH_2 diradical is created and a solid deposit is formed through polymerization of SiH_2. Hydrogen content of the deposit is in general, far less than 67 at.%, which is expected for $(SiH_2)_n$ solid. Therefore, hydrogen scavenging process from the solid surface should exist and it is likely that atomic hydrogen radicals created by reaction (3.2.2) attack SiH_2 diradicals adsorbed on the growing surface to eliminate hydrogen bonds through the following reaction:

$$SiH_2 + 2H \rightarrow Si + 2H_2 \ (\varDelta H = -2.37 \text{ eV}) \qquad (3.2.3)$$

This reaction is exothermic and may be very rapid. If the above reaction scheme holds even for direct photo-CVD, the low hydrogen content in the photo-CVD film compared to that of the thermal CVD sample could be attributed to the presence of atomic hydrogen in the photo-CVD system. Variation of dark- and photo-conductivity prior to and after light exposure at AM1 200 mW/cm^2 is seen in Fig. 3.2.15. There is very little change in dark-conductivity before and after illumination and photo-conductivity as high as 5.7×10^{-4} $\Omega^{-1} \cdot cm^{-1}$ for AM1 200 mW/cm^{-2} is hardly degraded during light exposure. Light soak degradation commonly observed in plasma-deposited a-Si:H[28] is absent also in thermal CVD films from Si_2H_6[8] and very little in HOMO-CVD films.[29] It is important to note that hydrogenated amorphous silicon produced by silent processes such as thermal- and photo-

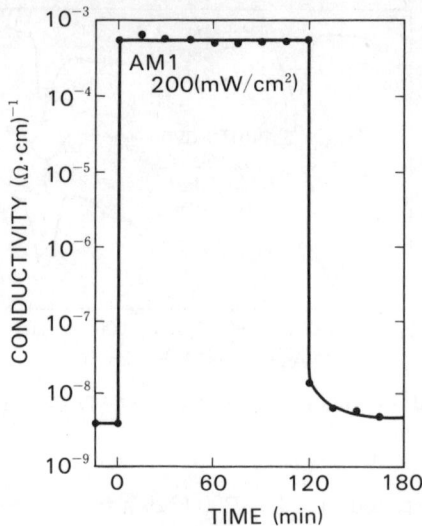

Fig. 3.2.15 Dark- and photo-conductivity variation prior to and after light exposure at AM1 200mW/cm^2. The sample was deposited at 300° C.

CVD or HOMO-CVD exhibit no significant Staebler-Wronski effect.

3.2.5 Summary

Preparation techniques using higher silanes and physical properties of resulting a-Si : H were described. The results can be summarized as follows :

(1) Deposition rate of a-Si : H is considerably increased by using Si_2H_6. Dimeric species such as Si_2H_y seem to play an important role in the deposition of a-Si : H films from Si_2H_6 glow discharge.

(2) Low temperature Si_2H_6 CVD can lead to improvement of material properties compared with SiH_4 CVD films while substitutional doping is possible.

(3) Photochemical deposition of a-Si : H either through mercury-photosensitized process or through direct photolysis of Si_2H_6 has been demonstrated. The electrical properties were found to be as good as SiH_4 glow discharge films.

References

1) D. E. Carlson and C. R. Wronski : Appl. Phys. Lett., *28* (1976) 671.
2) P. G. LeComber, W. E. Spear, and A. Ghaith : Electron. Lett., *15* (1979) 179.
3) B. A. Scott, M. H. Brodsky, D. C. Green, P. B. Kirby, R. M. Plinik, and E. E. Simonyi : Appl. Phys. Lett., *37* (1980) 727.
4) K. Ogawa, I. Shimizu, and E. Inoue : Jpn. J. Appl. Phys., *20* (1981) L 639.
5) A. Matsuda, T. Kaga, H. Tanaka, L. Malhotra, and K. Tanaka : Jpn. J. Appl. Phys., *22* (1983) L 115.
6) S. C. Gau, B. R. Weinberger, M. Akhtar, Z. Kiss, and A. G. MacDiarmid : Appl. Phys. Lett., *39* (1981) 436.
7) Y. Mishima, M. Hirose, and Y. Osaka : in Proceedings of the 7th International Conference on Vacuum Metallurgy (Tokyo, 1982), 10-5, 461.
8) V. L. Dalal : in Proceedings of the U. S.-Japan Joint Seminar on Technological Applications of Tetrahedral Amorphous

Solids (Palo Alto, 1982).
9) T. Saitoh, S. Muramatsu, S. Matsubara, and M. Migitaka: Jpn. J. Appl. Phys. Suppl. *22-1* (1982) 617.
10) Y. Mishima, M. Hirose, Y. Osaka, K. Nagamine, Y. Ashida, N. Kitagawa, and K. Isogaya: Jpn. J. Appl. Phys. *22* (1983) L 46
11) A. Matsuda, T. Kaga, H. Tanaka, and K. Tanaka: J. Non-Cryst. Solids, *59 & 60* (1983) 687.
12) T. Ohashi, J. Kenne, M. Konagai, and K. Takahashi: Appl. Phys. Lett., *42* (1983) 1028.
13) M. Hirose, M. Taniguchi, and Y. Osaka: Amorphous and Liquid Semiconductors, ed. W. E. Spear (CICL, University of Edinbrugh, 1977) 352.
14) Y. Ashida, Y. Mishima, M. Hirose and Y. Osaka: J. Appl. Phys., *55* (1984) 1425.
15) E. M. Maxted: Advances in catalysis (Academic, New York), III (1951) 129.
16) L. Pauling: The Nature of the Chemical Bond, 3rd ed. (Cornell University, Ithaca, 1960) 85.
17) Y. Yasuda, K. Hirabayashi, and T. Moriya: Jpn. Soc. Appl. Phys., *43* (1974) 400.
18) P. Rai-Choudhury and P. L. Hower: J. Electrochem. Soc., *120* (1973) 1761.
19) T. Nakashita, M. Hirose, and Y. Osaka: Jpn. J. Appl. Phys., *20* (1981) 471.
20) A. Matsuda, M. Matsumura, K. Nakagawa, T. Imura, H. Yamamoto, S. Yamasaki, H. Okushi, S. Iizima, and K. Tanaka: in Proceedings of the Topical Conference on Tetrahedrally Bonded Amorphous Semiconductors, AIP Conf. Proc., *73* (1981) 192.
21) C. C.Tsai: Phys. Rev., *B 19* (1979) 2041.
22) M. Hirose, M. Taniguchi, T. Nakashita, Y. Osaka, T. Suzuki, S. Hasegawa, and T. Shimizu: J. Non-Cryst. Solids, *35 & 36* (1980) 297.
23) R. F. Sarkozy: Technical Digest of 1981 Symposium on VLSI Technology, *68* (1981).
24) C. H. J. v. d. Brekel and P. J. Severin: J. Electrochem. Soc., *119* (1972) 372.
25) H. Ito, M. Hatanaka, K. Mizuguchi, K. Miyake, and H. Abe: Rroc. of 4th Int. Conf. on Solid State Devices (Tokyo, 1982), C-5-LN 5.
26) T. Inoue, M. Konagai, and K. Takahashi: Appl. Phys Lett., *43* (1983) 774.
27) T. L. Pollock, H. S. Sandhu, A. Jodhan, and O. P. Strausz: J. Am. Chem. Soc., *95* (1973) 1017.
28) D. L.Staebler and C. R. Wronski: Appl. Phys. Lett., *31* (1977) 292.
29) B. A. Scott, J. A. Reimer, R. M. Plecenik, E. E. Simonyi, and W. Reuter: Appl. Phys. Lett., *40* (1982) 973.

3.3 Microcrystalline Silicon (μc-Si) Prepared by Plasma-Chemical Techniques

Yukio OSAKA* and Takeshi IMURA*

Abstract

Preparation techniques of μc-Si, r. f. glow discharge of SiH_4 and reactive Si sputtering, are described with emphasis of the correlation of preparation conditions with grain and the film structures and properties. Effective medium theory is applied to explain the electrical properties of μc-Si film. Effects of doping B and P into a mixed-phase film composed of μc-Si and amorphous silicon are also given. Infrared absorption features are presented in detail for μc-Si film having high volume fraction of the crystallite, for which a structure model with hydrogen covalently bonding to the surface silicon is proposed. Finally, examples of μc-Si and amorphous silicon metamorphosis are shown.

3.3.1 Introduction

Amorphous-microcrystalline mixed-phase hydrogenated silicon (μc-Si) has received increasing attention due to its various potential applications such as solar cells.[1] The microcrystal was first prepared by chemical transport of silicon in a hydrogen plasma.[2] Since this initial work, several groups have reported preparing μc-Si by glow discharge of SiH_4[2~7] and reactive sputtering techniques.[8] The μc-Si film prepared by the latter technique is composed mostly of hydrogenated, fine Si crystallites and a small amorphous region.

To control μc-Si film properties, the preparation condition for μc-Si formation and its correlation with the μc-Si film structure must be defined. For further improvement of μc-Si characteristics, a suitable structural model is needed that explains the physical properties of μc-Si.

This paper presents recent progress in the investigation of μc-Si focussing on μc-Si prepared by glow discharge decomposition of SiH_4 in H_2 gas and reactive sputtering techniques. For μc-Si prepared by glow discharge plasma of SiH_4, preparation conditions for μc-Si formation, correlation between the preparation conditions and film structure, the electrical and optical properties of μc-Si (structural modeling), and the doping effects on μc-Si are described. Characterization of μc-Si prepared by reactive sputtering is presented, including the morphology, crystalline nature, hydrogen content, and infrared (IR) absorption features. A morphological model is proposed for the μc-Si film. Finally, examples of μc-Si and amorphous hydrogenated silicon (a-Si : H) transformation induced by adding

---

* Faculty of Engineering, Hiroshima University, Higashihiroshima 724.

nitrogen or by growth in film thickness are presented.

3.3.2 μc-Si Prepared by Glow Discharge Plasma of SiH₄

Preparation conditions for μc-Si formation

Veprek et al.[9] suggested that microcrystallite nucleation is formed under conditions whereby a quasi-chemical equilibrium is established at the plasma-substrate interface. Empirical conditions for μc-Si formation are strong dilution of SiH₄ by H₂ or inert gas and a somewhat higher r. f. power. Under these conditions, a large number of reactive hydrogen atoms exist in the silane plasma. Optical emission spectroscopy (OES) is useful in studying the behaviour of reactive species in a plasma during a-Si : H film growth.[10~12] Using OES, Mishima et al.[13] have found that a rapid decrease in the intensity ratio of SiH (414nm) to H (656nm), [SiH]/[H], occurs when microcrystallite nucleation takes place (Fig. 3.3.1).

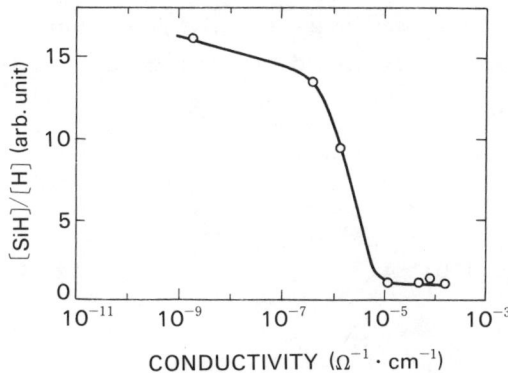

Fig. 3.3.1 The relative intensity ratio SiH (414nm) to H (656nm), [SiH]/[H] plotted against conductivity. Microcrystallization is initiated at $4.05 \times 10^{-7} \Omega^{-1} \cdot cm^{-1}$. [after M. Mishima et al.[13]]

Matsuda[14] has found that nucleation occurs when the ratio of [H] intensity to film growth rate is over a critical value. In light of Veprek's suggestion, a large number of reactive hydrogen atoms lower film growth rate, move the plasma-substrate interaction towards a quasi-equilibrium, and help to reconstruct the amorphous network.

Characterization of the film structure of μc-Si

Average grain size δ of microcrystallite and volume fraction f of microcrystalline phase are convenient quantities to characterize μc-Si film structure. X-ray diffraction peaks corresponding to the {111}, {220} and {311} planes have been observed in μc-Si. Usually, (111) diffraction peak half-width provides an average grain size from Scherrer's formulae (Azaroff[15]). Annealing effects on the X-ray-diffraction pattern give the microcrystallite volume fraction f.[16] Mishima[13] et al., with the aid of the effective medium theory, have shown that the Raman intensity ratio of 520cm⁻¹ mode to 480cm⁻¹ and the specimen conductivity determine volume fraction f.[17] Validity in the use of the effective medium theory will be discussed in "Electrical and optical properties (structural modeling)". Tsu et al.[18]) have proposed f determination method from Raman scattering only.

Shimada et al.[19]) have shown that the Raman peak at 150cm^{-1}, due to the TA-phonon-like mode, is useful to characterize the structural μc-Si properties.

Correlation between preparation conditions and film structure

According to Vepřek's interpretation on μc-Si growth it may be expected that there is a close correlation between the film growth rate and film structure. We have found the partial pressure of silane determines the growth rate R and that the following relation holds

$$R \propto [SiH_4]/([SiH_4] + [H_2]) \qquad (3.3.1)$$

Total flow rate of the reactant gas, total pressure, r.f. power (13.56 MHz) and substrate temperature were held constant at values of 300 sccm, 0.18 Torr, 60 W and 250°C, respectively. Deposition of μc-Si from glow discharge of $SiH_4 + H_2$ gas mixture was carried out in a conventional plasma CVD apparatus. Decrease in the growth rate results in simultaneous increases of both average grain size δ and volume fraction f in the microcrystalline phase (Fig. 3.3.2). As evidenced by small angle X-ray scattering measurements,

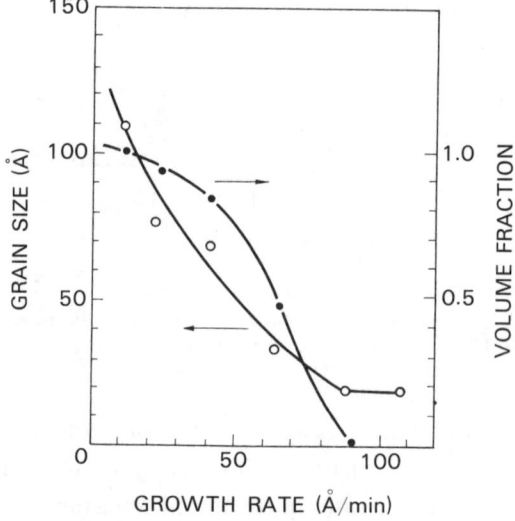

Fig. 3.3.2　An average grain size and a volume fraction of microcrystallite as a function of the growth rate of films.

this simultaneous increase of grain size δ and volume fraction f is accompanied with a reduction in structural inhomogeneities of morphological fluctuations.[20]) These facts could be explained by assuming that when the growth rate slows, discrepancy from the quasi-chemical equilibrium in the deposition system becomes smaller.

Matsuda[14]) has studied effects of ions in plasma on the growing surface of μc-Si (Fig. 3.3.3). Grain size δ and relative lattice expansion ($\Delta a/a_0$) of microcrystallite (a_0 is the lattice constant of c-Si) are strongly affected by hydrogen ions, which were measured by mass spectroscopy through a fine orifice. Matsuda et al.[21]) employed the triode configuration in glow discharge system to vary the amounts of ion species impinging on the surface. By

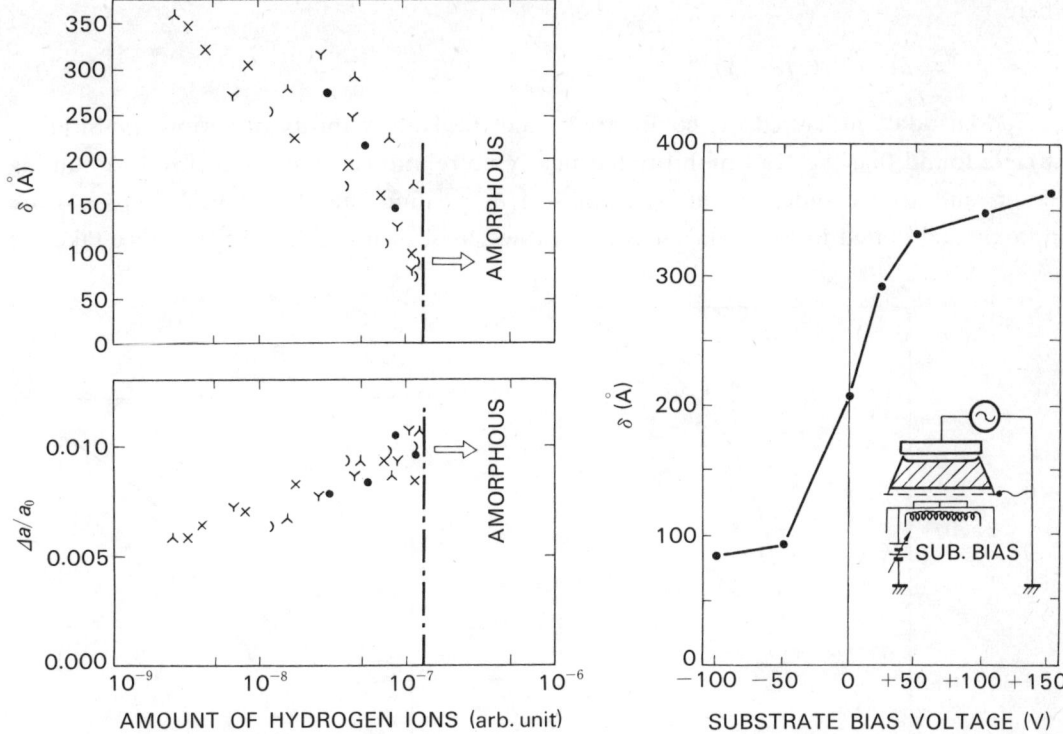

Fig. 3.3.3 An average grain size δ and a relative lattice expansion ($\Delta a/a_0$) plotted against the amount of hydrogen ions impinging on the growing surface. [after A. Matsuda[14)]]

Fig.3.3.4 The crystallite size control of μc-Si using the triode glow-discharge system. [after A. Matsuda[14)]]

changing the substrate bias under a fixed plasma condition, they have succeeded in controlling μc-Si grain size δ (Fig. 3.3.4).

Electrical and optical properties (structural modeling)

Based on the model proposed for explaining the electrical properties of polycrystalline Si,[22)] Spear et al.[23)] and LeComber et al.[24)] have analyzed the μc-Si specimen Hall mobility μ_H to be in the range of f exceeding 0.8. Their analysis holds μc-Si film electrical properties with $f \simeq 1$ prepared by a chemical transport method.[2)] This method is equivalent to the reactive sputtering technique.[8)] In this section, we shall discuss μc-Si electrical properties in the range of 0.15~0.8 in f. At first, Mishima et al.[25)] analysed measured μc-Si conductivity in terms of the effective medium theory on a random system in which the microcrystalline phase with the volume fraction f possesses a high conductivity σ_0 and a high mobility μ_0 with the remaining (amorphous) phase being of a low conductivity σ_1 and a low mobility μ_1. Under conditions where $\sigma_0 \gg \sigma_1$ and $\mu_0 \gg \mu_1$, this theory predicts conductivity σ and the Hall mobility μ_H of the material to be as follows; [17,26)]

$$\sigma = 0 \text{ and } \mu_H = 0 \qquad \text{for } f < 1/3 \qquad (3.3.2)$$
$$\sigma = \sigma_0 a(f) \qquad \text{for } f > 1/3 \qquad (3.3.3)$$
$$\mu = 4\mu_0 f a(f)/\{4a^2(f) + (1-f)(1+4a(f))\} \qquad \text{for } f > 1/3 \qquad (3.3.4)$$

where

$$a(f) = (3f - 1)/2 \tag{3.3.5}$$

Matsuda[14] measured the conductivity and the Hall mobility of various μc-Si ($f < 0.8$). He found that the Hall mobility has nearly no relationship to grain size δ. Assuming that σ_0 and μ_0 are independent of grain size δ, we find that Eqs. (3.3.2) ~ (3.3.5) are approximately fitted to Matsuda's masured values, as shown in Fig.3.3.5. Here we take the

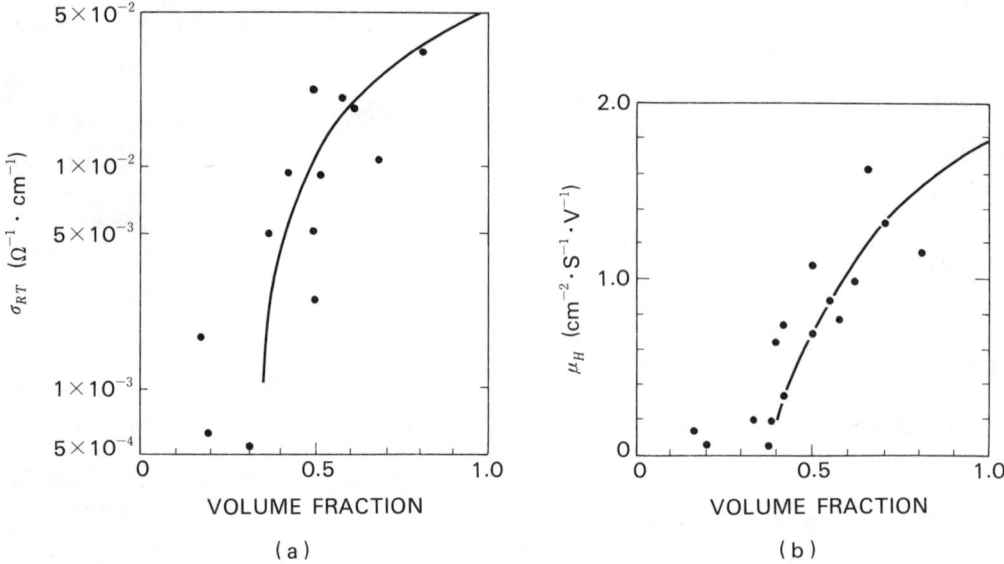

Fig. 3.3.5 Fitting of the measured values on dark conductivity σ_{RT} and Hall mobility μ_H (at a room temperature) of μc-Si[14] to the effective medium theory (solid lines).

values $\sigma_0 = 5 \times 10^{-2}$ (S·cm^{-1}) and $\mu_0 = 1.8$ (cm^2·V^{-1}·s^{-1}). Uchida et al.[1] have shown that the conductivity of phosphorus doped μc-Si as a function of f is in fair agreement with the effective medium theory. This applicability of the effective medium theory suggests the inadequacy of such a structural model where the internal surface of the microcrystalline phase is mainly covered with the amorphous phase and that the μc-Si network contains a smaller part of region composed of the microcrystalline phase. In this model, electrical transport would be dominated by conduction in the amorphous phase.

Data of IR absorption associated with Si-H stretching vibration shows that bonded hydrogen is distributed mainly in the amorphous phase and bonding is in the form of SiH$_2$.[5,13,43] A photoluminescence (PL) study of μc-Si[13,26] shows that PL is mainly contributed by the amorphous phase. Detailed analysis of the PL carried out by Hata et al.[27] leads to the fact that μc-Si consists of three different phases: microcrystalline phase, PL-emitting amorphous phase and non-PL amorphous phase. They suggests that the non-PL phase constitutes a boundary between the microcrystalline phase and the PL-emitting phase. Scanning electron microscopy studies on μc-Si suggest the presence of a columnar structure which is mutually connected with tissue containing SiH$_2$ bond.[28] These discussions seem to give support to a structural model on μc-Si proposed by Mishima et al.[13] This model is schematically shown in Fig. 3.3.6.

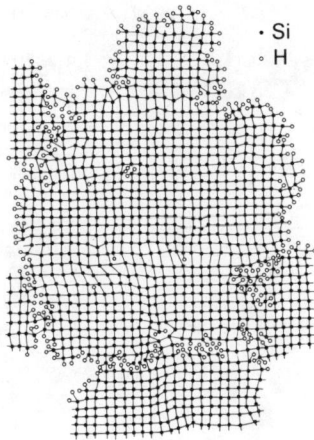

Fig. 3.3.6 A schematic model for the structure of μc-Si films. [after Y. Mishima et al.[13]]

Doping effects on μc-Si

Hamasaki et al.[29] have found preferential segregation of boron atoms into the amorphous phase of heavily B-doped μc-Si. In this section, we shall discuss this interesting effect. Boron doped μc-Si films were prepared by r. f. glow discharge deposition of a SiH_4/B_2H_6/H_2 gas mixture in a capacitively coupled reactor. Deposition conditions are summarized in Table 3.3.1. A low concentration of SiH_4 and a relatively high r. f. power favor the

Table 3.3.1 Deposition conditions.

Parameter	Range
Substrate Temperature	300°C
Electrode Spacing	4 cm
Diameter of Electrodes	12 cm
Gas Pressure	0.5 Torr
r. f. Power	50~70 W
Flow Rate	
SiH_4 (10% in H_2)	15 SCCM
B_2H_6 (1020 ppm in H_2)	40 SCCM
Doping Ratio (B_2H_6/SiH_4)	2.5%

formation of B-doped μc-Si. Characteristic values of B-doped μc-Si are summarized in Table 3.3.2. Real boron density in μc-Si and effective hole mobility were determined using SIMS and free carrier optical absorption measurements, respectively. Hole mobility shown in Table 3.3.2 is calculated by using light hole mass in crystalline Si or the free electron mass.

Segregation effect of boron atoms is studied by the impurity-induced IR absorption

Table 3.3.2 Characteristic values for specimens.

Conductivity (at 300K) (S·cm^{-1})	Activation Energy (meV)	Volume Fraction f	Grain Size (Å)	B Content by SIMS (cm^{-3})	Hole Mobility (cm^2/V·sec)	
					light hole	$m^* = m_0$
8.7	8.5	0.49	75	1.4×10^{21}	1.52	0.24
4.11	32.2	0.45	73	1.1×10^{21}	3.65	0.58
0.11	67.7	0.34	49	7.5×10^{20}	1.20	0.19

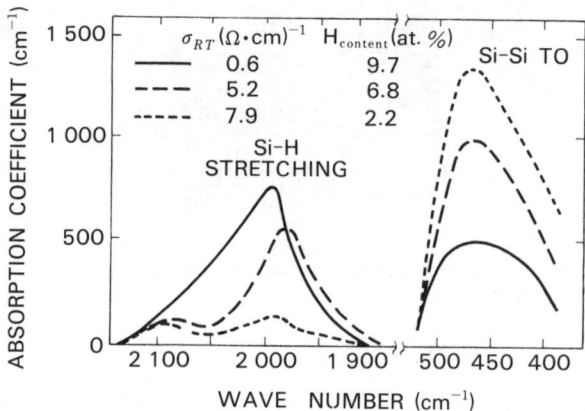

Fig. 3.3.7 IR absorption spectra of B-doped μc-Si. [after T. Hamasaki et al.[29]]

of the TO-lattice vibration mode. IR absorption of B-doped μc-Si is shown in Fig. 3.3.7. Impurity-induced lattice absorption near 480 cm^{-1} is very similar to that of B-doped amorphous Si:H and there is no such absorption in highly B-doped crystalline Si.[30,31] By assuming that most boron atoms in μc-Si are incorporated in the amorphous phase and very little in the microcrystalline phase, the boron concentration could be determined from the impurity-induced TO absorption, using the volume fraction obtained by Raman measurements. The obtained average boron concentration is in agreement with the result of SIMS measurements. This suggest that the boron atoms in μc-Si tend to segregate into the amorphous phase.

The conductivity σ of B-doped μc-Si is described as

$$\sigma = \sigma_0 \exp(-\Delta E/kT) \tag{3.3.6}$$

and a slight decrease in the activation energy ΔE is caused by a considerable increase in the average boron concentration. This gives support to the fact that electrical conduction in B-doped μc-Si is dominated by percolation transport through the microcrystalline phase. Spear et al.[23] have shown that most phosphorous sites act as donar in P-doped μc-Si and the conduction activation energy ΔE in P-doped μc-Si decreases from 75 meV to 6 meV corresponding with an increase of phosphorus density from 2×10^{17} to 2×10^{19} cm^{-3}. It must be noted that a similar change in the ΔE of B-doped μc-Si occurs through variation in boron density from 7.5×10^{20} to 1.4×10^{21} cm^{-3}. This pronounced difference in doping efficiency between B-doped and P-doped μc-Si may be due to boron atoms in B-doped μc-

Si segregating into the amorphous phase and phosphorus atoms in P-doped µc-Si are unlikely to result in such segregation. It may be speculated that defect creation in B-doped a-Si : H may be related with boron atoms segregation in a-Si : H two-phase structure.[32,33]

3.3.3 Reactive Sputtering µc-Si

Preparation conditions

The µc-Si film (or powder) is prepared in the usual sputtering apparatus mounted with Si target in a hydrogen atmosphere of 10^{-2} to about 3 Torr. Growth rate increases linearly with the logarithm of hydrogen atmospheric pressure. However, higher gas pressure conditions of, for example, about 1 Torr favor powder formation rather than µc-Si film deposition. Existence of a large amount of reactive hydrogen atoms look essential for the formation of µc-Si, as we have already noted in "Preparation conditions for µc-Si formation". Reactive sputtering in hydrogen gas diluted with a minor amount of other rare gases also produces µc-Si.

An example of growth rate versus input r. f. power is shown in Fig.3.3.8. Magnetron sputtering is effective in raising the growth rate. Higher r. f. power also tends to form µc-Si powder rather than µc-Si film.

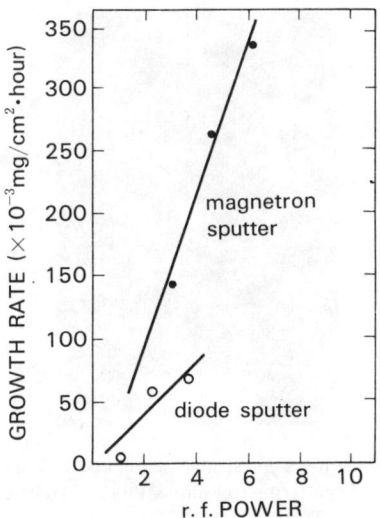

Fig. 3.3.8 Input r. f. power dependence on growth rate of µc-Si by reactive sputtering at a hydrogen pressure about 0.1 Torr and water cooled substrate. [after M. Haba et al.[37)]]

Substrate temperatures to get typical µc-Si film are usually 200°C~300°C. Samples prepared on a cooled substrate with liquid nitrogen (or even with water) sometimes show peculiar structural features especially in IR spectra and morphology,[34,35] as will be mentioned in "Crystalline nature and morphology" and "IR absorption due to Si-H bond vibrations".

This technique of using the reactive sputtering procedure in hydrogen seems essentially the same as the method developed by Vepřek et al.[2] Characteristic fine structures in

IR absorption spectra (see "IR absorption due to Si-H bond vibrations" in both specimens are similar to each other, in contrast with those in the μc-Si film deposited by glow discharge of SiH_4. The sputtering procedure in hydrogen plasma is not merely a physical process (or mass effect of the bombarded ion), but involves chemical Si target etching with reactive hydrogen atoms and chemical transport of some species through hydrogen plasma onto the substrate. Thus, in spite of a heavier atomic weight (4) of He rather than the atomic or molecular weight (1 or 2) of H or H_2, deposition rate in He has been far less than in H_2.[8]

Microcrystalline films of Ge have also been prepared by the same technique.[36] Grain size of Ge has been larger than that of Si. This may be ascribed to the weaker interaction of H with Ge than Si to from a covalent bond on the grain surface. This bond formation strongly suggests a role of surface hydrogen to stabilize microcrystal grain.

Crystalline nature and morphology

The X-ray diffraction line width for the {111} plane gives average dimensions of 10 ~20 nm for μc-Si prepared under various conditions. Distribution of diameters roughly from 5 to 50 nm, however, is observed in the transmission electron micrograph (TEM) of a film deposited at 250°C, 3.6 W/cm² and 0.5 Torr H_2 on KBr. Black masses of irregular shapes in Fig. 3.3.9 are grains of μc-Si, which seem to have some other internal structures. The

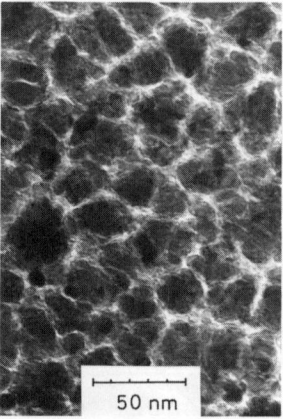

Fig. 3.3.9 TEM image for a μc-Si film of thickness about 0.5 μm deposited by reactive sputtering technique. Power : 3.6 W/cm², T_s : 250°C, H_2 Pressure : 0.5 Torr.

diffraction pattern for this μc-Si film is composed of sharp rings, showing the existence of fine particles of μc-Si. On the other hand, white contrasted regions may be those of smaller thickness. On the basis of these observations, we believe that the film has columnar structure, in which the aggregate of μc-Si grains grows normally to the film surface. Surface and sectional morphology revealed by SEM also supports μc-Si grain columnar stacking to develop normally to the film surface.

With preparation conditions of higher r. f. power, higher hydrogen pressure, and lower substrate temperature, spherical μc-Si grains which have diameters of a few tens of nm are sometimes obtained. These are usually powdery in appearance. An example of a

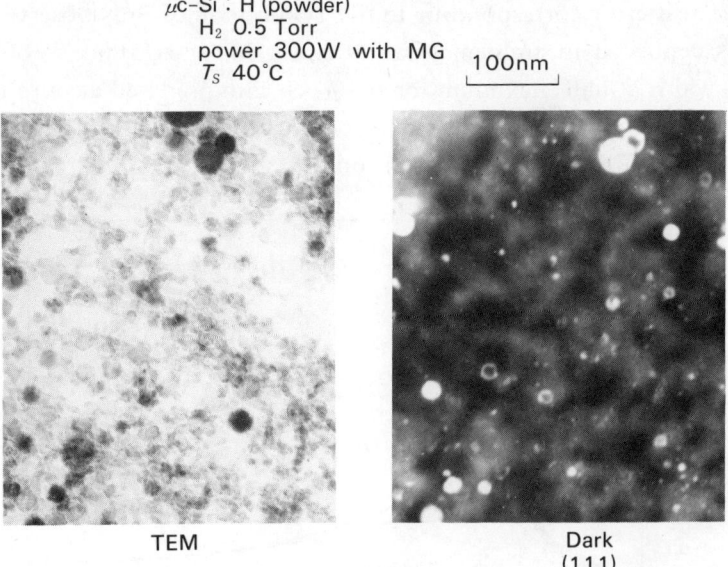

Fig. 3.3.10 Bright and dark field TEM images (at nearly a same field) of μc-Si powder fabricated by magnetron (MG) type reactive sputtering.

spherical grain is shown in Fig. 3.3.10. The dark field image on the right hand side, taken with a (111) diffracted beam, clearly shows that some of these spherical grains are composed of single crystals with diameters of a few tens of nm. Roughly spherical grains aggregation has also been displayed by SEM observation.[37]

A high resolution electron micrograph has been taken of a μc-Si film about 0.1 μm thick, prepared by magnetron sputtering technique at a hydrogen pressure of 10^{-2} Torr. Lattice images arising from the {111} plane are observed throughout the film with a spacing of about 0.3 nm in Fig. 3.3.11, showing that many {111} planes are normal to the film surface.

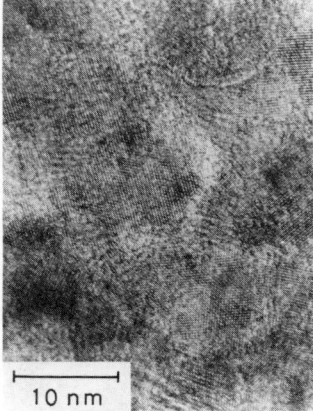

Fig. 3.3.11 Lattice image (111) for μc-Si grains prepared by magnetron reactive sputtering technique. Film thickness 0.1 μm. (kindly photographed at Japan Electron Optics Laboratory Co., Ltd., with JEOL 200 CX, direct magnification : × 220000)

Raman scattering corresponding to the TO phonon of Si is observed at 518.5 cm^{-1} for μc-Si films deposited in ambient gases of 0, 10, 30, and 75 mol. % of H$_2$ in He (Fig. 3.3.12). A full width at half maximum for the μc-Si film prepared at pure H$_2$ ($x = 100\%$) has been 9.6 cm^{-1}, in comparison with 4.5 cm^{-1} at 521 cm^{-1} for a crystal Si wafer measured under the same Raman measurement conditions.[8]

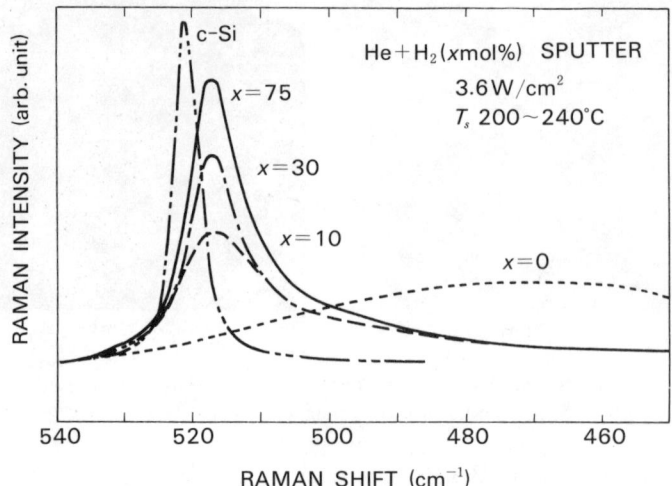

Fig. 3.3.12 Raman spectra of μc-Si films fabricated with $x = 0, 10, 30,$ and 75 mol. % of H$_2$ in He. The scattering intensity from amorphous film deposited in pure He ($x = 0$) is enlarged. The spectrum for TO phonon band in a single crystal of silicon is also shown. [after T. Imura et al.[8]]

The optical absorption spectrum of μc-Si deposited by the reactive sputtering technique in pure hydrogen is almost the same as that of crystalline Si at the observed photon energy range of 1.6 eV to 2.6 eV, but different from that of a-Si : H by roughly an order of magnitude in the absorption coefficient unit (cm^{-1}) at the higher photon energy range.

Very weak photoluminescence of μc-Si film deposited at 3.6 W/cm^2 and 240°C has been observed at 4.2 K upon Kr laser excitation (531 nm, 70 mW), having a broad peak at 0.96 eV. This differs much from those reported in μc-Si and a-Si by glow discharge of SiH$_4$,[27] but seems to correlate with the recombination luminescence in irradiated crystal Si.[38]

Elemental composition and impurities

Silicon and hydrogen are major elemental components in μc-Si. Si atomic density in a μc-Si film deposited at r. f. power of 3.6 W/cm^2, substrate temperature 250°C, and hydrogen gas pressure of about 0.5 Torr has been determined to be about 70% of the value in crystal Si, by means of Rutherford backscattering spectrometry. Relatively lower Si density is due to the film columnar structure, as observed by TEM (see Fig. 3.3.9) and SEM.

Gas chromatography shows the presence of about 30 at.% of H in μc-Si samples deposited under various conditions.[35] A nuclear elastic scattering technique has also been utilized to give a hydrogen content of 1×10^{22} H/cm^3 in a μc-Si film.[39] Some samples evolve hydrogen on heating with evolution peaks at 300°C and 450°C; these correspond to the decomposition of SiH$_3$ groups and SiH$_2$ groups, respectively, from IR absorption changes by annealing.[37]

A small amount (1~2at.%) of carbon has been detected as an impurity on the surface of the μc-Si film by means of electron probe microanalysis. Oxygen in form of Si-O-Si configuration is also sometimes perceived slightly in the μc-Si film IR spectra.

IR absorption due to Si-H bond vibrations

Several separated peaks are observed in the IR absorption bands due to Si-H bond stretching vibration in μc-Si prepared by reactive sputtering and chemical transport techniques in hydrogen plasma. Appearance of these well-separated fine structures is a striking feature that has hardly ever been observed in μc-Si, deposited by glow discharge decomposition of SiH$_4$. In addition, no absorption near 1990cm^{-1} assigned to the SiH monohydride configuration is present. This restricts the possible configuration of SiH within the combination of only SiH$_2$ and SiH$_3$, to which these peaks have been assigned.[40,41] Assignment is in principle based on the fact that the absorption peaks vary with electronegativity of substituted atoms connected to the SiH$_2$ or SiH$_3$ group, developed by Lucovsky.[42] This simple method, however, can neither distinguish (SiH$_2$) from -SiH$_2$- structure etc., nor can explain the appearance of no fewer than six peaks observed in μc-Si prepared under some condi-

Table 3.3.3 Stretching vibration frequencies (ν_{Si-H}) calculated and observed in μc-Si.

Si-H configuration*1)		sum of SR*2)	ν_{Si-H} (cm^{-1})	
			calculated	observed
=SiH$_2$	a.	5.24	2089	2089
	b.	5.67	2100	2104
	c.	6.10	2111	2118
	d.	5.91	2106	2108
	e.	6.34	2117	—
—SiH$_3$	f.	2.62	2145	2140
	g.	3.05	2155	2158

*1) [diagrams of Si-H configurations a–g; ○ Si, ● H]

*2) $X^{1/2} = 0.21$ SR $+ 0.77$, where X is Pauling's electronegativity. [after A. Hiraki et al.[40]]

tions. Therefore, influence due up to the third nearest neighbor atoms from the H atom should be considered. As a matter of convenience, it has been done by taking the geometrical average of stability-ratio (SR) electronegativities (given in note ∗2 in Table 3.3.3) of the second and third nearest neighbor atoms. Oscillation frequencies for calculated values are shown in Table 3.3.3, along with revised values measured by means of a high resolution prism spectrometer.

Instead of Lucovsky's three equations,[42] a unified equation has been proposed to give Si-H stretching frequencies in Si compounds and proton implanted Si crystals, taking a different value of SR electronegativity for H atom from the usual value.[43,44] They have also discussed the correlation between Si-H bond vibration frequencies and hydrogen-related defects in the proton implanted Si crystal.

The μc-Si sample with strong IR absorption bands due to the-SiH_3 groups has been prepared on the lower temperature substrate.[34] By cooling down the substrate to about $-60°C$ the content of the-SiH_3 group increases extremely. In this sample the μc-Si grain shows a uniform polyhedron or spherical shape with a uniform diameter of 10nm. IR absorption spectra in the Si-H stretching region are shown in Fig. 3.3.13 along with the effect of annealing on the spectra and those of other reference samples. A peak near 2150 cm^{-1} is responsible for-SiH_3 groups, and peaks near 2100 cm^{-1} for =SiH_2 groups. Peaks near 2260 and 2200 cm^{-1} denote oxygenation.

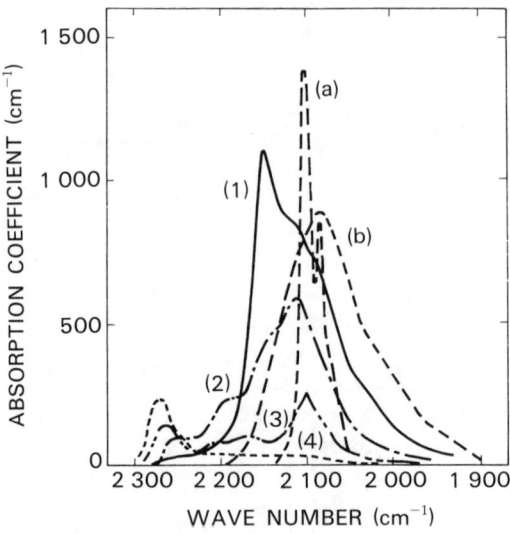

Fig. 3.3.13 IR absorption spectra of various μc-Si films at Si-H stretching region. (a) T_S: 250°C, (b) T_S: 50°C, (1) T_S: $-60°C$, (2)~(4): successive annealing of (1) at 300, 400, and 500°C, respectively, in a vacuum of 1×10^{-6} Torr. [after T. Miyasato et al.[34]]

A model for the μc-Si grain

The lattice image implies that hydrogen does not exist inside the μc-Si grain, but may instead be on the grain surface or among the columns (Fig. 3.3.11). The crystalline phase of Si can never contain $10^{22}H/cm^3$ of hydrogen by nature. This is quite different from a-Si:H where hydrogen is distributed inside the amorphous network at such high concentrations as 10^{22} H/cm^3.

Roughly spherical μc-Si grain models with diameters of about 0.5 nm and 1.5 nm have been made out of plastic balls. The surface of these microcrystrallite models has been covered with covalently bonded SiH$_2$ and SiH$_3$ structure units so exactly that no dangling bond remains on the surface. Following this Si and H balls have been counted. The number of both these atoms in the sphere of 10 nm diameter, on the other hand, has been computed from the sphere volume and surface area, where Si density is 5.0×10^{22} Si/cm^3 and an average area density of H is assumed to be 2×10^{15} H/cm^2. The result upon counting the plastic models, shown in Table 3.3.4, agrees closely with that from elemental analyses.

Table 3.3.4 Numbers of Si and H atoms in spherical models of Si crystal covered with SiH$_2$ and SiH$_3$ groups.

diameter (nm)	number of atom		H/(Si + H)		
	Si	H	calculated	elemental analyses[35]	IR[35]
0.5	14	24	0.63	0.65	0.026
1.5	74	84	0.53	—	—
10	2.6×10^4	6.3×10^3	0.20	about 0.3	$\lesssim 0.05$

In these models the presence of non-bonded hydrogen is not postulated. NMR measurements for hydrogen in μc-Si have also shown the presence of a large amount of non-bonded hydrogen such as hydrogen molecules of atomic hydrogens to be unlikely.[45] Hydrogen contents obtained from IR absorption intensity, however, compared to the calculated results and those from elemental analyses, are extraordinarily small. This is probably due to the decrease in the absorption cross-section of Si-H stretching oscillator, located on the μc-Si surface.

On this model no chain structure such as $(SiH_2)_2$ and $-SiH_2SiH_3$ is attached, although some of these structures are experimentally observed in IR spectra of the μc-Si film having grain size of about 10 nm. Therefore, the actual sample may contain more hydrogen atoms than those counted in the model. In actual μc-Si film, grain boundaries may be connected by these chain structures.

One or two atom layers on this model surface must be different from the regular diamond structure in order to end and cover the surface with $=SiH_2$ and $-SiH_3$ structures. The surface structure has been easily deformed here by introducing amorphous structure or wurtzite structure to stabilize the strain on the surface.

Metamorphism between μc-Si and a-Si : H

Structure change from a-Si : H to μc-Si with 10 to 100 nm film thickness has been observed by TEM and IR absorption.[46] When the film is thin enough ($\lesssim$ 30 nm), IR spectra do show the presence of a SiH monohydride configuration (Fig 3.3.14). At the same time TEM indicates the film is amorphous with uniform morphology rather than microcrystalline. With increments of the film thickness, the SiH monohydride configuration gradually disappears in the IR spectrum (Fig 3.3.14). Also the film morphological structure changes with the thickness from uniform amorphous to a non-uniform μc-Si columnar structure. μc-Si grain stacking is clarified to be mainly of (110) orientation parallel to the film surface, by RHEED pattern analysis. The monohydride configuration of SiH looks necessary to stabilize the uniform amorphous network of Si. Since the tetrahedral structure

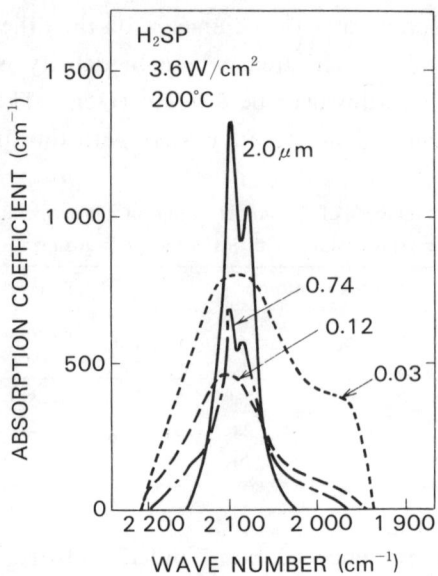

Fig. 3.3.14 Film thickness dependence on the IR absorption spectrum in Si-H stretching region of μc-Si films prepared by the reactive sputtering technique. [after T. Imura et al.[46)]]

of Si is very rigid, construction of the random network induces a high strain. Some connecting bonds may be broken to attach hydrogen atom and result in the SiH monohydride configuration. This reduction of strain by SiH, however, becomes insufficient when uniform amorphous film thickness exceeds the order of 10 nm under preparation conditions where μc-Si would form in thick film. Consequently, with the increments of the film thickness the structure gradually changes to a less strained one and finally into μc-Si.

The same structure change with film thickness has also been observed in the μc-Si film prepared by the glow discharge deposition from SiH_4.[46)]

Reverse transformation of μc-Si to the amorphous state is caused by gradual addition of 0 to 5 mol% of N_2 into H_2 atmosphere by magnetron reactive sputtering.[47)] Nitrogen in the Si network is a strain-relieving element with a lower coordination number and higher electronegativity than Si. The phase transition from μc-Si to a-Si:H containing nitrogen has been clearly observed as a function of the ratio of N/Si in the film by TEM. Characteristic μc-Si columnar morphology has become uniform at N/Si = 0.5. The change in IR spectra with the N/Si ratio in the film is shown in Fig. 3.3.15. Here, the N/Si ratio is determined from XPS signal intensity (areal) standardized with stoichiometric Si_3N_4. The SiH monohydride configuration appears in slightly nitrogenized films. Broad absorption bands are also observed in these films with an increase in N/Si ratio.

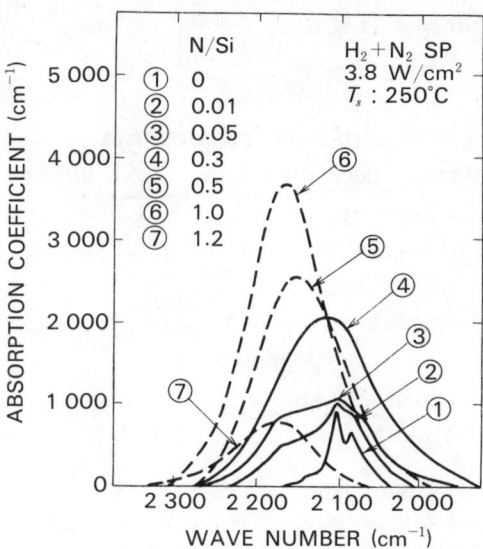

Fig. 3.3.15 Change in IR absorption spectra in Si-H stretching region induced by addition of nitrogen into μc-Si films deposited by magnetron reactive sputtering. [after A. Hiraki et al.[47]]

3.3.4 Conclusions

Findings for μc-Si prepared by a glow discharge are summarized as follows:

(1) The key condition for μc-Si formation is the amount of reactive hydrogen atoms in the silane plasma.

(2) There is a close correlation between the growth rate and μc-Si film structure. The ionic species reaching the substrate deteriorate the film structure.

(3) The electrical transport in μc-Si is reasonably explained by the effective medium theory.

(4) μc-Si electrical and optical properties give support to a structural model on μc-Si proposed by Mishima et al.[13]

(5) The pronounced difference in doping efficiency between B-doped μc-Si and P-doped μc-Si is related to boron atom segregation into the amorphous phase on B-doped μc-Si.

The above mentioned results (1) and (2) may apply also to μc-Si prepared by the reactive sputtering technique. Characteristic results of μc-Si by this technique are summarized as follows:

(6) A large amount of hydrogen is contained in μc-Si.

(7) Fine structures in IR absorption at Si-H stretching are assigned to structure units composed of SiH_2, SiH_3, etc.

(8) Morphological models have been proposed from TEM and compositional data for μc-Si film and grain, where the surface of the grain is covalently covered with $=SiH_2$ and SiH_3 groups, and the film is composed of these grains connected weakly to each other and fairly oriented in columnar stacks.

(9) The film to become microcrystalline is amorphous when it is very thin, or when a

small amount of nitrogen is added.

Acknowledgements

The photoluminescence spectrum of the μc-Si film by reactive sputtering was measured at the Institute for Solid State Physics, The University of Tokyo, in a cooperative project with Prof. Kazuo Morigaki, Prof. Akio Hiraki, and Dr. Izumi Hirabayashi.

References

1) Y. Uchida, T. Ichimura, M. Ueno, and M. Ohsawa: J. Phys., *42* (1981) Suppl. C4-265.
2) S. Veprek and V. Mareček: Solid State Electronics, *11* (1968) 683; Z. Iqbal, A. P. Webb, and S. Veprek: Appl. Phys. Lett., *36* (1968) 163.
3) S. Usui and M. Kikuchi: J. Non-Cryst. Solids, *34* (1979) 1.
4) T. Hamasaki, H. Kurata, M. Hirose, and Y. Osaka: Appl. Phys. Lett., *37* (1980) 1084.
5) A. Matsuda, S. Yamasaki, K. Nakagawa, H. Okushi, S. Iizima, M. Matsumura, and H. Yamamoto: Jpn. J. Appl. Phys., *19* (1980) L305.
6) T. Hamasaki, H. Kurata, M. Hirose, and Y. Osaka: Jpn. J. Appl. Phys., *20* (1981) L84.
7) A. Matsuda, M. Matsumura, H. Yamamoto, T. Imura, S. Yamasaki, H. Okushi, S. Iizima, and T. Tanaka: Jpn. J. Appl. Phys., *20* (1981) L183.
8) T. Imura, K. Mogi, A. Hiraki, S. Nakashima, and A. Mitsuishi: Solid State Commun., *40* (1981) 161.
9) S. Veprek, Z. Iqbal, H. R. Oswald, F. A. Sarott, and J. J. Wagner: J. Phys., *42* (1981) Suppl. C4-251.
10) M. Taniguchi, M. Hirose, T. Hamasaki, and Y. Osaka: Appl. Phys. Lett., *37* (1980) 787.
11) A. Matsuda, K. Nakagawa, K. Tanaka, A. Matsumura, S. Yamasaki, H. Okushi, and S. Iizima: J. Non-Cryst. Solids., *35 & 36* (1980) 183.
12) F. J. Kampas and R. W. Griffith: J. Appl. Phys., *52* (1981) 1285.
13) Y. Mishima, S. Miyazaki, M. Hirose, and Y. Osaka: Philos. Mag., *B 46* (1981) 1.
14) A. Matsuda: J. Non-Cryst. Solids, *59 & 60* (1983) 767.
15) L. V. Azaroff: Element of X-Ray Crystallography (McGrow-Hill, N. Y., 1968) 552.
16) A. Matsuda, T. Yoshida, S. Yamasaki, and K. Tanaka: Jpn. J. Appl. Phys. *20* (1981) L439.
17) R. Landauer: J. Appl. Phys., *23* (1952) 779.
18) R. Tsu, J. Gonzalez-Hernancleg, S. S. Chao, S. C. Lee, and K. Tanaka: Appl. Phys. Lett., *40* (1982) 534.
19) T. Shimada, Y. Katayama, K. Nakagawa, H. Matsubara, M. Migitaka, and E. Maruyama: J. Non-Cryst. Solids, *59 & 60* (1983) 783.
20) S. Miyazaki, Y. Mishima, M. Hirose, and Y. Osaka: J. Non-Cryst. Solids, *59 & 60* (1983) 787.
21) A. Matsuda, T. Kaga, H. Tanaka, L. Malhotra, and K. Tanaka: Jpn. J. Appl. Phys., *22* (1983) L115.
22) J. Y. W. Seto: J. Appl. Phys., *46* (1975) 5247.
23) W. E. Spear, G. Willeke, P. G. LeComber, and A. G. Fitzgerald: J. Phys., *42* (1981) Suppl., C4-257.
24) P. G. LeComber, G. Willeke, and W. E. Spear: J. Non-Cryst. Solids, *59 & 60* (1983) 795.
25) Y. Mishima, T. Hamasaki, H. Kurata, M. Hirose, and Y. Osaka: J. Appl. Phys., *20* (1981): L121.
26) M. H. Cohen and J. Jortner: Phys. Rev. Lett., *30* (1973) 699.
27) N. Hata, S. Yamasaki, H. Oheda, A. Matsuda, H. Okushi, and K. Tanaka: Jpn. J. Appl. Phys., *20* (1981) L793.
28) M. Hirose: Jpn. J. Appl. Phys., *21* (1982) Suppl. 21-1, 275.
29) T. Hamasaki, M. Ueda, Y. Osaka, and M. Hirose: J. Non-Cryst. solids, *59 & 60*

(1983) 811.
30) S. C. Shen and M. Cardona : Phys. Rev., *B 23* (1981) 5322.
31) M. Cardona, S. C. Shen, and S. P. Varma : Phys. Rev., *B 23* (1981) 5329.
32) J. A. Reimer, R. W. Vaughan, and J. C. Knights : Phys. Rev. Lett., *44* (1980) 193.
33) S. G. Greenbaum, W. E. Carlos, and P. C. Taylor : Solid State Commun., *43* (1982) 663.
34) T. Miyasato, Y. Abe, M. Tokumura, Y. Imura, and A. Hiraki : Jpn. J. Appl. Phys., *22* (1983) L 580.
35) T. Imura and A. Hiraki : Japan Annual Reviews in Electronics, Computers & Telecommunications, ed. J. Nishizawa (Ohmsha/North-Holland) (Tokyo/Amsterdam, 1983) 155.
36) T. Imura, M. Tashiro, T. Ohbiki, H. Terauchi, A. Hiraki, K. Tsuji, and S. Minomura : Jpn. J. Appl. Phys., *22* (1983) L 505.
37) M. Haba, T. Imura, and A. Hiraki : Solid State Physics[Kotai-Butsuri], *17* (1982) 581 [in Japanese].
38) R. J. Spry and W. Dale Compton : Proc. of Santa Fe Conf. on Radiation Effects in Semiconductors, ed. S. L. Vook (Plenum Press)(New York, 1968) p. 421.
39) N. Fukada, T. Imura, A. Hiraki, T. Itahashi, T. Fukuda, and M. Tanaka : Jpn. J. Appl. Phys., *21* (1982) L 532.
40) A. Hiraki, T. Imura, K. Mogi, and M. Tashiro : J. Phys., *43* (1981) Suppl. C 4-277.
41) T. Imura, H. Terauchi, M. Tashiro, and A. Hiraki : Preceedings of 6th Symposium on Ion Sources and Ion-Assisted Technology (Kyoto, 1982) 273.
42) G. Lucovsky : Solid State Commun., *29* (1979) 571.
43) S. N. Sahu, T. S. Shi, P. W. Ge, J. W. Corbett, A. Hiraki, T. Imura, M. Tashiro, and V. A. Singh : J. Chem. Phys., *77* (1982) 4430.
44) T. S. Shi, S. N. Sahu, G. S. Oehrlein, A. Hiraki, and J. W. Corbett : Phys. Stat. Sol., (*a*) *74* (1982) 329.
45) M. Kumeda, Y. Yonezawa, K. Nakazawa, S. Ueda, and T. Shimizu : Jpn. J. Appl. Phys., *22* (1983) L 194.
46) T. Imura, H. Kaya, H. Terauchi, H. Kiyono, A. Hiraki, and M. Ichihara : Jpn. J. Appl. Phys., *23* (1984) 179.
47) A. Hiraki, Y. Fukushima, T. Sato, H. Kiyono, H. Terauchi, and T. Imura : J. Non-Cryst. Solids, *59 & 60* (1983) 791.

3.4 Photo-CVD of a-Si : H

Katsumi AOTA*†, Yasuo TARUI* and Tadashi SAITOH**

Abstract

Photo-chemical vapor deposition of a-Si films at a high rate using SiH_4 gas and a 185-nm low-pressure mercury lamp is described. A maximum rate of 1-nm/s is attained using the 185-nm lamp. This is approximately ten times higher than when using a 254-nm lamp. Assuming that there is no interaction between the effects of the two wavelengths, the deposition rate per light output power for 185-nm light is 160 times higher than for 254-nm light. Additionally, the absorption cross-section of 185-nm light is ten times greater than that of 254-nm light.

Microcrystalline silicon prepared by photo-CVD under low gas pressure at a low deposition rate is also described. From Raman scattering and X-ray diffraction, it was found that such silicon films have a mixed-phase structure that includes both microcrystalline and amorphous regions.

3.4.1 Introduction

Low-temperature epitaxial silicon growth utilizing photon energy has been attempted by Nishizawa et al.[1] as one approach to low-temperature processing. Subsequent to that work, reactant gases to which mercury atoms were added for sensitization were utilized in deposition of silicon nitride, [2,3] silicon oxide[4]~[6] and amorphous silicon[7,8] under low-pressure mercury lamp irradiation. Laser beam was also used to deposit amorphous silicon films.[9,10] Photo-dissociation of disilane gas will be described in 3.2.

The photosensitized process has enabled actualization of low-temperature processing with various incident photon energies. Thus, this method might well be regarded as a way to cause specific reactions and reduce such damage as commonly occurs in plasma chemical vapor deposition.

Although a low-pressure mercury discharge emits light at wavelengths of both 185 nm and 254 nm, most reports on photo-CVD have been concerned only with the latter. This is due primarily to the fact that the wall material of the lamp, i. e. quartz, absorbs most of the 185-nm wavelength light. If 185-nm wavelength light emission is utilized, another reaction process can be expected. Accordingly, the first half of this paper pursues this

---

* Department of Electronic Engineering, Tokyo University of Agriculture and Technology, Koganei, Tokyo 184.
** Central Research Laboratory, Hitachi Ltd., Kokubunji, Tokyo 185.
† Permanent Address : Technical Laboratory, Citizen Watch Co., Ltd., Tokorozawa, Saitama 359.

possibility for a-Si,[11] and the last half discusses microcrystalline silicon film prepared by photo-CVD.

3.4.2 Experiment with a 185-nm Lamp

Photo-chemical vapor deposition of a-Si film was performed using a low-pressure mercury lamp with a Suprasil (syntheses quartz) wall, developed by Ushio Inc., to verify the effect of 185-nm light on the reaction process. This low-pressure lamp will hereafter be referred to as a 185-nm lamp. The quartz-tubed "ozone-free type" will hereafter be called a 254-nm lamp.

Figure 3.4.1 shows the light output power at the two different wavelengths as a function of distance from the low-pressure 185-nm mercury lamp. The light output power of the 254-nm lamp is the same as that of the 185-nm lamp at a wavelength of 254 nm.

A schematic diagram of the photo-CVD apparatus is shown in Fig. 3.4.2. The reactor is made of stainless steel and has a window at the top. The reactant gas adopted was 100% SiH_4, and the elementary reaction that occurred in the reactor is

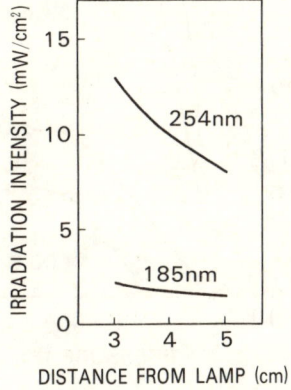

Fig. 3.4.1 Light intensities from a 185-nm lamp. [after Y. Tarui et al.[11]]

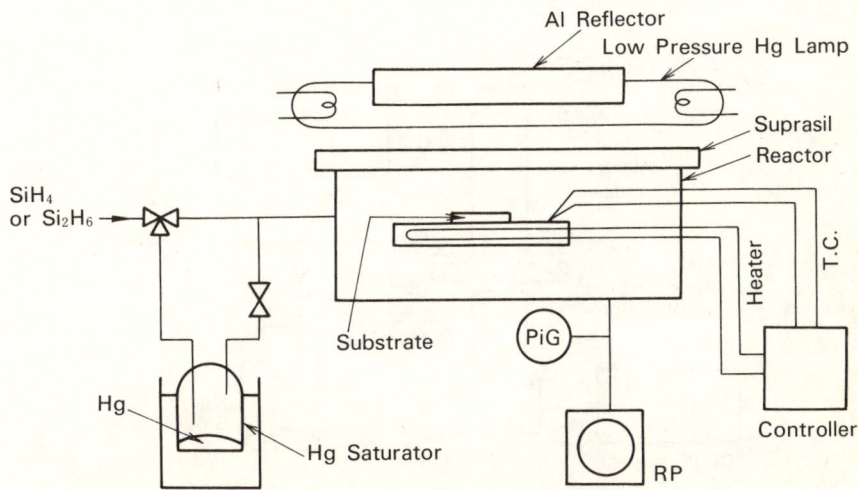

Fig. 3.4.2 Schematic diagram of the photo-CVD apparatus.

$$\text{SiH}_4 \xrightarrow[\text{Hg}]{h\nu} \text{SiH}_4^* \longrightarrow \text{SiH}_x + \text{H}$$

SiH$_4$ gas was transferred through a "saturator", held at a constant temperature, in which mercury atoms were added, prior to the gas being fed into the reactor.

The deposition rate of a-Si film on a glass substrate (Corning 7059) at 250°C is shown in Fig. 3.4.3 as a function of the reactant pressure. This pressure was controlled by varying the conductance of the vacuum system. The light output power of the 185-nm lamp on the substrate was 0.6 mW/cm^2 at 185 nm and 10 mW/cm^2 at 254 nm. These values have been corrected for reflection effect from the Al reflector on the lamp, as well as for absorption by the reactor quartz window.

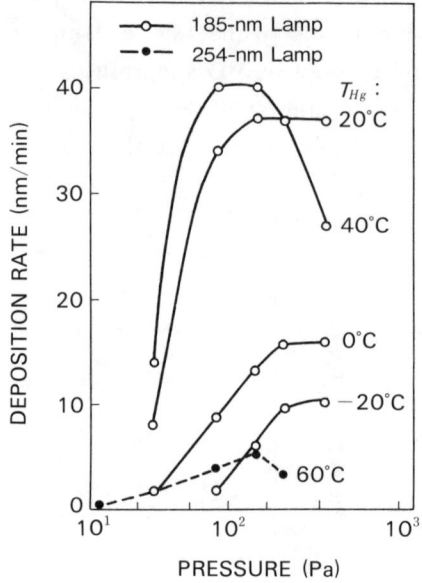

Fig. 3.4.3 Comparison of deposition rates at various Hg saturator temperatures (T_{Hg}) for 185-nm and 254-nm lamps. [after Y. Tarui et al.[11]]

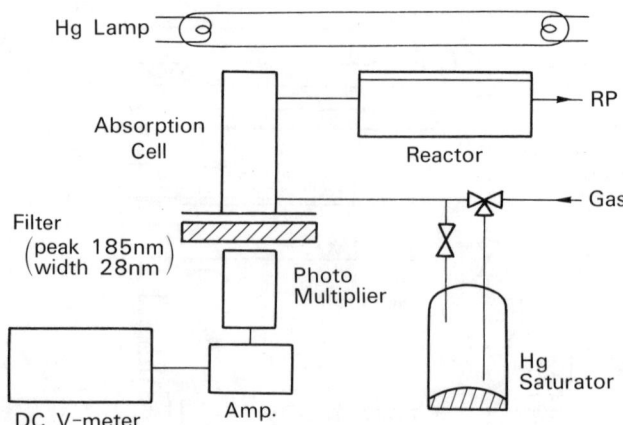

Fig. 3.4.4 Schematic diagram of the system for measuring the absorption coefficient at 185 nm.

Absorption of light at 185 nm and 254 nm by the reactant gas was measured additionally by means of an absorption cell. Figures 3.4.4 and 3.4.5 show schematic diagrams of the measuring systems. Unfortunately, irradiating Hg-containing SiH_4 gas fed into the absorption cell, deposited a-Si film on the inside of the window, causing measurement error. The absorption coefficients of Hg atoms picked up by SiH_4 and N_2 gas were therefore measured and compared with special care by cleaning the cell window frequently. The absorption coefficients derived are shown in Fig. 3.4.6. It is evident that the absorption coefficients for Hg atoms carried by the two gases at 185 nm are almost the same. This suggests that SiH_4 can be replaced by N_2 gas for this kind of measurement. The photo (at AM1) and dark conductivities of a-Si films deposited under various reactant pressures were also measured. Photo conductivity was 10^4 times greater than dark conductivity as shown in Fig. 3.4.7.

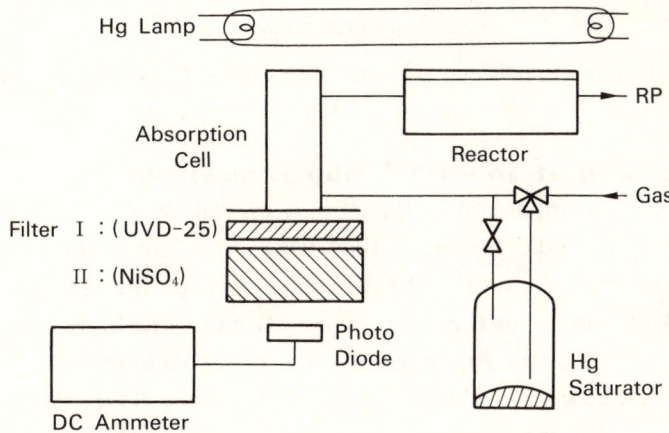

Fig. 3.4.5 Schematic diagram of the system for measuring the adsorption coefficient at 254 nm.

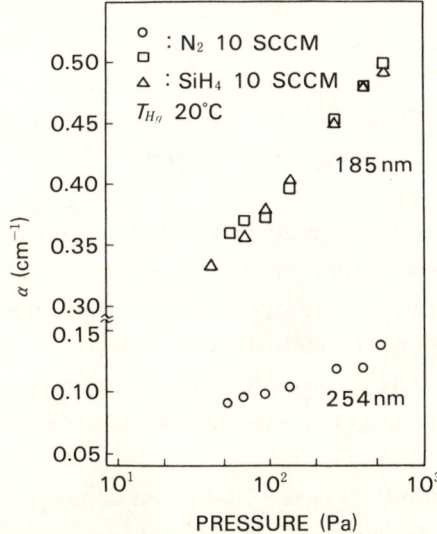

Fig. 3.4.6 Comparison of absorption coefficients for Hg atoms added to N_2 and SiH_4 gases at various pressures. [after Y. Tarui et al.[11]]

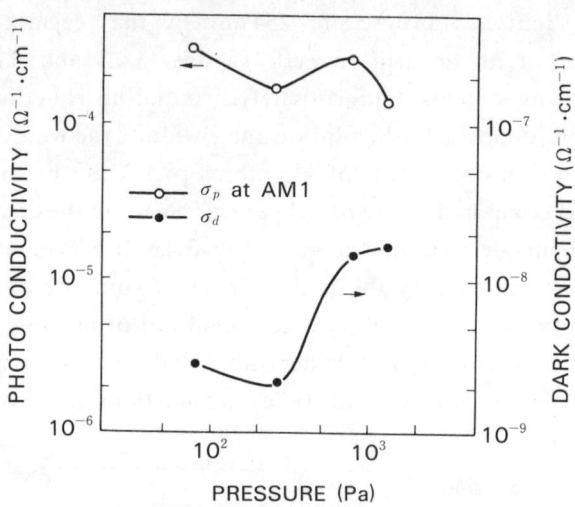

Fig. 3.4.7 Variation in photo and dark conductivities with reactant pressure.

3.4.3 Discussion of 185-nm Lamp Deposition

Figure 3.4.2 is a comparison of the deposition rate at $T_{Hg} = -20°C \sim 40°C$ for the 185-nm lamp, and $T_{Hg} = 60°C$ for the 254-nm lamp. The former shows a value one order of magnitude greater than the latter. The difference between the absorption coefficients at 185 nm and at 254 nm, shown in Fig. 3.4.6, is one reason for the higher deposition rate with the 185-nm lamp in spite of the lower temperature of the mercury saturator. However, the difference between the maximum deposition rates of the two lamps cannot presumably be explained only as the difference between absorption coefficients.

It is assumed that there is no multiplier effect for simultaneous illumination of 185-nm and 254-nm wavelength light emitted from the 185-nm lamp. Considering that the ratio of 185-nm light output to that of 254-nm light is 0.6/10 and that the maximum deposition rate for the 185-nm lamp is about 10 times greater than that for 254-nm light, as shown in Fig. 3.4.3, the deposition rate per unit output of 185-nm wavelength light is 160 times greater than that for 254-nm light. This indicates that the a-Si deposition process caused by illumination with 185-nm wavelength light, which probably leads to excitement of Hg ($^1P_1$), is quite efficient.

Figure 3.4.8 shows the pressure-dependence of deposition rates of a-Si films and the transmissivity of 185-nm wavelength light on the substrate at $T_{Hg} = 20°C$. It is evident from the figure that deposition rate is disproportional to transmissivity. At low pressure, the decrease in deposition rate may be caused by mass transfer limitation. At high pressure, saturation of the deposition rate disproportional to the transmissivity decrease is caused by an increase in the monosilane quantity and probably also by excited molecules between the window and substrate.

Figure 3.4.9 shows film thickness for a-Si film deposited on a sloping substrate, i. e. at various distances from the window, on the pressure at which the deposition rate is proportional to transmissivity. The maximum growth rate obtained is about 1 nm/s. Both positive and negative gradients were used to determine the effect of different gradients on

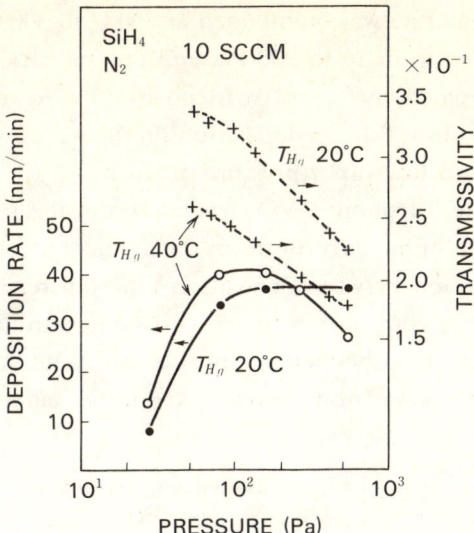

Fig. 3.4.8 Variations in deposition rate and light transmission with reactant pressure at Hg saturator temperature $T_{Hg} = 20°C$ and $40°C$.

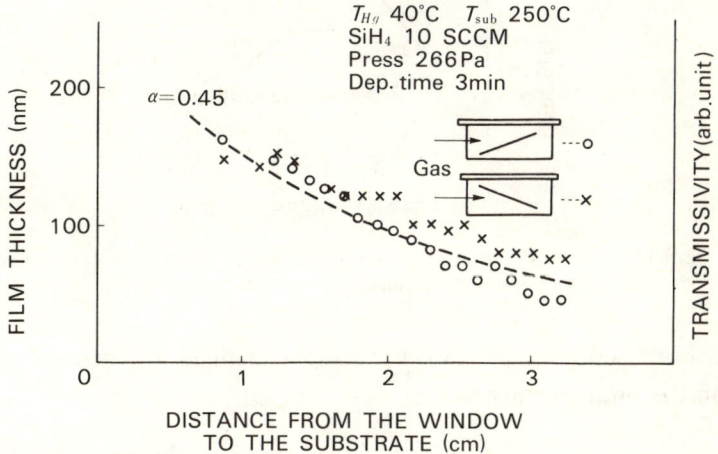

Fig. 3.4.9 Thickness of a-Si film deposited at different gas absorption layer thicknesses. Dashed line represents the transmissivity calculated for the absorption coefficient $\alpha = 0.45$. [after Y. Tarui et al.[11]]

gas flow direction. The dashed line in Fig. 3.4.9 shows the transmissivity of light calculated for $\alpha = 0.45$, where α is the absorption coefficient of reactant gas. The value is derived from the result of light absorption experiments with Hg shown in Fig. 3.4.6.

3.4.4 Microcrystalline Silicon Film Prepared by Photo-CVD

Since another section (3.3) of this paper gives details on microcrystalline silicon films, especially those prepared by plasma-CVD,[12)13)] this section only describes the preparation of microcrystalline silicon films using photo-CVD.[14)] The optical, electrical and structural properties have also been previously characterized.

Silicon films were deposited by Hg ($^3P_1$) photosensitized reaction of SiH_4 gas on glass and crystalline Si. The substrate was maintained at 200°C to 300°C with the pressure of the SiH_4-H_2 gas system ranging from 26 to 200 Pa, and the mercury container was kept at 60° C. Thick films were prepared by repetitive deposition following plasma etching of Si deposits on the quartz window. Silicon deposition on the window was otherwise effectively prevented by application of low-vapor-pressure oil to it.

The deposition rate for photo-CVD depends upon the reactant pressure, substrate temperature, concentration of mercury atoms in the ambient area and photon energy of the irradiating light. The relation between deposition rate and reactant pressure is shown in Fig. 3.4.10. The lower curve corresponds to repetitive deposition of Si explained above. A maximum rate of 2 nm/min was obtained at a pressure of about 90 Pa, but the rate increased to about 6 nm/min for the deposition shown in the upper curve.

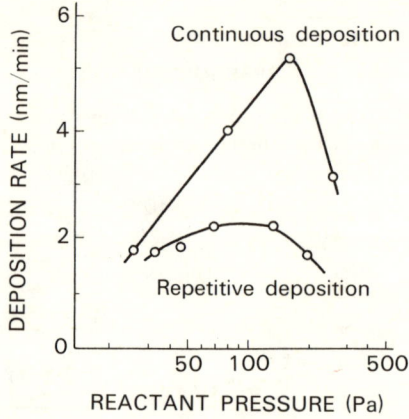

Fig. 3.4.10 Effect of reactant pressure on deposition rate for photoproduced silicon. [after T. Saitoh et al.[14]]

Deposition efficiency in photo-CVD can be defined by analogy to the quantum efficiency within the photochemistry:

$$\text{Deposition efficiency} = \frac{\text{Number of atoms in the deposited film}}{\text{Number of photons in the irradiation}}$$

The maximum rate in Fig. 3.4.10 corresponds to $4.9 \times 10^{14} cm^{-2} \cdot s^{-1}$ due to the number of atoms in the film. Using a photon flux of $1.3 \times 10^{16} cm^{-2} \cdot s^{-1}$, equivalent to a 254-nm light energy of $10 mW/cm^2$, deposition efficiency was calculated to be 0.038. This value seems to be relatively high if light absorption in the reactant is taken into account and reactor design is not optimized.

Both photo and dark conductivity, σ_p and σ_d, tend to decrease with increased reactant pressure. At pressure in the vicinity of 100 Pa, typical values of σ_p and σ_d are in the order of 10^{-3} and 10^{-8} mho·cm^{-1}, respectively, and at lower pressure they are in the order of 10^{-2} and 10^{-4} mho·cm^{-1}, respectively.

Mobility-lifetime product obtained from photo current measured in gap-cell structure is shown in Fig. 3.4.11 as a function of photon energy. Typical values for amorphous films are 1 to $2 \times 10^{-5} cm^2/V$, while values of 10^{-3} to $10^{-4} cm^2/V$ are obtained

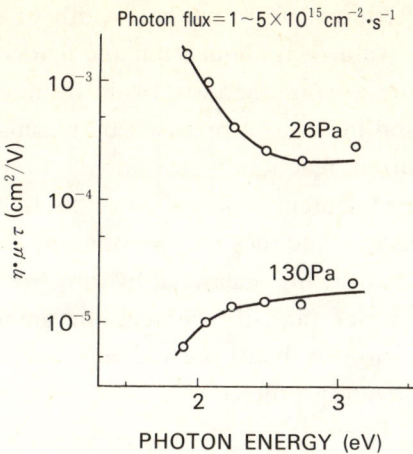

Fig. 3.4.11 $\eta \cdot \mu \cdot \tau$ product as a function of photon energy. η is the generation efficiency, μ the mobility and τ the lifetime of free carriers. [after T. Saitoh et al.[14]]

for films deposited at 26 Pa.

The activation energy of dark conductivity tends to decrease with decreased reactant pressure, becoming 0.08 eV for film deposited at 26 Pa. This low value suggests that the Fermi level is near the bandgap edge for electronic conduction. The mean optical gap is about 1.8 eV but the optical gap tends to increase slightly with decreased reactant pressure.

Raman scattering was employed to characterize the structural properties of photo-CVD films. A typical Raman spectrum is shown in the lower curve of Fig. 3.4.12 for Si film

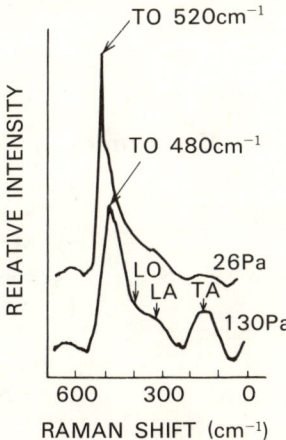

Fig. 3.4.12 Raman spectra for amorphous and microcrystalline silicon prepared by photosensitization. [after T. Saitoh et al.[14]]

deposited at a pressure of 130 Pa. The main Raman peak lies at 480 cm^{-1}, which corresponds to the TO-phonon peak for amorphous silicon. For silicon films deposited at 26 Pa of 14% SiH$_4$, a sharp Raman peak is located near 520 cm^{-1}, indicating that microcrystallites exist. There is an additional lower peak at 480 cm^{-1}. Thus, it was concluded that such silicon films have a mixed-phase structure that includes crystalline and amorphous regions.

Crystallite size was determined from the half-width of the X-ray (111) diffraction peak. For amorphous film, the value was about 2 nm and tended to increase with decreased reactant pressure. The crystallite size obtained was 14 nm for microcrystalline silicon, which is equal to that for microcrystalline silicon prepared by plasma-CVD.

Mercury-photosensitization reaction is considered to be caused by collision of reactant molecules with excited mercury atoms.[15] For SiH_4 diluted with H_2, hydrogen radicals are thought to be generated and they may play an important role in deposition. The hydrogen content of microcrystalline film measured by infrared absorption was about 10%, or approximately two times lower than the content for amorphous film. These results suggest that hydrogen elimination by hydrogen radicals from the surface or reactants is important in the microcrystallization process.

3.4.5 Conclusion

High-rate deposition of a-Si employing Hg ($^1P_1$) photosensitized reactions of SiH_4 was performed for the first time. A high deposition rate per light output power was attained using 185-nm light. This value was 160 times that for 254-nm light. These results suggest that the method presented here is a promising candidate for fabrication of such devices as solar cell, where high-speed deposition is essential.

Acknowledgements

The authors wish to express their sincere thanks to Asst. Prof. Yoshitaka Takubo and Mr. Takashi Sugiura of the Tokyo University of Agriculture and Technology, as well as Citizen Watch Co., Nippon Sanso K. K., Ushio Inc., Nippon Tylan Co. and Hitachi Ltd., for their cooperation.

References

1) M. Kumagawa, H. Sunami, T. Terasaki, and J. Nishizawa : Jpn. J. Apple. Phys., 7 (1968) 1332.
2) C. Brekel and P. Severin : J. Electrochem. Soc., 119 (1972) 372
3) J. Peters, F. Gebhart, and T. Hall : Solid State Technology Sept., (1980) 121.
4) J. Peters : IEEE IEDM Tech. Digest, (1981) 240.
5) R. F. Sarkozy : Tech. Digest of 1981 Symp. on VLSI Tech., (1981) 69.
6) H. Kim, S. Tai, S. Groves, and K. Schuegraf : CVD Conf. (1981) 258.
7) H. Ito, K. Mizuguchi, and H. Abe : Proc. of the 22nd Symp. on Semiconductor and Integrated Circuits, (1982) 90 [in Japanese].
8) T. Saitoh, S. Muramatsu, S. Matsubara, and M. Migitaka : Jpn. J. Appl. Phys., 22 Suppl. (1983) 617.
9) M. Meunier, J. H Flint, D. Adler, and J. S. Haggerty : J. Non-crystalline Solids, 59 ~60 (1983) 699.
10) M. Hanabusa, S. Moriyama, and H. Kikuchi : J. Non-crystalline Solids, 59 ~60 (1983) 703.
11) Y. Tarui, K. Sorimachi, K. Fujii, and T. Saitoh : Abstract of the 10th Int. Conf. on Amorphous and Liquid Semiconductors (1983) 272 ; J. Non-Crystalline Solids, 59 ~60 (1983) 711.
12) S. Usui and M. Kikuchi : J. Non-Crystalline Solids, 34 (1979) 1
13) A. Matsuda, S. Yamasaki, K. Nakagawa, H. Okushi, K. Tanaka, S. Iijima, M. Matsumura, and H. Yamamoto : Jpn. J.

Appl. Phys., *19* (1980) L 305.

14) T. Saitoh, T. Shimada, and M. Migitaka: Abstract of the 10th Int. Conf. on Amorphous and Liquid Semiconductors (1983) 273; J. Non-Crystalline Solids, *59-60* (1983) 715.

15) H. Okabe: Photochemistry of small molecules, (John Wiley & Son, New York, 1978) p. 144.

3.5 Tetrahedral Alloys

Yukinori KUWANO* and Shinya TSUDA*

Abstract

In this section, recent developments in amorphous silicon based alloys are reviewed. A brief history of each a-Si based alloy is described. Then, some recent approaches for obtaining good quality a-Si based alloys are presented. A number of materials have been investigated, with emphasis on reducing the density of states. Preparation methods focusing on reducing damage or improving deposition speed are discussed. Some new characterization methods and new device applications are also described.

3.5.1 Introduction

Amorphous semiconductors do not have fixed material parameters, so their material parameters can be varied by changing composition and fabrication conditions. Especially, a-Si based alloys which were mixed with other atoms in a-Si : H have been extensively investigated. These alloy materials are very important for obtaining good characteristic a-Si solar cells, image pick up devices, and so on. Therefore, a-SiC : H and a-SiN : H were investigated as materials having wider bandgaps than a-Si, and a-SiGe : H and a-SiSn : H were investigated as materials having narrower bandgaps.

3.5.2 Preparation and Properties of a-SiC : H Films

Brief history

Since Spear and Anderson reported that a-SiC : H could be prepared by a glow discharge method,[1] many of its structural, optical and electrical properties have been investigated, including optical bandgap (E_{opt}), photoluminescence (P_L), photoconductivity (σ_{ph}), dark conductivity (σ_d). A number of preparation methods, such as the GD-CVD method (Glow Discharge Chemical Vapor Deposition), sputter method, thermal-CVD method, and so on, have also been investigated.

Recently Hamakawa et al. reported valence controllability[2] in a-SiC : H prepared by glow discharge with ($SiH_4 + CH_4$) gases. They also showed its applicability as a window material[3] for a-Si solar cells, and have reported high efficiency solar cells[4] using an a-SiC : H/a-Si : H hetero-junction structure.

In glow discharge methods many material gases have been studied such as $SiH_4 +$

---

* Research Center, Sanyo Electric Co., Ltd., 1-18-13, Hashiridani, Hirakata City, Osaka 573.

CH_4, $SiH_4 + C_2H_6$, $Si(CH_3)_4$, and so on. The addition of each these gases resulted in different properties in a-SiC : H film. $SiH_4 + CH_4$ gases have generally been used for fabricating a-SiC : H fims, because of their low state density.

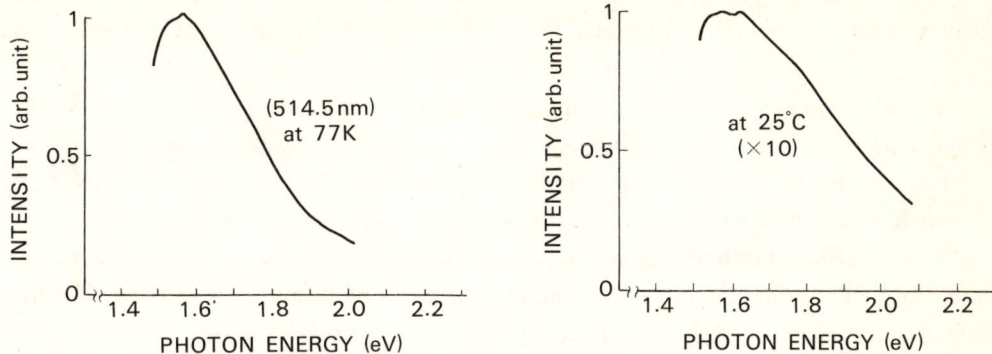

Fig. 3.5.1 Photoluminescence spectra of a-SiC : H fabricated with Si_2H_6 and C_2H_6 gases. [after A. Hatano et al.[12]]

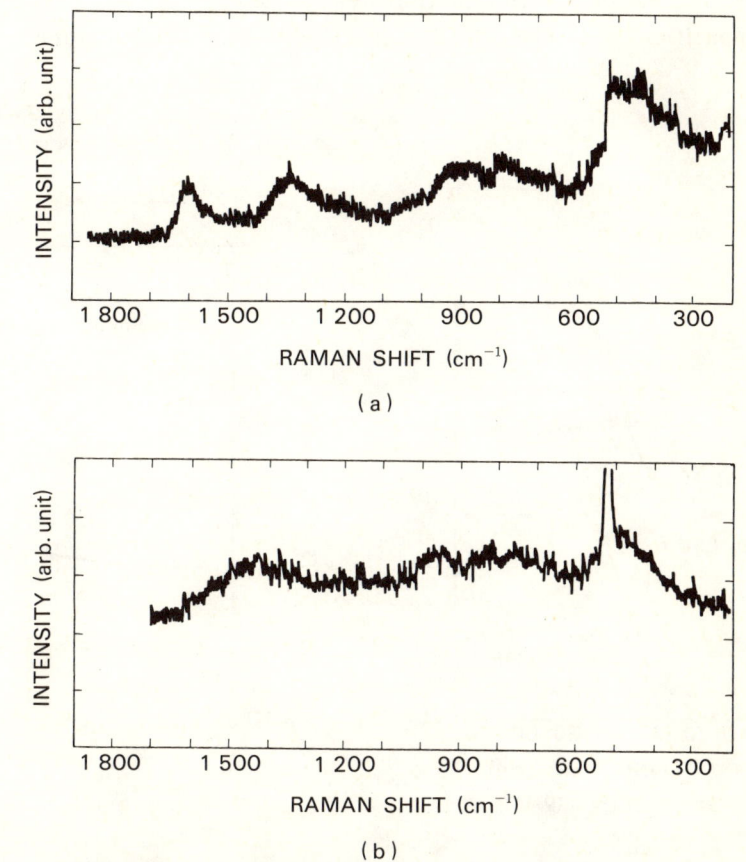

Fig. 3.5.2 Raman shift spectra of a-SiC : H fabricated by
(a) sputter method
(b) glow discharge method. [after Y. Inoue et al.[14]]

Recent topics

Preparation methods   New a-SiC : H fabrication methods, such as reactive evaporation,[8] ICB (Ion Cluster Beam)[9] and IBS (Ion Beam Sputter),[10] have been tried. As for the glow discharge method, various material gases have been tried. New gas materials used for fabricating a-SiC : H were TMG $((CH_3)_3 Ga) + SiH_4$[11] for Ga doped a-SiC : H and $Si_2H_6 + C_2H_6$.[12]

A comparison of various methane group gases, aiming at good property a-SiC : H films, has also been reported.[13] Figure 3.5.1 shows photoluminescence spectra of a-SiC : H films fabricated by a glow discharge method using a new gas combination : $Si_2H_6 + C_2H_6$.[12] The photoluminescence was visible, and was observable at room temperature.

Characterization methods and new application   The new characterization methods applied to a-SiC : H films were the measurement of Raman spectroscopy,[14] NMR,[15] Hydrogen quantification by nuclear reaction,[16] and so forth. Figure 3.5.2 shows Raman spectra of a-SiC : H films[14] fabricated by the glow discharge method (a) and the sputter method (b). Structural information for a-SiC can be obtained from these spectra.

Recently, a-SiC : H films have found new applications in exprimentally fabricated electro-luminescence devices[17] and quantum well devices.[18]

Effects of impurities   The diffusion effects of indium and tin atoms[5,6] from TCO to

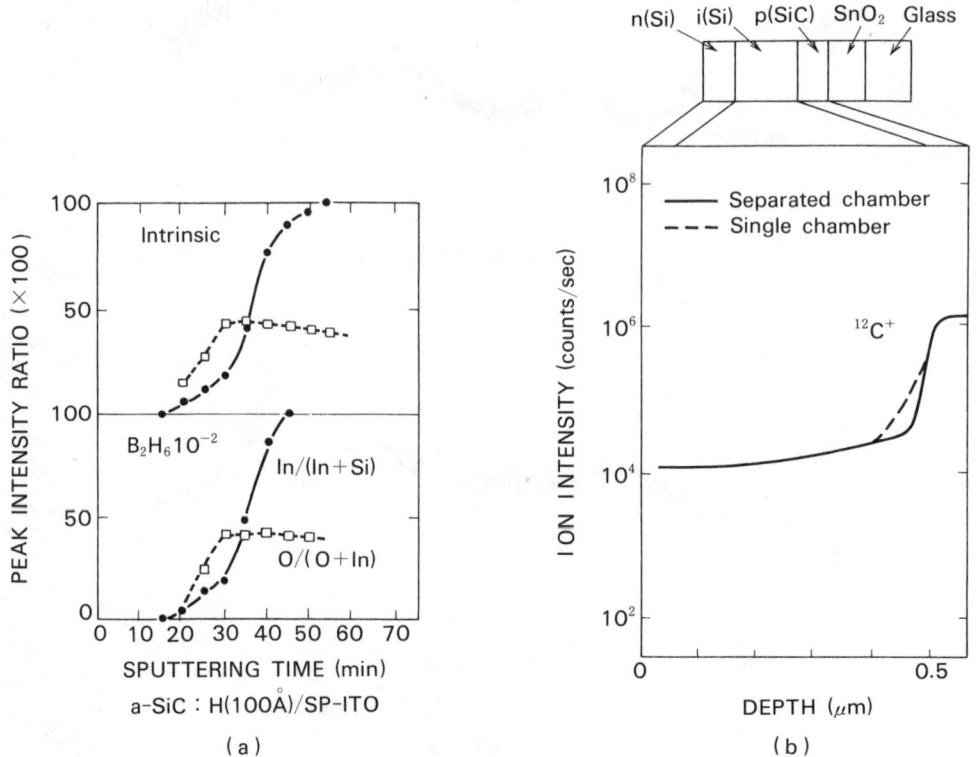

Fig. 3.5.3   Diffusion profiles of impurities
(a) In and O in a-SiC : H from ITO [after N. Fukada et al.[5]]
(b) C in a-Si : H from a-SiC : H. [after K. Enomoto et al.[7]]

a-SiC : H and of carbon atoms[7] from a-SiC : H to a-Si were investigated in order to determine the feasibility of using a-SiC : H for the window material in a-Si solar cells.

Figure 3.5.3 shows diffusion profiles of indium and oxygen atoms from ITO to a-SiC : H and carbon atoms from a-SiC : H to a-Si : H. These atoms diffuse to a-SiC : H during the fabrication process. Indium atoms, in particular, were known to act as acceptors and caused a decrease in the conversion efficiency of a-Si : H solar cells.[5] Furthermore, the diffusion of carbon atoms into the i layer decreases the electrical properties.[7]

3.5.3 Preparation and Properties of a-SiN : H Films

Brief history

a-SiN alloys fabricated by the CVD (chemical vapor deposition) method or glow discharge method have been studied as insulating materials. In recent years, a-SiN : H has been investigated as a wide-bandgap semiconductor material.[19]

When a small quantity of nitrogen atoms were introduced to a-Si : H, the optical bandgap of a-SiN : H films increased, while their photoconductivity and dark conductivity remained within the range of semiconductor properties.

In this semiconductor region, the coordination number of nitrogen atoms was studied. It was found that some nitrogen atoms were fourfold coordinated and others were threefold coordinated. As the nitrogen content increases, the threefold coordinated nitrogen becomes dominant. Recently, many other application studies have been reported; for example, impurity doping, thermal stability,[20] film quality improvements,[21] and so on.

Recent topics

Impurity effects   Slight boron doping was studied in order to improve the electrical properties of a-SiN : H.[21] As some of the nitrogen atoms act as donors in a-SiN : H films, boron atoms were used as a compensator. The Fermi level therefore shifted to near the

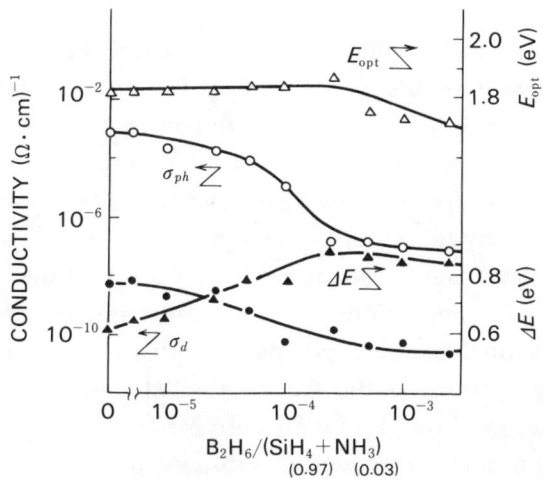

Fig. 3.5.4 Fundamental properties of slightly boron doped a-SiN : H films as functions of gas ratio. [after H. Nishiwaki et al.[21]]

midgap. Figure 3.5.4 shows the fundamental properties of slightly boron doped a-SiN : H films as functions of the gas ratio. As the boron doping ratio increases, activation energy (ΔE) increases and the Fermi level shifts toward the midgap. But in the case of heavy boron doping ($> 10^{-4}$), σ_{ph} and E_{opt} decrease, and these decreases are not suitable for solar cells. To improve the electrical properties, therefore, an optimum value exists for the gas ratio, as shown in Fig. 3.5.4.

Characterization methods   Figure 3.5.5 shows changes in the infrared absorption spectra of a-Si : H and a-SiN : H films for different annealing conditions.[20] The decrease in

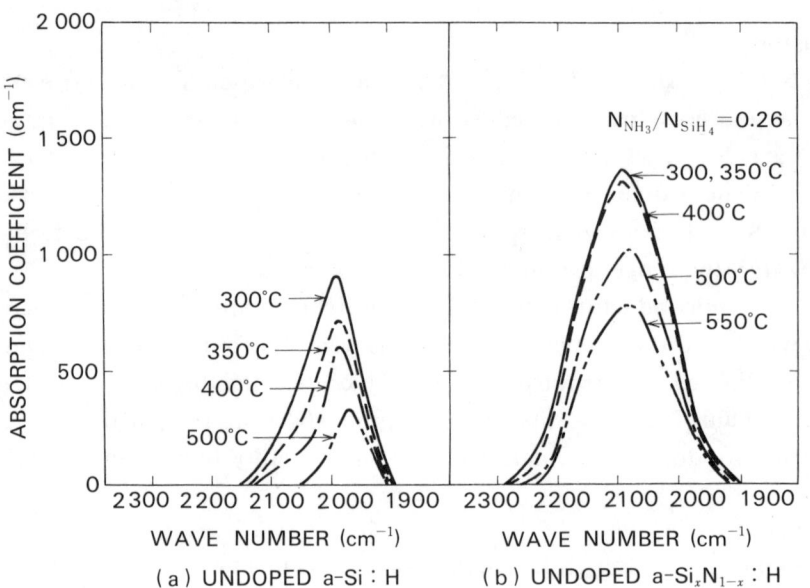

Fig. 3.5.5 Changes in infrared absorption spectra by Si-H stretching mode in a-Si : H and a-SiN : H films for different annealing temperatures. [after H. Kurata et al.[20]]

the Si-H stretching mode of a-SiN : H films was smaller than that of a-Si : H films. Hydrogen evolution in a-Si : H generally occurs above 350°C.[22] The Si-H$_2$ stretching absorption due to dihydride was reduced above that temperature. In contrast to this, the Si-H bonds in a-SiN : H were stable even at 400°C. This result indicates that the hydrogen atoms which are bonded to silicon atoms in a-SiN : H films were more stable than those in a-Si : H. So a-SiN : H was a good possibility for a thermostable material.

New preparation methods and application   Recent studies of a-SiN : H have focused mainly on practical applications. As a new preparation method for a-SiN : H, photo-CVD yields higher quality film,[23] compared with the glow discharge method, by reducing damage during fabrication. In the photo-CVD method, reaction gases were decomposed by ultraviolet rays from a low pressure mercury lamp. a-SiN : H films could therefore be fabricated on a low temperature substrate and there was minimal atomic damage to the substrate, compared to the conventional plasma CVD method. Quantum well structure devices were investigated as a new application of the a-Si : H/a-SiN : H heterostructure.

3.5.4 Preparation and Properties of a-SiGe : H Films

Brief history

a-SiGe : H was expected to become a new material for improving the long wavelength sensitivity of a-Si solar cells. This was because germanium could easily produce the Si-Ge fourfold coordinated structure, and the E_{opt} of a-SiGe : H could be varied from 1.0 eV to 1.7eV by varing the germanium content.[25] The state density in the forbidden band of a-SiGe could be reduced by hydrogen, the same as in a-Si : H. But the binding energy of Si-H and Ge-H was different. As for a-SiGe : H, hydrogen atoms were more likely to bind silicon atoms than germanium atoms. So the characteristics of a-SiGe : H were therefore inferior to those of a-Si : H. Many methods have been studied for improving a-SiGe : H and a number of physical characterization methods that had been studied for a-Si : H were applied to a-SiGe : H.

Now, a-SiGe : H is used for the 2nd or 3rd i layer of multi-bandgap a-Si solar cells.[26]

Recent topics

Characterization methods   Figure 3.5.6 shows the results of EXAFS[27] (Extended X-ray Absorption Fine Structure), from which we can obtain information about the coordination number and bond length of the germanium atom. From these results, Si-Ge alloys have been found to be a fourfold coordinated structure, and each atom distributes completely randomly. The bond length between silicon and germanium in the a-SiGe alloy was larger than that of the crystal Si-Ge alloy, and if hydrogen atoms are present, the bond length of a-SiGe : H is shorter than that of a-SiGe. This is because hydrogen atoms reduce stress in silicon-germanium networks.

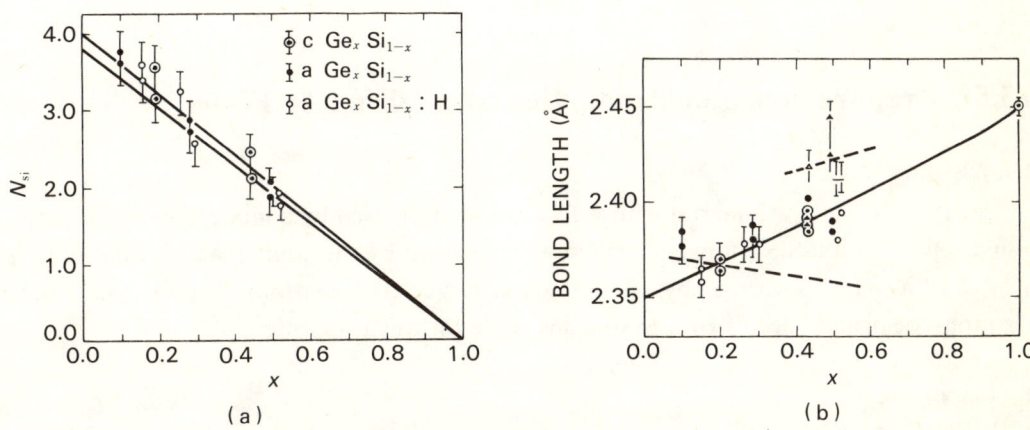

Fig. 3.5.6 Structural analysis by EXAFS
(a) Coordination number of Si atoms around Ge atoms
(b) Bond length between Si atoms and Ge atoms around Ge atoms. [after S. Minomura et al.[27]]

Preparation methods   Some new preparation techniques have been reported; for example, glow discharge methods with various frequencies[28] and with new combinations of gas material, such as $Si_2H_6 + GeH_4$[29] and $SiH_4 + GeF_4$.[30] Figure 3.5.7 shows the photoconductivity of a-SiGe:H film fabricated by $SiH_4 + GeF_4$ gases. It was found that the photoconductivity of a-SiGe:H film was improved by using $SiH_4 + GeF_4$ gases. The improvement was caused by the decrease of dangling bonds which were terminated by fluorine atoms.

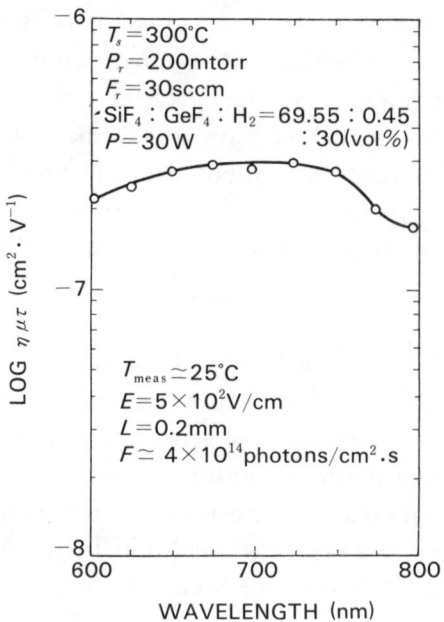

Fig. 3.5.7   Photoconductivity of a-SiGe:H fabricated with GeF_4 and SiH_4 gases. [after K. Nozawa et al.[30]]

3.5.5 Preparation and Properties of a-SiSn:H Films

Brief history

a-SiSn:H is a new material whose E_{opt} can be decreased by a mixture with tin atoms. The first report[31] on a-SiSn films described a fabrication by the sputtering method and the E_{opt} of a-SiSn:H was varied, by changing the target compositions. Figure 3.5.8 shows absorption coefficient spectra of a-SiSn films with various tin content.

Recent topics

Preparation methods   Recently a new preparation method was reported[33] for a-SiSn films. It is a glow discharge method with $SiH_4 + Sn(CH_3)_4$[32] or $SiH_4 + SnH_4$[33] gases. In this method, the tin atom ratio and E_{opt} of a-SiSn:H films can be varied by changing gas compositions. Figure 3.5.9 shows the fundamental properties, such as E_{opt}, σ_{ph}, and σ_d,

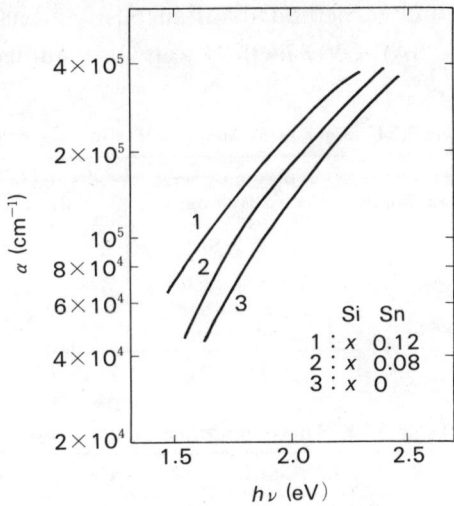

Fig. 3.5.8 Absorption coefficient of a-SiSn films fabricated by sputtering method. [after C. Verie et al.[31]]

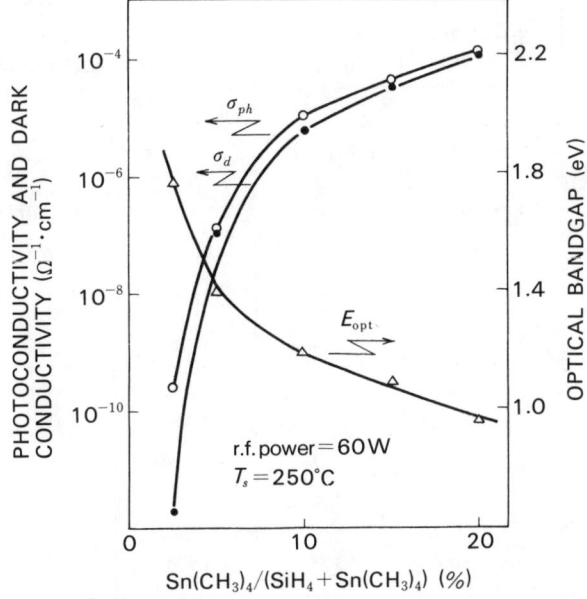

Fig. 3.5.9 Fundamental properties of a-SiSn:H films fabricated by glow discharge method as functions of gas ratio. [after Y. Kuwano.[32]]

as functions of the gas ratio. When the flow ratio of $Sn(CH_3)_4$ increases, the E_{opt} of a-SiSn: H films decreases and both σ_{ph} and σ_d increase. But these electrical properties were not suitable for solar cells and improvement was required. Fabrication by a glow discharge with SiH_4 and SnH_4 gases was studied to improve the electrical properties.

Another a-SiSn:H fabrication method was also proposed.[34] This was called SAP-CVD (Sputter Assisted Plasma Chemical Vapor Deposition) which used both the glow discharge method and the sputter method simultaneously. Preparations of a-SiGe : H and a-SiPb : H produced by the SAP-CVD method were also studied.[35]

Table 3.5.1 Some properties of a-Si alloy elements.

Element	Inclination of Dangling Bonds	Binding Energy to Hydrogen	$\dfrac{dN_s \text{ (spin density)}}{dx \text{ (content)}}$	Thermal Stability
Carbon	C > Si	C > Si	large	C > Si
Nitrogen	N < Si	N ≃ Si	small	N > Si
Germanium	Ge > Si	Ge < Si	small	Ge < Si

Table 3.5.2 Tipical properties of a-Si alloys.

Alloys	a-SiC	a-SiN	a-SiGe	a-SiSn	remarks
Reaction condition					
Fabrication method	GD-CVD	GD-CVD	GD-CVD	GD-CVD	representative value
Pressure (Torr)	0.3	0.3	0.3	0.3	
r. f. power (W)	30	20	10	20	
Substrate temperature (°C)	250	300	200	250	
Gas combination	$SiH_4(0.8)$ + $CH_4(0.2)$	$SiH_4(0.8)$ + $NH_3(0.2)$	$SiH_4(0.8)$ + $GeH_4(0.2)$	$SiH_4(0.8)$ + $Sn(CH_3)_4(0.2)$	
Electrical properties					
Dark conductivity ($\Omega^{-1}\cdot cm^{-1}$)	1×10^{-10}	1×10^{-8}	9×10^{-9}	1×10^{-4}	
Photoconductivity ($\Omega^{-1}\cdot cm^{-1}$)	8×10^{-7}	1×10^{-5}	2×10^{-4}	1×10^{-7}	AM-1 100 mW/cm²
Activation energy (eV)	0.9	0.65	0.85	0.3	
Electron mobility (cm²/vs)			0.05		
Hole mobility (cm²/vs)			1×10^{-3}		
Optical properties					
Optical band gap (eV)	1.85	2.0	1.68	0.95	
Absorption coefficient (cm⁻¹)	2.2×10^4	8×10^3	8×10^4	5×10^5	(= 2.0eV)
Structual properties					
IR absorption X-H (cm⁻¹)	2800 3000	3340	1970		Stretching mode
Hydrogen content					
H_{Si} (%)	35		5		
Hx (%)	15		38		
H total (%)	50		43		
Photoluminescence					
Peak energy (eV)	2.0	1.65	1.2		
Intensity ratio			0.7		a-Si alloy/a-Si
Electrion spin resonance					
Spin density (cm⁻³)	8×10^{19}	3×10^{17}	1×10^{17}		
Raman shift (cm⁻¹)	750	850	370		Si-X

3.5.6 Conclusion

a-Si based alloys were prepared by various methods, such as the GD-CVD method, thermal-CVD method, photo-CVD method, sputter method, sputter assisted plasma CVD method, ICB method, and ISB method. These preparation methods exert different influences on internal stress, hydrogen content, plasma damage to films, etc. The most important feature in these preparation methods was that different atoms could be introduced to a-Si networks without resulting in stress or dangling bonds. Glow discharge methods are used mainly for preparation of a-Si based alloys, because they don't cause extensive damage and the content of each atom can be varied by changing gas compositions. Typical properties of a-Si alloys at various gas compositions are shown in Table 3.5.1.

In a-Si based alloys, carbon and germanium atoms (group IV elements) can be easily introduced in a-Si networks to make alloys. Nitrogen atoms (group V elements) vary their coordination number from 4 to 3 as the nitrogen content increases. Tin atoms increase stress in a-Si network because of their large atomic radius and because oxygen is easily introduced into a-SiSn films.

Many characterization methods were applied in an attempt to estimate the tendency to create dangling bonds and to measure the binding energy to hydrogen, when carbon, nitrogen and germanium atoms were mixed in a-Si networks. Some of these are shown in Table 3.5.2 with spin densities (N_s) of ESR and thermal stabilities. In the case of tin, even though there was not so much data, it was determined that tin atoms tend to create a large number of dangling bonds and that they do not bind very well with hydrogen atoms, as compared with germanium atoms.

In order to obtain a-Si alloys with good properties, it is important to choose reaction conditions carefully. Therefore, the reaction process and damages to films, should be considered, and in some cases, the addition of a bias voltage, the gas combination, the reaction frequency at the glow discharge, etc., must also be taken into consideration. Recently, investigation has begun into the photo-CVD method for both the reduction of plasma damage and the improvement of the deposition rate.

References

1) D. A. Anderson and W. E. Spear: Philos. Mag., 35 (1977) 1.
2) Y. Tawada, H. Okamoto, and Y. Hamakawa: Appl. Phys. Lett., 39 (1981) 237.
3) Y. Tawada, M. Kondo, H. Okamoto, and Y. Hamakawa: Jpn. J. Appl. Phys., 21 (1981) Suppl. 21-1, 273.
4) Y. Tawada, K. Tsuge, K. Nishimura, M. Kondo, H. Okamoto, and Y. Hamakawa: Proc. 3rd Photovoltaic Science and Engineering Conf. in Japan.
5) N. Fukada et al.: Jpn. J. Appl. Phys. 21 (1982) Suppl. 21-2.
6) Y. Fukada, M. Fukushima, K. Imura, A. Hiraki, K. Nishimura, and Y. Kowada: Preprint of Spring Meet. Jpn. Soc. Appl. Phys., (1983) 376.
7) K. Enomoto, H. Nishiwaki, K. Watanabe, Y. Nakashima, S. Tsuda, M. Ohnishi, and Y. Kuwano: Jpn. J. Appl. Phys. 21 (1982) Suppl. 21-2, 265.
8) K. Yasui, M. Sato, T. Yokobori, K. Miyazaki, and S. Kameda: Preprint of Spring Meet. Jpn. Soc. Appl. Phys., (1983) 374.
9) M. Yoshitake, K. Sato, H. Kataoka, K. Matsubara, and T. Takagi: Preprint of Autumn Meet. Jpn. Soc. Appl. Phys.,

(1983) 326.
10) J. Saraie, et al.: Thin Solid Films *80* (1981) 189.
11) K. Maruyama, T. Hama, K. Ichimura, H. Sakai, and Y. Uchida: Preprint of Autumn Meet. Jpn. Soc. Appl. Phys., (1983) 329.
12) A. Hatano and S. Nitta: Preprint of Autumn Meet. Jpn. Soc. Appl. Phys., (1983) 334.
13) A. Hatano and S. Nitta: Preprint of Autumn Meet. Jpn. Soc. Appl. Phys., (1983) 333.
14) Y. Inoue, S. Nakashima, A. Mitsuishi, S. Tabata, Y. Hamakawa, and H. Kukimoto: Preprint of Spring Meet. Jpn. Soc. Appl. Phys., (1982) 503.
15) Y. Yonesawa, K. Yamada, K. Nakazawa, S. Ueda, M. Kumeda, and T. Simizu: Preprint of Spring Meet. Jpn. Soc. Appl. Phys., (1982) 504.
16) F. Fujimoto, A. Ohtsuka, K. Komaki, Y. Hashimoto, T. Yamame, H. Yamashita, K. Ozawa, Y. Tawada, M. Kondo, H. Okamoto, and Y. Hamakawa: Preprint of Spring Meet. Jpn. Soc. Appl. Phys., (1982) 500.
17) H. Munakata and H. Kukimoto: Appl. Phy. Lett. *42*, 432.
18) H. Munakata, A. Hiroe, and H. Kukimoto: Preprint of Spring Meet. Jpn. Soc. Appl. Phys., (1983) 377.
19) M. Hirose: Jpn. J. Appl. Phys., *21* (1981) Suppl. 21-1, 275.
20) H. Kurata, H. Miyamoto, M. Hirose, and Y. Osaka: Jpn. J. Appl. Phys., *21* (1982) Suppl. 21-1, 205.
21) H. Nishiwaki, H. Tarui, K. Enomoto, Y. Nakashima, N. Nakamura, T. Takahama, S. Tsuda, M. Ohnishi, and Y. Kuwano: Preprint of Autumn Meet. Jpn. Soc. Appl. Phys., (1982) 334.
22) D. A. Anderson and W. E. Spear: Philos. Mag., *35* (1976) 935.
23) H. Ito, M. Hatanaka, K. Miyake, K. Nizuguchi, and K. Abe: Preprint of Autumn Meet. Jpn. Soc. Appl. Phys., (1982) 394.
24) S. Miyazaki and M. Hirose: Preprint of Autumn Meet. Jpn. Soc. Appl. Phys., (1983) 332.
25) G. Nakamura, K. Sato, and Y. Yukimoto: Proc. 4th E. C. Photovoltaic Solar Energy Conf., (1982).
26) Y. Yukimoto: Jpn. Ann. Rev. in Electronics Computers and Telecommun., (OHM ＊North-Holland, 1983).
27) S. Minomura and K. Morigaki: Latest Amorphous Si Hand BooK, Science Forum, (1983) 206.
28) K. Sato, T. Ishihara, H. Yakushiji, G. Nakamura, and Y. Yukimoto: Preprint of Autumn Meet. Jpn. Soc. Appl. Phys., (1983) 321.
29) K. Nozawa, S. Hanna, and I. Simizu: Preprint of Spring Meet. Jpn. Soc. Appl. Phys., (1983) 367.
30) K. Nozawa, S. Hanna, and I. Simizu: Preprint of Autumn Meet. Jpn. Soc. Appl. Phys., (1983) 347.
31) C. Verie, J. E. Rochette, and J. P. Rebouillat,: Proc. of 9th Int. Conf. on Amorphous and Liquid Semiconductor.
32) Y. Kuwano: Latest Amorphous Si Hand Book, Science Forum, (1983) 222.
33) H. Nishiwaki, H. Tarui, H. Doujyo, Y. Nakashima, K. Enomoto, T. Takahama, S. Tsuda, M. Ohnishi, and Y. Kuwano: Preprint of Spring Meet. Jpn. Soc. Appl. Phys., (1983) 365.
34) H. Itosaki, H. Kawai, Y. Fujita, K. Igarashi, and H. Ichiyanagi: Preprint of Spring Meet. Jpn. Soc. Appl. Phys., (1983) 381.
35) H. Itosaki, H. Kawai, Y. Fujita, K. Igarashi, and H. Ichiyanagi: Preprint of Autumn Meet. Jpn. Soc. Appl. Phys., (1983) 342.

CHAPTER 4

TOPICS IN DEVICE PHYSICS FIELD

4.1 Photogenerated Carrier Transport

Hiroaki OKAMOTO*

Abstract

The current status of studies on photogenerated carrier transport in amorphous semiconductors, particularly amorphous silicon, is reviewed. Several characteristics of transport phenomena observed in time-of-flight, photoconductivity and photovoltaic mode operations are interpreted in a unified manner on the basis of a simple model for carrier trapping and recombination associated with exponentially distributed band-tail states and correlated dangling bonds.

4.1.1 Introduction

Long-range disorder in amorphous solids has significant influence on the electronic density of the states. The sharp conduction and valence band edges of the periodic system are replaced by localized tail states extending into the forbidden gap. The absence of steric constraints imposed by the necessity of long-range periodicity results in a variety of structural defects, which tend to create localized states within the gap. The presence of these localized states characterizes the electronic properties of amorphous semiconductors, particularly their transport property. Band-tail states control macroscopic carrier transport and, provided they are sufficiently extensive, give rise to the phenomenon of trap-controlled dispersive transport through multiple-trapping.[1] Structural defects, on the other hand, potentially play a central role in deep trapping and recombination of photogenerated carriers,[2] and limit the performance of opto-electronic devices made of amorphous semiconductors; e. g. solar cells, photosensors, imaging devices and devices used in electrophotography.

In order to optimize such materials for possible device applications, it is necessary to understand the nature of localized gap states and their role in the transport and recombination processes. At present, there are several methods for investigating these processes. The time-of-flight (TOF) transient photocurrent technique provides direct information about transport and deep-trapping of both electrons and holes.[3] Details of the carrier recombination mechanism are studied through the excitation intensity and temperature dependences of photoconductivity (PC).[4] The collection efficiency spectra of photovoltaic devices (PV) can also be used to deduce the mobility-lifetime ($\mu\tau$) product of minority carriers.[5] Physical parameters, for example $\mu\tau$, evaluated by these methods often

---

* Faculty of Engineering Science, Osaka University, Toyonaka, Osaka, 560.

can not be reconciled with each other because different sample configurations and measuring conditions result in different distributions of photogenerated carriers and occupations of gap states. Therefore, such physical parameters should be translated into more universal quantities, such as gap-state distributions and relevant carrier capture cross sections, using appropriate theoretical analyses.

It is widely accepted, due to various experimental results, that amorphous semiconductors contain structural defects characterized by the effective correlation energy, U,[6] and band-tail states exponentially distributed from the conduction and valence band edges.[7] Section 4.1.2 describes a simple model for the carrier trapping and recombination associated with these localized electronic states in the gap.[8] Based on this model, recent experimental results of TOF, PC and PV mode measurement are interpreted in a unified manner, and from this, capture cross sections of well-defined defects (dangling bonds) are deduced for the amorphous silicon (a-Si) system. Section 4.1.3 briefly describes Kagawa's interpretation of transient photocurrent behavior, taking account of multi-phonon emission for weak electron-phonon coupling.[9]

4.1.2 Carrier Recombination and Trapping

Figure 4.1.1 schematically illustrates the gap state densities in amorphous semiconductors.[8] Band tail states D_{ct} and D_{vt}, which extend from the conduction band edge (ε_c) and the valence band edge (ε_v), are shown as shaded areas. It is assumed here that an amorphous semiconductor includes a single set of isolated defects characterized by the effective correlation energy, U.[10] Following Street and Mott,[6] these defects are labeled as

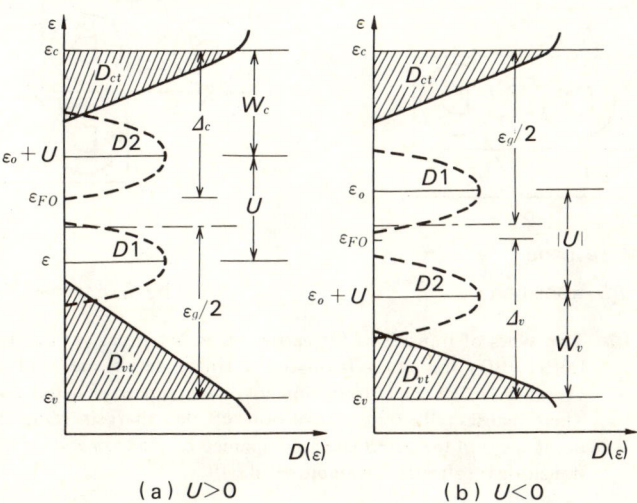

Fig. 4.1.1 Schematic illustrations of the density of states $D(\varepsilon)$ in amorphous semiconductors: (a) a-Si containing positively correlated defects and (b) chalcogenide glass containing negatively correlated defects. Here, ε_g is the energy band gap defined by $\varepsilon_c - \varepsilon_v$, and ε_{F0} denotes the Fermi level in equilibrium. [after H. Okamoto et al.[8]]

D^+, D^0 and D^-, to denote an unoccupied positive site, a singly occupied neutral site and a doubly occupied negative site, respectively. In a one-electron picture,[11] the single defect creates energy levels centered at ε_0 ($D1$) and $\varepsilon_0 + U$ ($D2$), as indicated in Fig. 4.1.1.[8] U is considered to be positive in a-Si[12] and negative in chalcogenide glasses.[13] Thus, Fig. 4.1.1 (a) and (b) correspond to the density of gap states in a-Si and chalcogenide glasses, respectively.

The model for carrier trapping and recombination adopted here assumes that photogenerated carriers propagating in transport states ($\varepsilon > \varepsilon_c$ or $\varepsilon < \varepsilon_v$) with a multiple trapping mode directly interact with correlated defects. The occupation of band tail states in a non-equilibrium steady state can be treated in terms of Schokley-Read statistics, as verified by Simmons and Taylor.[14] Carrier trapping and recombination associated with correlated defects have been investigated by Frye and Alder,[15] and Okamoto et al.[8] Figure 4.1.2 shows two types of transitions for electron-hole recombination through correlated defects.[8] Recombination with band-to-defect (BD) transition (a) is straightforward. For example, free electrons are captured by a particular D^+ sites at the net rate R_n^+, which converts the state of the defect from D^+ to D^0. Then free holes are captured by these D^0 sites at the net rate R_n^0, and recombination is completed. That is, carrier recombination due to BD transition involves only single defect sites. On the other hand, a pair of defect sites is required for defect-to-defect (DD) transition, as indicated in Fig. 4.1.2 (b). Moreover, DD transition can not complete carrier recombination by itself, but must be accompanied by BD transition.

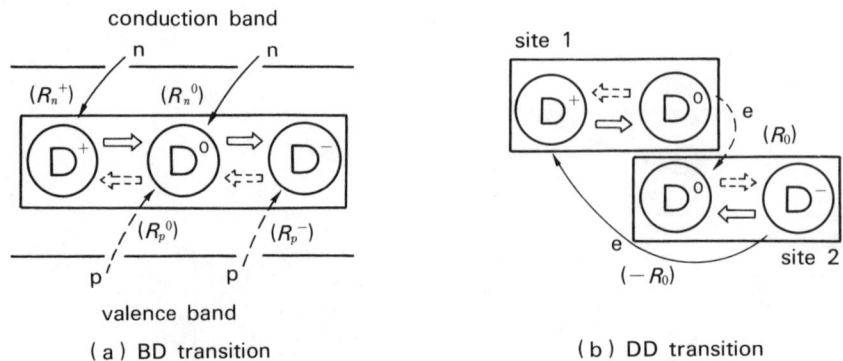

Fig. 4.1.2 Twp types of transitions for carrier recombination: (a) Band-to-Defect (BD) transition involving a single defect site, and (b) Defect-to-Defect (DD) transition involving a pair of defect sites. These figures illustrate conversion of the charged states (or occupancy) of the defect site accompanied by electron and/or hole transitions. [after H. Okamoto et al.[8]]

Taking into account the five elemental transitions shown in Fig. 4.1.2, Okamoto et al. have derived distribution functions F^+, F^0 and F^-, which represent the existence probabilities of D^+, D^0 and D^- sites, respectively, in a non-equilibrium steady state.[8] When the effect of DD transition is ignored, they can be written as

$$F^+ = P^-P^0/(N^0N^+ + N^+P^- + P^-P^0)$$
$$F^0 = N^+P^-/(N^0N^+ + N^+P^- + P^-P^0) \quad (4.1.1)$$
$$F^- = N^0N^+/(N^0N^+ + N^+P^- + P^-P^0)$$

where

$$\begin{aligned}
N^0 &= C_n^0 n + C_p^- p_1^-; & p_1^- &= (N_v/2)\exp[(\varepsilon_v - \varepsilon - U)/kT] \\
N^+ &= C_n^+ n + C_p^0 p_1^0; & p_1^0 &= (2N_v)\exp[(\varepsilon_v - \varepsilon)/kT] \\
P^0 &= C_p^0 p + C_n^+ n_1^+; & n_1^+ &= (N_c/2)\exp[\varepsilon - \varepsilon_c)/kT] \\
P^- &= C_p^- p + C_n^0 n_1^0; & n_1^0 &= (2N_c)\exp[(\varepsilon - \varepsilon_c + U)/kT]
\end{aligned} \quad (4.1.2)$$

C_n^0 and C_n^+ denote the free electron-capture probabilities of D^0 and D^+ sites, respectively, and C_p^0 and C_p^- are the free hole-capture probabilities of D^0 and D^+ sites, respectively, and C_p^0 and C_p^- are the free hole-capture probabilities D^0 and D^- sites. Here ε is the energy position of the Dl (singly-occupied electron state) level, k denotes Boltzman's constant and T is the absolute temperature. N_c and N_v are the effective density of states (N_{eff}) of the conduction and valence bands, respectively. Thus, the recombination rate, R, for the single-level case can be given by[8]

$$R = (np - n_i^2)(C_n^0 C_p^- N^+ + C_p^0 C_n^+ P^-)N/(N^0N^+ + N^+P^- + P^-P^0) \quad (4.1.3)$$

where n_i is the intrinsic carrier density and N is the total defect density.

4.1.3 Time-of-Flight Measurement

Time-of-Flight (TOF) transient photocurrent measurement is a very useful technique for understanding the transport and trapping processes in amorphous semiconductors. In the transit regime of the TOF trasient characteristics, where all of the injected carriers cross the sample, the mobility associated with shallow trapping can be directly determined from the transit times. In the range-limited regime, where carriers are trapped at deep gap states before completing transit, the mobility-lifetime product ($\mu\tau$) of the carriers is estimated from the field dependence of the collected charge, so far as the effect of geminate recombination can be neglected. Thus, TOF transient photocurrent independently identifies the type of transport, mobility and deep trapping lifetime.

Range limited regime

Figure 4.1.3 shows a plot of $\mu\tau$ for electrons vs. holes for a series of undoped and doped a-Si samples taken by Street.[16] For an intrinsic sample, he suggests, due to the constancy of the quantity $\mu\tau N_{s(i)}$, where $N_{s(i)}$ is the spin density corresponding to the lowest estimate of neutral dangling bonds (D^0), that the D^0 center controls deep trapping of both electrons and holes. For doped samples, as shown in Fig. 4.1.3, the $\mu\tau$ of minority carriers decreases inversely with the squre root of the doping concentration, whereas that of majority carriers is essentially unchanged. This observation is in direct agreement with the results of experiments using ESR,[17] photoluminescence,[18] photothermal deflection spectroscopy[19] and isothermal-capacitance-transient-spectroscopy (ICTS),[20] which all show that the density of defects increases with the squre root of the doping concentration. That is, the result of Fig. 4.1.3 can be readily explained in terms of doping-induced charged dangling bonds (D^+ or

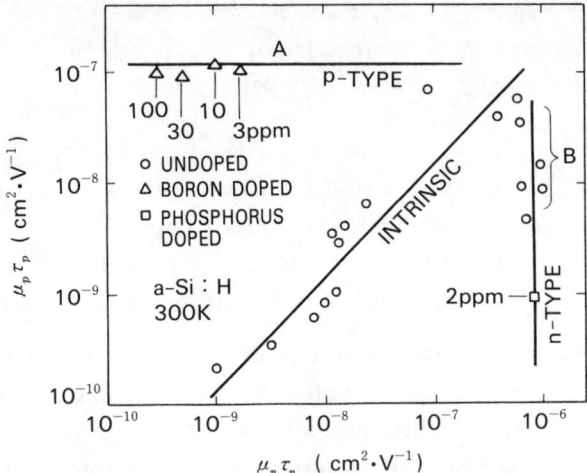

Fig. 4.1.3 Plot of $\mu\tau$ for electrons vs. holes for a series of undoped and doped a-Si samples. The groups of samples labeled A and B are contaminated with boron and phosphorus, respectively. [after R. A. Street et al.[16]]

D^-) that associate with deep trapping of minority carriers possessing charge opposite to the introduced dangling bonds.

Consider, for example, n-type material doped with a donor additive having atomic fraction F. According to Street's doping mechanism,[21] negatively charged dangling bonds (D^-) with a concentration proportional to the square of F are induced.[22] Thus, the total density of dangling bonds can be approximated by[23]

$$N \sim N_{s(i)} + AF^{1/2} \qquad (4.1.4)$$

where A is a constant on the order of 10^{19}cm^{-3}.[24] When deep trapping at correlated dangling bonds occurs from the transport state, the TOF deep trapping lifetime of free carriers can be formulated as

$$\tau_{n(\text{TOF})} \simeq [N(C_n^+ F_0^+ + C_n^0 F_0^0)]^{-1}$$
$$\tau_{p(\text{TOF})} \simeq [N(C_p^- F_0^- + C_p^0 F_0^0)]^{-1} \qquad (4.1.5)$$

where F_0^+, F_0^0 and F_0^- are distribution functions of D^+, D^0 and D^- centers in equilibrium[11,22] which can be obtained by replacing n and p with their equilibrium densities, n_0 and p_0, in Eq. (4.1.1). With appropriate simplification, Eq. (4.1.5) is reduced to

$$\tau_{n(\text{TOF})} \sim [C_n^0 N_{s(i)}]^{-1} \qquad (4.1.6)$$

$$\begin{aligned}\tau_{p(\text{TOF})} &\sim [C_p^0 N_{s(i)}]^{-1} \quad \text{(for the intrinsic case)} \\ &\sim [C_p^- AF^{1/2}]^{-1} \quad \text{(for the moderately doped case)}\end{aligned} \qquad (4.1.7)$$

For derivation of Eqs. (4.1.6) and (4.1.7), the correlation energy, U, is assumed to be much greater than kT. Similar results are easily obtained for p-type material. Eqs. (4.1.6) and (4.1.7) successfuly explain the variation of $\mu\tau$ products with spin density (intrinsic case) and doping, as long as the mobilities are not significantly influenced by doping. Assuming that trapping is from free carriers by a ballistic mechanism, the capture cross section, σ, can be given by the relevant capture probability, C, divided by the velocity, v ($\sim 10^7$ cm/s). Then, also assuming the free carrier mobilities of electrons and holes to be 13 cm²·V/s and 0.7 cm²/V·s,[7] a combination of the above theoretical formula and the experimental data in Fig. 4.1.3 leads to $\sigma_n^0 = 2 \sim 5 \times 10^{-15}$ cm², $\sigma_n^+ = 2 \sim 4 \times 10^{-14}$ cm², $\sigma_p^0 = 1 \sim 2 \times 10^{-15}$ cm² and $\sigma_p^- = 3 \sim 5 \times 10^{-15}$ cm² at 300 K. The reason for the large discrepancy between the magnitude of σ_n^0 thus evaluated and that directly measured by ICTS[20] ($10^{-17} \sim 10^{-19}$ cm²) is not yet clear. A key point for resolving this inconsistency might be a difference in the internal electric field under which each measurement was carried out.

Transit regime

TOF transient photocurrent behavior in the transit regime is usually analysed on the basis of the multiple trapping model developed by Tiedje and Rose,[25] in which the attempt-to-escape frequency is assumed to be independent of trap-state energy. Their analyses, when applied to secondary transient photocurrent, leads to physical inconsistencies:

(1) the electron occupation probability of trap states exceeds unity, and

(2) the demarcation level, which separates the thermalized and frozen states, moves with time from the conduction band to beyond the equilibrium Fermi level for moderately doped a-Si. In order to overcome these difficulties, Kagawa et al.[9] proposed a new model for transient photocurrent by taking account of the energy dependent attempt-to-escape frequency arising from multiphonon emission in a weak phonon-electron coupling limit.

It is assumed that photogenerated electrons in the conduction band are non-radiatively trapped by continuously distributed trap states with multiphonon emission. For weak electron-phonon coupling, the attempt-to-escape frequency, $\nu(\varepsilon)$, of trap states at energy ε is expressed as[9]

$$\nu(\varepsilon) = \nu_0 \exp[-(\varepsilon_c - \varepsilon)/kT_p] \qquad (4.1.8)$$

with

$$kT_p = \hbar\omega/[\gamma - \ln(1 + n_{ph})]$$

where n_{ph} denotes phonon density determined by the Bose-Einstein distribution, γ is constant and $\hbar\omega$ is phonon energy. At time t, trap states above the demarcation level, $\varepsilon_d(t)$, are already thermalized by the detrapping process. Thus, $\varepsilon_d(t)$ can be defined from

$$\varepsilon_c - \varepsilon_d(t) = [kTT_p/(T + T_p)] \ln(\nu_0 t) \qquad (4.1.9)$$

Through the procedure made by Tiedje et al., the photocurrent behavior can be represented as

$$J_{ph} \propto t^{-(1-\alpha^*)} \qquad (4.1.10)$$

with

$$\alpha^* = T(T_c + T_p)/T_c(T + T_p)$$

where T_c is the characteristic temperature of the exponential tail-states profile. For $T_p \to \infty$, Eq. (4.1.10) coincides with the result for the conventional MT model, where $\alpha = T/T_c$. For transient TOF photocurrent, the transit time, t_T, is given by

$$t_T = \nu_0^{-1+1/\alpha^*}[L^2\alpha^*(1-\alpha^*)\pi/\mu_0 V\sin(\alpha^*\pi)]^{1/\alpha^*} \qquad (4.1.11)$$

where L is sample thickness, V is applied voltage, and μ_0 is the microscopic mobility in the conduction band. The drift mobility, μ_d, is defined by $\mu_d = L^2/Vt_T$.

Table 4.1.1 summarizes several parameters tentatively estimated[9] from the experimental data of Tiedje et al.[7] The microscopic mobility, μ_0, which is assumed to be temperature independent, is quite different from those evaluated using the conventional MT model. Also, the coupling strength of holes is much greater than that of electrons, and a ν_0 of 4×10^{13} cm^{-1} for holes is too large for weak coupling. Hence, strong coupling may be more appropriate for holes. It is thought that the great difference between the drift mobilities of electrons and holes is due to the difference in strengths of coupling with phonons rather than to the difference in microscopic mobilities.

Table 4.1.1 Parameters obtained from analysis of TOF[7] assuming weak coupling. T_p for a temperature of 0K is presented. Parameters presented in ref. 7) are shown for comparison. [after T. Kagawa et al.[9]]

		T_c(K)	T_p(K)	ν_0(s^{-1})	μ_0(cm^2/V·s)
Weak Coupling	Electron	400	570	5×10^{11}	4.4
	Hole	750	730	4×10^{13}	3.2
Reference 7)	Electron	312	∞	4.6×10^{11}	13
	Hole	500	∞	1.6×10^{12}	0.67

Kagawa et al. have also made detailed numerical analysis of the time evolution of occupation probability, and pointed out that difficulties (1) and (2) described above can be successfuly removed by assuming weak electron-phonon coupling for multiphonon emission in electron capture.[9] This conclusion is directly supported by the capture cross section obeying the energy gap law, as experimentally verified by Ohkushi et al. using ICTS.[20]

4.1.4 Photoconductivity

Photoconductivity is a simple means of obtaining detailed information about transport and recombination mechanisms. The photoconductivity characteristics of a-Si, where electrons are assumed to be the main charged carriers for photoconduction, can be analyzed by solving the simultaneous equations arising from the recombination equation,[8]

$$G = \frac{C_n^+ N}{C_p^0 n_1^0}[(1-\xi)n^2 - (1+\xi)n_c^2 - \xi n_1^0 n]\frac{C_p^- n + C_p^0 n_1^0}{2n + n_1^0} \qquad (4.1.12)$$

and charge neutrality equation,

$$\xi = F^- - F^+ = \frac{n_0^2 - n_{c0}^2}{n_0^2 + n_0 n_1^0 + n_{c0}^2} - \frac{N_c^{1-T/T_c}}{N}(n^{T/T_c} - n_0^{T/T_c}) \qquad (4.1.13)$$

where n_{c0} is defined as the electron density that appears in the conduction band when the Fermi level is located at the center of the $D1$ and $D2$ levels.[8] Similar equations are obtained for chalcogenide glasses in which photogenerated holes dominate photoconduction. The results of such analysis are summarized in Tables 4.1.2 and 4.1.3.[8] In these tables, T_c and T_v denote the characteristic temperature (T_0) representing the widthes of the conduction and valence band tail states, respectively. Notations Δ_c and Δ_v are the energy separations between the transport states and the equilibrium Fermi level (Δ). The other notations are defined in Fig. 4.1.1 and in previous sections.

Table 4.1.2 Photoconductivity characteristics of a-Si with $n_a = N_c(n_1^0/N)^{T_c/(T_c-T)}$, $n_b^2 = (n_1^0/N)N_c^{1-T/T_c}n_0^{1+T/T_c}$ and $n_c = N_c[Nn_1^0/N_c(n_0+n_1^0)]^{T_c/T}$ [after H. Okamoto et al.[8]]

regimes	I$_1$	I$_2$	II$_1$
definition	$\delta n < n_0$		$n_0 < \delta n < n_a$
condition	$n_0 \gg n_b$	$n_0 \ll n_b$	$n_0 \ll n_b$
excitation-intensity dependence: G^γ	1	1	$T_c/(T+T_c)$
temperature dependence	$\exp[(\Delta_n - W_c)/kT]$	$T_c/(T+T_c)$	$(G/N_c^2 C n^+)^{T_c/(T+T_c)}$

regimes	II$_2$	III$_1$	III$_2$
definition	$n_a < \delta n < n_1^0$	$n_1^0 < \delta n < n_c$	$\delta n > n_c$
condition	$n_a < n_1^0$	$n_1^0 < n_c$	—
excitation-intensity dependence: G^γ	1/2	1/2	$T_c/(T+2T_c)$
temperature dependence	$\exp[-W_c/2kT]$	$(n_0+n_1^0)^{1/2}$	$\exp[-W_c T_c/kT(T+2T_c)]$

Table 4.1.3 Photoconductivity characteristics of chalcogenide glass with $p_a = N_v(p_1^0/N)^{T_v/(T_v-T)}$ and $p_b^2 = (p_1^0/N)N_v^{1-T/T_v}p_0^{1+T/T_v}$. [after H. Okamoto et al.[8]]

regimes		I$_1(p_0 \gg p_b)$	I$_2(p_0 \ll p_b)$
definition		$\delta p < p_0$	
excitation-intensity dependence: G^γ		1	1
temperature dependence	BD	$\exp[(\Delta_v - W_v)/kT]$	$T_v/(T+T_v)$
	DD	$\exp[(\Delta_v - 2W_v)/kT]$	$\sim \exp[(|U|-W_v)/kT]$

regimes		II$(p_0 \ll p_b)$	III$(p_0 \gg p_b)$
definition		$p_0 < \delta p < p_a$	$\delta p > p_a$
excitation-intensity dependence: G^γ		$T_v/(T+T_v)$	1/2
temperature dependence	BD	$\sim \exp[|U|T_v/kT(T+T_v)]$	$\exp[-W_v/2kT]$
	DD	$\sim \exp[(|U|-W_v)T_v/kT(T+T_v)]$	$\exp[-W_v/kT]$

It is obvious that the excitation intensity dependence, G^γ ($0.5 < \gamma < 1$), can appear when the following condition is satisfied:

$$T[\Delta - kT_0 \ln(2N_{\text{eff}}/N)] < (U/2)T_0 \tag{4.1.14}$$

The left side of this equation is usually positive for both a-Si and chalcogenide glasses. Hence, Eq. (4.1.14) can be satisfied for a positive U case if temperature T is lower than a certain critical temperature, but it can never be satisfied for a negative U case. That is, the excitation intensity dependence, G^γ ($0.5 < \gamma < 1$), indicates the positive-U nature of correlated defects associated with carrier recombination in an a-Si system.

The photoconductivity characteristics of a-Si under moderately high excitation and/or low temperature ($\delta n \gg n_0$) can be approximated as

$$n^\delta \propto \frac{G}{\hat{C}N}(n + n_1^0/2) \tag{4.1.15}$$

with

$$\hat{C} = \begin{cases} C_n^+ & (\text{for } n < n_1^0/2) \\ C_n^+ C_p^-/2C_p^0 & (\text{for } n > n_1^0/2) \end{cases}$$

Exponent δ takes $1 + T/T_c$, 2, 3 or $3 + T/T_c$, depending on the excitation intensity (G) and temperature (T). In other words, exponent γ of the excitation intensity dependence of photoconductivity, G^γ, changes successively from $T_c/(T + T_c)$ to $1/2$ and then to $T_c/(T + 2T_c)$ with increased G and/or lowered T. Wronski et al. report transitions of the γ-value that are consistent with the above discussion.[26] Eq. (4.1.13) suggests that the main recombination channel will tend to change when the quasi-Fermi level moves through a doubly-occupied dangling bond state ($D2$) located W_c below the conduction band edge, resulting in transition of the photoconductivity characteristics. This theoretical prediction is verified by the experimental observation of Spear et al.[27] It should be pointed out here that transition of the recombination channel does not always accompany change of the γ-value. This only happens when the relation

$$T_c \gtrsim W_c/k\ln[N_c/N] \tag{4.1.16}$$

is satisfied. Otherwise, transition of the recombination channel occurs without any distinct change in the γ-value.

A detailed examination of the temperature dependence of photoconductivity[8] has led to the conclusion that the maximum activation energies observed in the high- and low-temperature regimes (I_1 and II_2) provide rough estimates of $U/2$ and $W_c/2$, respectively. By applying experimentally obtained values, the magnitude of positive correlation energy U has been estimated to be about 0.4 eV, and W_c to be about 0.5 eV. In other words, singly-occupied ($D1$) and doubly-occupied dangling bond states ($D2$) locate approximately 0.9 eV and 0.5 eV below the conduction band edge, respectively. These values are identical with those deduced by other, more direct measurements.[28,29]

Kagawa et al. investigated the variation in photoconductivity characteristics with the Fermi level shift by utilizing a metal-oxide field effect transistor (MOSFET) configuration.[30]

Figure 4.1.4 shows the variation in γ-value with Fermi level shift induced by gate voltage. Detailed numerical calculation of the photoconductivity characteristics was made on the basis of modified Schokley-Read statistics.[30] Carrier recombination is assumed to occur via single-particle dangling bond states deep in the gap. As shown in Fig. 4.1.5, their calculation can well reproduce variation in γ-value with Fermi level shift when $\sigma_p^0 = 5 \times 10^{-16}$ cm^2 and the relation $\sigma_n^+ \gg \sigma_p^0$ are assumed. An important consequence of this good coincidence is that the γ-value can be varied by the Fermi level shift without any change in the recombination mechanism. Based upon this argument, they have concluded that carrier recombination is dominated by hole capture at neutral dangling bonds (D^0), and negatively charged dangling bonds (D^-) play no significant role. However, this conclusion may not always hold true because, as previously discussed, a change in recombination mechanism without any distinct change in the γ-value is possible in a correlated dangling-bond system.

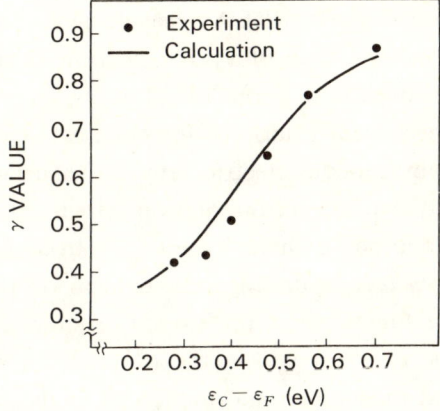

Fig. 4.1.4 Fermi-level-position dependence of exponent γ. The closed circles and solid line represent experimental and calculated results, respectively. [after T. Kagawa et al.[30]]

Lifetimes of free electrons and holes in undoped a-Si can be roughly estimated, using steady state photoconductivity and the delayed collection field technique,[31] as $\tau_{n(PC)} = 8 \times 10^{-6}$ s and $\tau_{p(PC)} = 3 \times 10^{-7}$ s. These photoconductivity lifetimes are given by

$$\tau_{n(PC)} = n_1^0 / 2 \sigma_n^+ v n_0 N$$
$$\tau_{p(PC)} = 1 / \sigma_p^0 v N \qquad (4.1.17)$$

Assuming $N = 10^{15}$ cm^{-3} and $\Delta_c - W_c = U/2 = 0.2$ eV, Eq. (4.1.17) yields $\sigma_n^+ = 3 \times 10^{-14}$ cm^2 and $\sigma_p^0 = 3 \times 10^{-16}$ cm^2. These values match Kagawa's postulation, and agree with those deduced from TOF measurement.

4.1.5 Photovoltaic Effect

The macroscopic situation of photogenerated carrier transport in photovoltaic mode operation is much more complicated than that encountered in the TOF and PC mode configurations. Photogenerated carrier transport occurs due to both diffusion and drift.[32] On transit towards either side of collecting electrodes, photogenerated electrons and holes are trapped at the gap states and recombine, affecting the occupation of those gap states, and thereby the space charge distribution. Redistribution of the space charge, then, violates the original internal electric field, which, in turn, results in modification of the photogenerated carrier transport. In such a situation, a formal procedure for the understanding of photovoltaic characteristics must be based upon simultaneous solutions to the electron and hole continuity equations and Poisson's equation. Moreover, not equilibrium but non-equilibrium statistics must be used for occupation of the gap states. Such a formal procedure, however, involves considerable difficulty because it requires detailed information about the gap state profile and relevant carrier capture cross sections.

Instead, Okamoto et al. developed a simple model for photogenerated carrier transport to give insight into the physics of transport in photovoltaic mode operation.[33] This model can be used not only to explain the photovoltaic characteristics but also to deduce several physical parameters related to carrier recombination.[5] According to back-surface-reflected-electroabsorption (BASREA) measurement,[34] the internal electric field is almost constant within the active i-layer of a p-i-n junction of practical dimensions, as long as the excitation intensity is lower than about 10^{15} photons/s·cm². Hence, a uniform internal electric field approximation is usually adopted for practical analysis.[5] The effects of carrier recombination and an inhomogeneous electric field in the surface and interface regions are renormalized into the effective surface recombination velocity at the p/i and i/n interfaces.[5] In general, photovoltaic lifetimes defined by $\tau_{n(PV)} = \delta n/R$ ($\delta n \ll \delta p$) and $\tau_{p(PV)} = \delta p/R$ ($\delta n \gg \delta p$), where R is the net recombination rate, depend on the distribution of electrons and holes, and thereby are functions of position. For simplicity, however, photovoltaic lifetimes are assumed to be constant within the i-layer. The validity of this assumption should be critically checked through formal numerical analysis.

Figure 4.1.5 shows mobility-lifetime product as a function of spin density, N_s, at $g = 2.0055$. The mobility-lifetime product ($\mu\tau$), which now denotes $\mu_n\tau_n + \mu_p\tau_p$, was evaluated from the bias voltage dependence of the carrier collection efficiency spectrum on the basis of the variable minority-carrier transport model.[5] As shown in this figure, $\mu\tau$ is inversely proportional to N_s. Since N_s gives the lowest estimate of dangling bond density N in the case of intrinsic material, the relation $\mu\tau \propto N_s^{-1}$ suggests that carrier recombination takes place via the dangling bond states. Fig. 4.1.5 also shows how defect density N_A, which is responsible for optical absorption around 1.0 eV, correlates with dangling bond density. The proportionality of N_A to N_s indicates that optical absorption around 1.0 eV originates from dangling bond states in the case of undoped intrinsic a-Si.

What influence the impurity inclusion into the active i-layer has on photogenerated carrier transport is of practical interest. A preliminary study on variation in the $\mu\tau$ products with boron and phosphorus doping revealed that both $\mu_n\tau_n$ and $\mu_p\tau_p$ attain their maxima for a-Si in which a small number of boron atoms are introduced and the Fermi level locates near the mid gap, as seen in Fig. 4.1.6.[5] Such a tendency is quite different from the variation in photoconductivity characteristics with Fermi level shift.[35] The $\mu\tau$ product responsible for photoconductivity is that of the majority carriers, while, the $\mu\tau$ product of minority

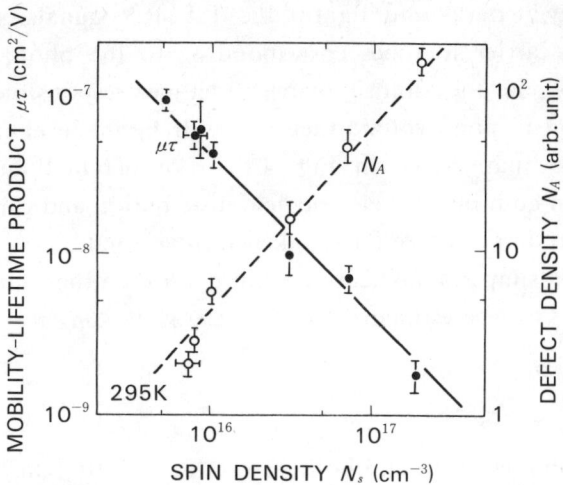

Fig. 4.1.5 Summation of $\mu_n\tau_n$ and $\mu_p\tau_p$ ($\mu\tau$) and defect density (N_A) evaluated from the absorption spectra below 1.4 eV as a function of spin density (N_s) at $g = 2.0055$. Here, $\mu\tau$ and N_A were measured in a series of p-i-n junctions, while N_s was measured on corresponding undoped i-type films.

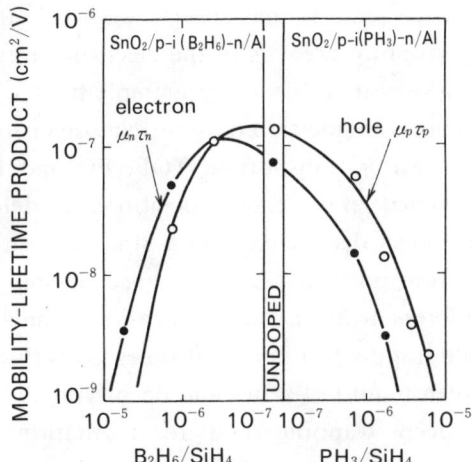

Fig. 4.1.6 Mobility-lifetime products of electrons ($\mu_n\tau_n$) and holes ($\mu_p\tau_p$) at 293K as a function of gaseous composition during plasma deposition of the i-layer. [after H. Okamoto et al.[5]]

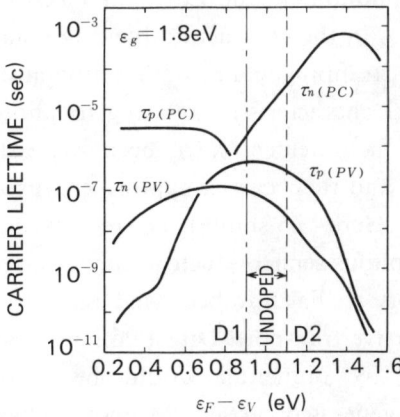

Fig. 4.1.7 Calculated results of carrier lifetimes related to the photoconductivity and photovoltaic properties. [after H. Okamoto et al.[37]]

carriers predominately limits the photovoltaic characteristics.

From this point of view, the variation in $\mu\tau$ products dominating both characteristics with Fermi level shift has been calculated on the basis of carrier recombination kinetics associated with positively correlated dangling bonds and band tail states, as presented in Section 4.1.2. The result is shown in Fig. 4.1.7. The physical parameters, e.g. gap state profile and relevant carrier capture cross sections were chosen to fit the experimental data on photoconductivity,[35] dark- and light-induced ESR[36] signals. Here, $\tau_{n(PC)}$ and $\tau_{p(PC)}$ denote the majority carrier lifetimes corresponding to the photoconductivity lifetimes, while, $\tau_{n(PV)}$ and $\tau_{p(PV)}$ are the minority carrier lifetimes responsible for the photovoltaic effect.[37] The variation in photovoltaic lifetimes with Fermi level shift is qualitatively in good agreement with that shown in Fig. 4.1.6. For detailed, quantitative discussion, however, variations in both internal electric field distribution and gap states profile induced by impurity incorporation must be fairly taken into account.

According to a simple analytical procedure, on the other hand, the upper limits of photovoltaic lifetimes can be estimated for the case $\sigma_p^- p_1^0 \ll \sigma_n^+ n_1^0$ as

$$\tau_{n(PV)} \sim p_1^0 / 2 n_i N \sigma_n^+ v$$
$$\tau_{p(PV)} \sim 1 / N \sigma_p^0 v \qquad (4.1.18)$$

By setting the maximum electron and hole $\mu\tau$ products to 10^{-7} cm^2/V, the upper limits of the capture cross sections related to positively correlated dangling bonds are estimated to be $\sigma_n^+ \sim 8 \times 10^{-14}$ cm^2 and $\sigma_p^0 \sim 7 \times 10^{-16}$ cm^2. The magnitudes of these cross sections do not contradict those evaluated by TOF and PC mode measurements.

4.1.6 Summary

Photogenerated carrier transport in the time-of-flight, photoconductivity and photovoltaic mode configurations has been discussed. It has been suggested that multiphonon-emission in weak electron-phonon coupling prevails in the electron capture process in the conducting band tail states in an a-Si system. Based on a simple model for carrier trapping and recombination at structural defects characterized by correlation energy, several characteristics relating to photogenerated carrier transport in TOF, PC and PV mode measurement have been successfully interpreted and, energy positions of defect states and their carrier capture cross sections have been deduced for an a-Si system.

Here, it should be remarked that the transport and capture mechanisms in amorphous semiconductors may be essentially different from those in crystalline semiconductors. As has now been well established, multiple trapping in band tail states gives rise to dispersive transport. Direct communication between band tail states and deep-lying defect states by tunneling would be involved in deep trapping and recombination of photogenerated carriers. Moreover, the possibility of diffusion-limited capture can not be excluded. In addition to isolated defects, defects spatially correlated with each other or with impurities may contribute to the carrier trapping and recombination processes. These problems must be critically checked from both the experimental and theoretical standpoints to accurately understand photogenerated carrier transport in amorphous semiconductors.

Acknowledgements

The author wishes to thank Dr. T. Kagawa for sending useful transcripts of his work prior to publication. He also thanks Professor Y. Hamakawa and Mr. H. Kida for their useful suggestions and comments.

References

1) J. Noolandi : Phys. Rev., *B 16* (1977) 4466.
2) D. Adler : J. Non-cryst. Solids, *35 & 36* (1980) 819.
3) R. A. Street : Appl. Phys. Lett., *41* (1982) 1060.
4) A. Rose : "Concept in Photoconductivity and Allied Problems", Interscience Publishers, New York, 1963.
5) H. Okamoto, H. Kida, S. Nonomura, and Y. Hamakawa : J. Appl. Phys., *54* (1983) 3236.
6) R. A. Street and N. F. Mott : Phys. Rev. Lett., *35* (1975) 1293.
7) T. Tiedje, J. M. Cebuka, D. L. Morel, and B. Abels : Phys. Rev. Lett., *46* (1981) 1425.
8) H. Okamoto, H. Kida, and Y. Hamakawa : Phil. Mag., *B49* (1984) 231.
9) T. Kagawa and N. Matsumoto : J. Non-cryst. Solids, *59 & 60* (1983) 477.
10) P. W. Anderson : Phys. Rev. Lett., *35* (1975) 1293.
11) H. Okamoto and Y. Hamakawa : Solid State Comm., *24* (1977) 23.
12) R. A. Street and D. K. Biegelsen : J. Non-cryst. Solids, *35 & 36* (1980) 651.
13) M. A. Kastner, D. Adler, and H. Fritzsche : Phys. Rev. Lett., *37* (1976) 1504.
14) J. G. Simmons and G. W. Taylor : Phys. Rev., *B 4* (1971) 502.
15) R. C. Frye and D. Adler : Phys. Rev., *B 24* (1981) 5485.
16) R. A. Street, J. Zesch, and M. J. Thompson : Appl. Phys. Lett., *43* (1983) 672.
17) H. Dersch, J. Stuke, and J. Beichler : Phys. Status Solidi, *B 105* (1981) 265.
18) R. A. Street, D. K. Biegelsen, and J. C. Knights : Phys. Rev., *B 24* (1981) 969.
19) W. B. Jackson and N. M. Amer : Phys. Rev., *B 25* (1982) 5559.
20) H. Ohkushi, M. Miyakawa, T. Okuno, S. Yamasaki, Y. Tokumaru, and K. Tanaka : J. Non-cryst. Solids, *59 & 60* (1983) 437.
21) R. A. Street : Phys. Rev. Lett., *49* (1982) 1187.
22) H. Okamoto and Y. Hamakawa : J. Non-cryst. Solids, *33* (1979) 225.
23) H. Okamoto, H. Kida, and Y. Hamakawa : Solid State Comm., *49* (1984) 731.
24) C. R. Wronski, B. Ables, T. Tiedje, and G. D. Cody : Solid State Comm., *22* (1982) 1423.
25) T. Tiedje and A. Rose : Solid State Comm., *37* (1980) 49.
26) C. R. Wronski and R. E. Daniel : Phys. Rev., *B 23* (1981) 794.
27) W. E. Spear : J. Non-cryst. Solids, *59 & 60* (1983) 1.
28) H. Ohkushi, T. Kakahama, Y. Tokumaru, S. Yamasaki, H. Oheda, and K. Tanaka : Phys. Rev., *B 27* (1983) 5184.
29) W. B. Jackson : Solid State Comm., *44* (1982) 477.
30) T. Kagawa, N. Matsumoto, and K. Kumabe : Phys. Rev., *B 28* (1983) 4570.
31) J. Mort, I. Chen, A. Troup, and M. Morgan : Phys. Rev. Lett., *45* (1980) 1348.
32) H. Okamoto, T. Yamaguchi, and Y. Hamakawa : J. de Physique, *42* suppl. 10 (1981) C4-507.
33) H. Okamoto, H. Kida, S. Nonomura, and Y. Hamakawa : Solar Cells, *8* (1983) 317.
34) S. Nonomura, H. Okamoto, and Y. Hamakawa : Appl. Phys., *A 32* (1983) 31.
35) S. Oda, Y. Saito, I. Shimizu, and E. Inoue : Phil. Mag., *43* (1981) 1079.
36) J. C. Knights, D. K. Biegelesen, and I. Solomon : Solid State Comm., *22* (1977) 133.
37) H. Okamoto, H. Kida, S. Nomura, K. Fukumoto, and Y. Hamakawa : J. Non-cryst. Solids, *59 & 60* (1983) 1103.

4.2 Impurity Effects

Akio HIRAKI*

Abstract

Impurities tend to solve or invade into the amorphous film by 1 or 2 orders of magnitude higher in concentration than into the corresponding crystalline one. Some examples of the impurity-invasion are demonstrated and at the same time necessity of protection and monitoring against such invasion for dependable and reproducible a-Si : H films for solar cell devices is stressed. Also a positive effect of some impurities such as N, O and H for stabilizing the a-Si phase is described.

4.2.1 Introduction

Generally speaking, amorphous (a-) materials such as hydrogenated amorphous silicon (a-Si : H) have far higher solubility for impurity atoms than do crystalline (c-) materials. Therefore, in the study of a-Si : H, a determination of the types, concentrations, and distributions of impurity atoms in the host lattice is prime importance to study the amorphous material. Also, for the fabrication of reproducible and reliable solar cells from these a-Si : H films, establishment of dependable characterization methods is of course important. In other words, proper characterization methods for monitoring the fabrication process are highly valued. Unfortunately, however, many measurements of electrical and optical properties and their application to various devices have sometimes been performed in the absense of such crucial methods. For the compositional characterization of a-Si : H films, several methods, such as infrared absorption (IR), electron spectroscopy [X-ray photoelectron (XPS), Auger electron (AES) and so on], and secondary ion mass spectroscopy (SIMS) have been employed. But these methods have the disadvantages of producing only moderate information on film depth, being destructive to the samples and measuring only relative concentrations of impurities. In this respect, as has been pointed out already in JARECT Vol. 2 (1982) [P. 52~67] by Hiraki and Imura[1], the Rutherford backscattering (RBS) using MeV He$^+$ ion beams technique provides better results in the characterization of solid films. This technique is essentially non-destructive and can reveal the distribution depths of almost all impurity atoms in solid films. RBS analysis can also give the atomic density of a target film. In principle the RBS method is insensitive to impurities with low atomic numbers and cannot detect atoms with lower atomic numbers than incident ions, such as hydrogen. However, if the hydrogen content is higher than a few at. %, as is

---

* Department of Electrical Engineering, Osaka University, Osaka 565.

presently the case with a-Si : H films, obtaining information on the hydrogen concentration is possible. We have also shown that the usage of ultra-high energy (100 MeV) can be a more precise detection method of hydrogen concentration, as explained in JARECT Vol. 8 (1983) [P. 155~171] by Imura and Hiraki.[2]

In the present article, some recent topics in this field are described; first the usage of the RBS method for the film characterization with respect to impurity is demonstrated in 4.2.2, then in 4.2.3 some new knowledges and the positive role of invading impurities such as nitrogen (N), oxygen (O) and also hydrogen (H) for stabilizing the amorphous phase are described, and finally in 4.2.4 one example of impurity invasion from an electrode into the i-layer of a-Si in solar cell structure is shown.

4.2.2 Rutherford Backscattering (RBS) Spectrometry for Film Characterization

Besides the well known ability of RBS to reveal the depth distribution of impurities in the specimen films, this method has been found to be useful in the detection of hydrogen atoms. Present author and the coworkers first tested this possibility for a-Si : H films using a 1.5 MeV He$^+$ ion beam.[1] The principle is based on the reduction of RBS yields of Si in a-Si : H due to the energy loss by H.

The RBS method was also employed to determine the atomic density of silicon, the amounts of subcomponents, and the concentrations of impurities in films of a-Si : H (and a-Si). Details of the detection of H atoms as well as the density determination of a-Si : H are explained in Ref. 1).

Though the RBS method has an uncertainity of a few atomic % for the impurity

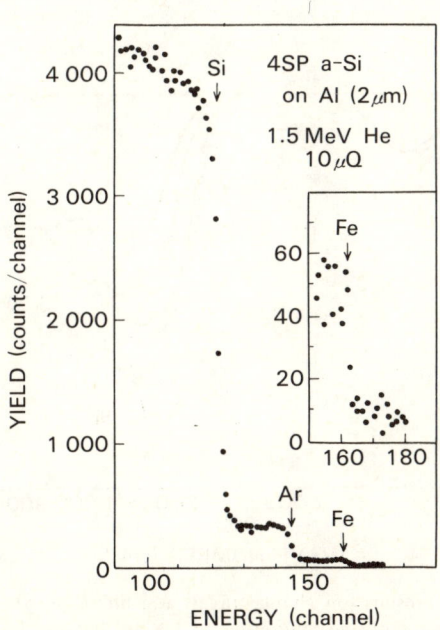

Fig. 4.2.1 Backscattering spectrum from a-Si film deposited on Al in Ar gas. Impurity (Fe) is clearly seen.

concentration due to statistical error, this method is simple and has the advantage of providing information on, in principle, all impurities and their depth distribution simultaneously. So, since amorphous films exhibit higher solubility for impurities than crystalline films, as mentioned already in 4.2.1, the chances of unfavorable impurities invading into a-Si : H films during the fabrication process are very good. One example is shown by the backscattering spectrum of an a-Si film (Fig. 4.2.1), fabricated by reactive sputtering in argon (Ar) plus H_2 (Ar + H_2) gas, to reveal a presence of about 0.5 at. % of Fe. The Fe atoms came from the stainless steel wall of the sputtering chamber and caused the film to be very highly conductive (or metallic). Please note that the solubility of Fe into c-Si is less than 10^{-2} at. %, even at 1000°C. Invasion of oxygen into a-Si films during and after the fabrication procedure can also be clearly determined by RBS as explained below.

This oxygen invasion was incidentally found in the course of studying a-Si : H film fabricated by reactive sputtering.[3] Namely, oxygen content in the film, placed in an air atmosphere, increased with the passage of time.

An r. f. diode sputtering apparatus supplying 140W at 13.56MHz was used. The pressure of Ar gas during sputtering varied between 3×10^{-2} and 1.5×10^{-1} Torr. The spacing between electrodes was 45mm. The sputtering chamber was evacuated to less than 5×10^{-7} Torr before introducing Ar gas. The gas was more then 99.99% pure, with the major impurities of O_2 and H_2O at less than 10ppm each. The deposition rate was about 2 ~3Å/s.

A series of infrared (IR) transmittance spectra of an a-Si film as a function of exposure time to air after fabrication are shown in Fig. 4.2.2. A film with a thickness of about 1.7μm was deposited on an intrinsic Si wafer at a substrate temperature of about 55°C (water cooled) in an Ar atmosphere of 1.5×10^{-1} Torr. The IR spectrum of the as-

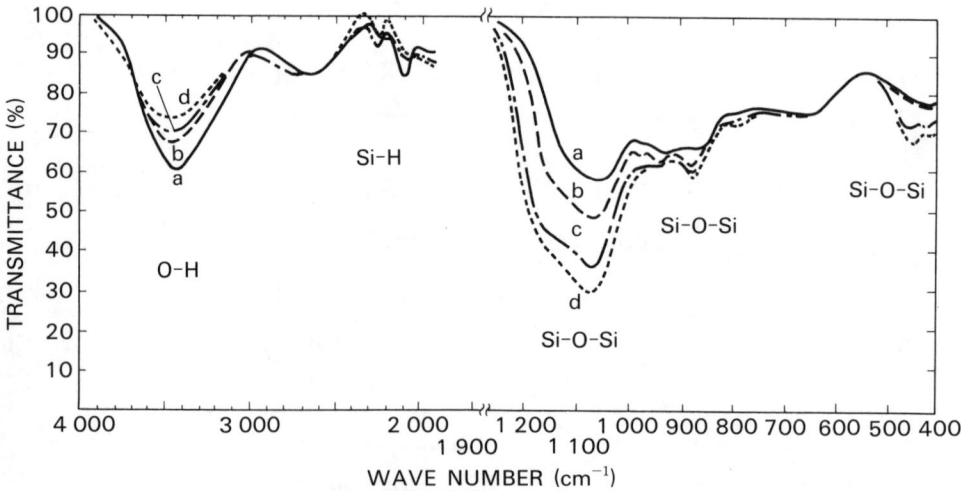

Fig. 4.2.2 IR transmission change in an a-Si film exposed to air. Ar gas pressure during sputtering: 1.5×10^{-1} Torr. It takes 0.2h to perform an IR measurement. a : 0h (as-deposited), b : 3h, c : 26h, d : 115h. The small transmission minimum near 2650cm^{-1} is due to interference.

deposited film shows the presence of Si-O-Si structural units from as early as the very beginning of deposition. Absorption bands near 450, 880 and 1070cm^{-1} have been assigned to the bending, symmetric stretching, and asymmetric stretching vibrations, respectively, of Si-O-Si configuration, on the basis of the standard spectra of organo-silicon molecules and oxygenated silicon with the Si-O-Si units.[4]

The absorbance around 450, 880 and 1070cm^{-1}, corresponding to the relative amounts of oxygen, increased nearly in proportion to the square root of the elapsed time in the initial 20 hours or so. Consequently, the process of oxygen inclusion or invasion may be a diffusion limited reaction. A diffusion constant, presumably for oxygen molecules in a-Si, is estimated from the time and the thickness to be of the order of 10^{-13}cm$^2 \cdot$s^{-1}, a reasonable value for the migration process of a gas molecule in a solid at room temperature.

Immediately after the film was taken from the evacuated desiccator which it had been stored in for a few days following preparation, the IR absorption spectrum presented hardly any difference from that of the as-deposited film. Gradually, however, it absorbed more and more oxygen from the air, as shown in Fig. 4.2.2. These results clearly show that a part of the oxygen in the film is taken from the air after removal from the sputtering chamber, and that the film has already been partially oxidized on deposition.

However, with Ar gas pressure during sputtering of 3×10^{-2} Torr or less, the as-deposited films of a-Si did show no indication of oxygen invasion in IR bands. Moreover, no oxygen absorption was observed during storage in air.

Similar gas pressure dependence on oxygen invasion was found[3] in a-Si : H films deposited by reactive sputtering under an Ar + H$_2$ (10mol%) atmosphere at a total pressure of 1.0×10^{-1} Torr, as shown in Fig. 4.2.3. The oxygen content in the as-deposited film of a-Si : H appears to be minimized by deposition in an atmosphere containing H$_2$, which can reduce O$_2$. In the as-deposited film of a-Si : H in Fig .4.2.3 (curve : a), Si-H bonding modes

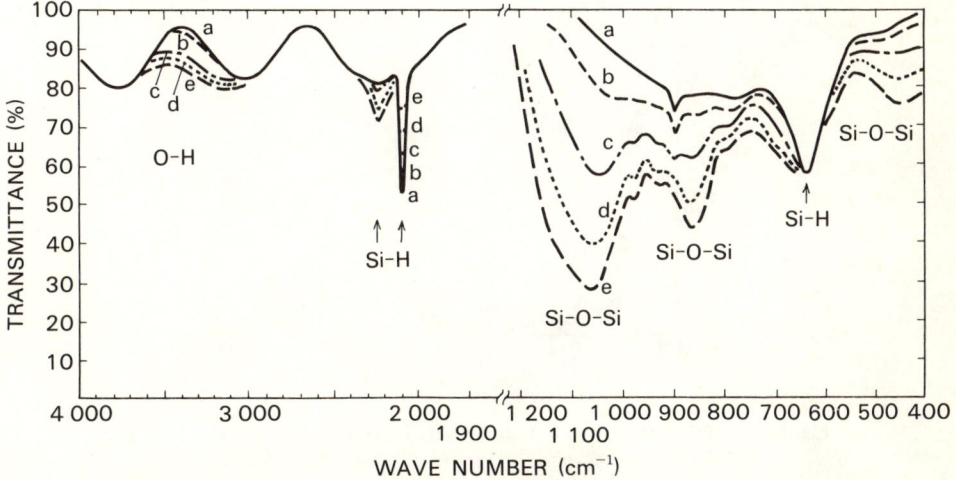

Fig. 4.2.3 IR transmission change in an a-Si : H film exposed to air. Total Ar + H$_2$ (10mol%) gas pressure during reactive sputtering : 1.0×10^{-1} Torr. Thickness : 2.7 μm. a : 0h (as-deposited), b : 2h, c : 15 h, d : 41h, e : 115h.

have been distinguished. They are a stretching mode near $2100\,cm^{-1}$, a bending mode near $895\,cm^{-1}$, and a rocking or wagging mode near $640\,cm^{-1}$. The hydrogen content in the film was estimated to be about 25 at. % from the integrated intensity of the stretching bands. Although the structure of the stretching bands (near 2100 and $2250\,cm^{-1}$) changed with the passage of time, the integrated intensity of the bands remained nearly constant, suggesting that the total content of hydrogen covalently bonded to Si is unaffected by the oxygen inclusion or invasion process. Only the environment surrounding the Si-H structures or the force constant of the Si-H bond appears to be modified by the invasion of electronegative oxygen. Also, while this oxygen absorption process resulted in undesirable changes in the film photocurrent, the dark (unilluminated) conductivity of a-Si : H was almost unchanged.

The dark conductivity of a-Si, however, decreased by two orders of magnitude, from about 10^{-3} to $10^{-5}\,\Omega^{-1}\cdot cm^{-1}$, upon sufficient oxygen inclusion, whereas optical band gaps of both a-Si and a-Si : H broadened by about 0.04 eV.[3]

RBS spectra of two double-layered specimens of a-Si and a-Si : H, are shown in Fig. 4.2.4. These specimens were deposited alternately with thicknesses of about $0.1\,\mu m$ each on sintered graphite plates, at a substrate temperature of about 55°C. The a-Si layers were

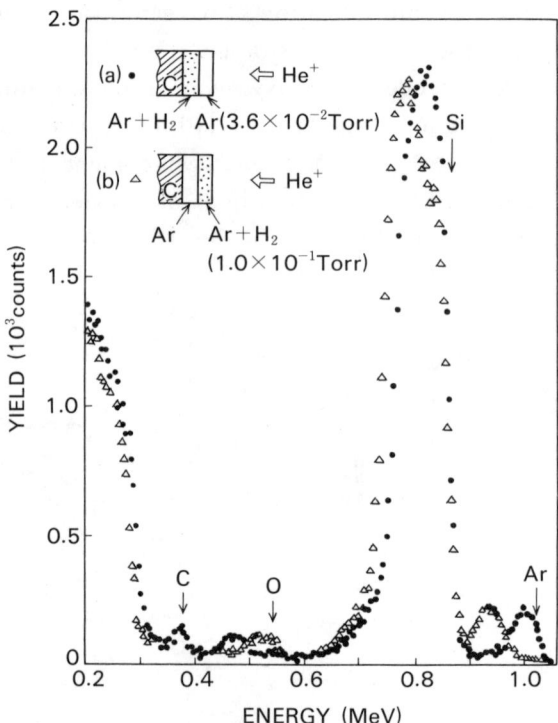

Fig. 4.2.4 RBS spectra of two double-layered specimens of a-Si and a-Si : H deposited on graphite substrates at pressures of Ar only 3.6×10^{-2} Torr for Ar and 1.0×10^{-1} Torr for Ar + H_2 (10 mol%). The measurements were carried out three days after the preparation. The small background of RBS yield in the region corresponding to oxygen in the film can be assumed to arise from multiple scattering.

deposited at such low pressures (i. e. 3.6×10^{-2} Torr) that no oxygen was absorbed, whereas the a-Si : H layers were deposited at the higher pressure of 1.0×10^{-1} Torr, as in Fig. 4.2.3. The two specimens were exposed to air for about three days after preparation. In specimen (a) oxygen was observed only in the inner layer of a-Si : H. This suggests that the oxygen present came through the outer layer of a-Si, where, however, no oxygen was observed. This outer layer of a-Si seems to allow the free migration of oxygen, while having no active sites for combination.

The oxygen and argon contents were determined from RBS spectra as a function of the argon gas pressure between 1.5×10^{-1} and 3×10^{-4} Torr, as shown in Fig. 4.2.5. No indication of the presence of oxygen was seen in the spectra of films fabricated at less than 3×10^{-2} Torr, except for slight surface contamination and multiple scattering background. On the other hand, the films fabricated at pressures of 5×10^{-2} Torr and above did show oxygen inclusion. A clear correlation was recognized between the Ar content, characteristics of oxygen invasion or inclusion, and the pressure during deposition of the films, with a critical value of $3 \sim 5 \times 10^{-2}$ Torr. The mean free path of sputtered Si atoms becomes much shorter than the electrode distance of 45mm in this pressure range. Apparently the kinetic energy of the sputtered Si atoms is of crucial importance in determining the structure of the film. In other words, structures such as voids and grain boundaries in the deposited film might change in this pressure range; at least, atomic arrangements of Si on these inner surfaces seem to be critically altered by the gas pressure on deposition. For more detailed discussion of this oxygen invasion, please refer to Ref. 3).

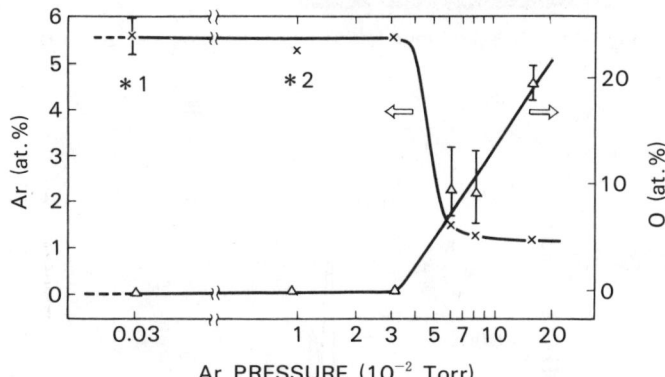

Fig. 4.2.5 Contents of oxygen and Ar, determined by RBS, in a-Si films sputter-deposited under Ar various pressures. The measurements were carried out two days after preparation. Rather large error-bars on some points for oxygen are due to the uncertainty in determination of low atomic number elements. *1: This sample was deposited by tetrode sputtering. Power: 170W. Distance between substrate and target: 65mm. Deposition rate: $1 \sim 2$ Å/s. *2: This sample was fabricated in Electrotechnical Lab. Headquarters. Distance between electrodes: 55mm. Power: 150W.

4.2.3 Positive Role of Some Impurities for Stabilization of Amorphous State Studied from Transformation of Microcrystalline Si : H to Amorphous One Due to Presence of Nitrogen

As has already been described in Ref. 1), r. f. -sputtering of Si in H_2-gas atmosphere produces films composed of microcrystalline hydrogenated Si (μc-Si : H) at substrate temperatures as low as ~130°K⋯ the size of the microcrystals ranging from ~50 nm to ~5 nm in diameter as clearly shown for example, by the high resolution image photograph in Fig. 4.2.6. The μc-Si-films thus produced show several interesting features.[1,5] One of them is infra-red (IR) property. A remarkable feature indicated by the IR spectra is that SiH monohydride configurations responsible to stretching near 1990 cm^{-1} are completely absent in the μc-Si-films,[1] as also seen in Fig. 4.2.12 ①.

The cause of this complete absence of SiH configuration has been pursued as a function of the film thickness through IR study (Fig. 4.2.7) correlated with film morphology

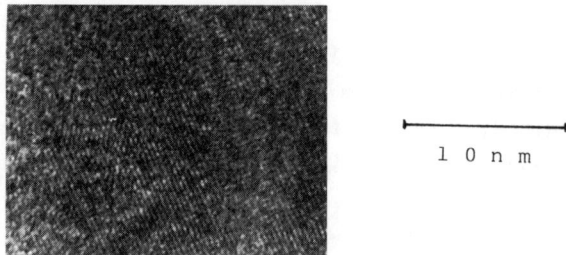

Fig. 4.2.6 High resolution lattice image photograph of μc-Si : H.

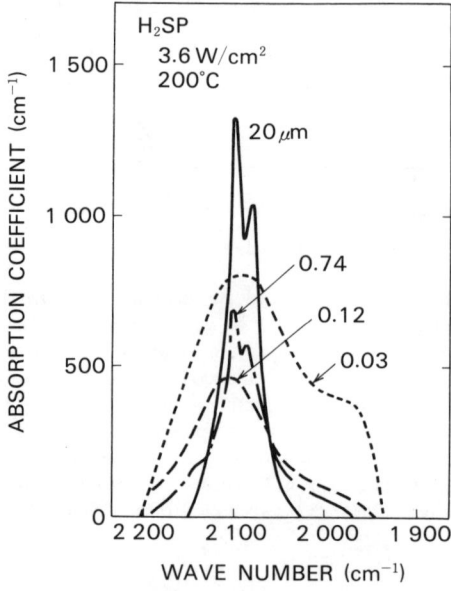

Fig. 4.2.7 Change of IR spectra of μc-Si : H with film thickness.

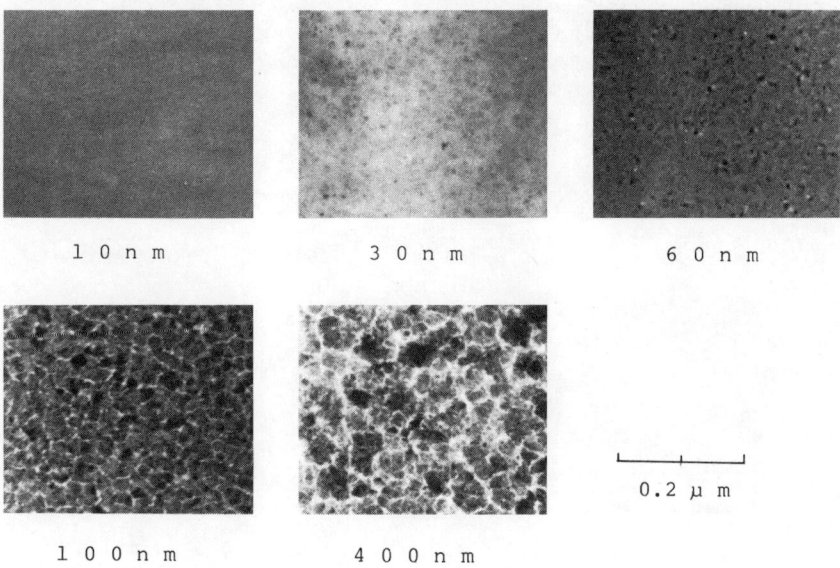

Fig. 4.2.8 Morphological change of μc-Si : H with film thickness by TEM.

and crystallinity by TEM observations (Fig. 4.2.8), and the following information was obtained. (a) When the film is sufficiently thin ($\lesssim 30$nm), IR spectra do show the presence of SiH configuration and at the same time the TEM indicates that the film is amorphous with uniform morphology rather than microcrystalline. (b) With the increment of the film thickness, the SiH configuration gradually disappears in the IR spectra and no more SiH configuration can be seen at $\sim 1\mu$m-thickness. (c) The morphology of the film also changes with the thickness from uniform amorphous to non-uniform columnar structure and finally microcrystalline structure with (110) orientation.

From (a), (b) and (c), a simple conclusion can drawn that the SiH configuration is at least necessary to stabilize uniform amorphous structure against nonuniform columnar or microcrystalline structure.

This is understood as follows. Since the tetrahedral unit of Si is very rigid, construction of amorphous structure by random connection of these units induces high strain. So, to reduce this strain some connecting bonds must be broken and hydrogen atoms attach to these broken bonds giving rises to the SiH configurations schematically shown in Fig. 4.2.9. But this reduction of strain by the SiH becomes insufficient when a uniform amorphous region exceeds 10nm in dimension. Consequently, with the increment of film thickness the

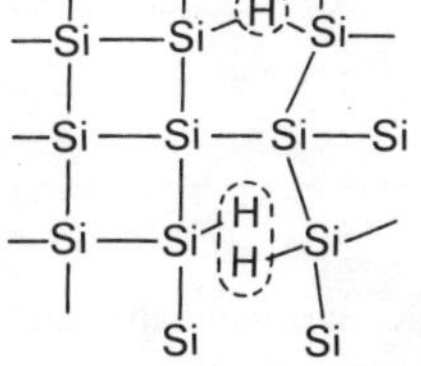

Fig. 4.2.9 Schematic of the network of a-Si : H.

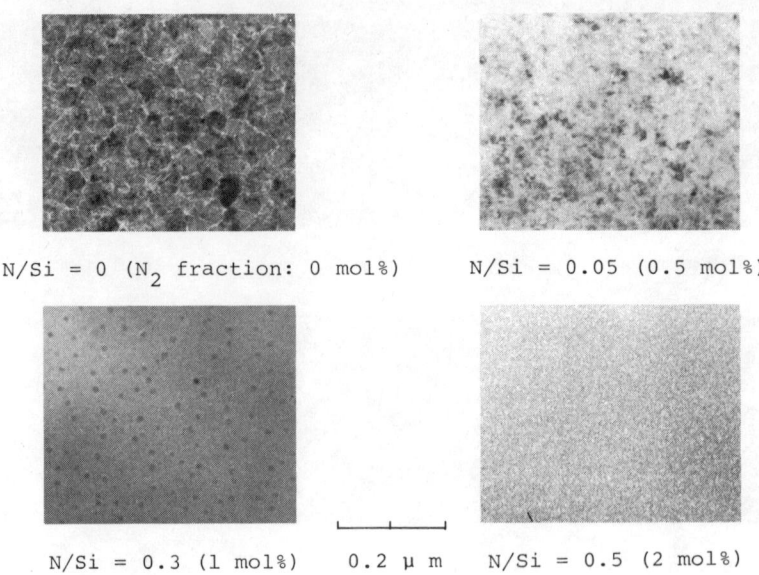

N/Si = 0 (N$_2$ fraction: 0 mol%)   N/Si = 0.05 (0.5 mol%)

N/Si = 0.3 (1 mol%)   0.2 μm   N/Si = 0.5 (2 mol%)

Fig. 4.2.10 Morphological change from microcrystal to amorphous due to addition of nitrogen by TEM.

structure gradually changes to a less strained one like columnar and finally microcrystalline.

This statement seems to be evidenced by the observation of microcrystalline to amorphous phase transition of the film through putting a small fraction of N$_2$ into H$_2$-gas for the sputtering atmosphere, since in this case nitrogen (N) is a strain-relieving element with a lower coordination number and higher electron negativity than Si (fabrication conditions: see Table 4.2.1). The phase transition as a function of N$_2$ fraction in H$_2$-gas is clearly seen from TEM (transmission electron micrograph) in Fig. 4.2.10,[6] ⋯ characteristic columnar morphology of the microcrystalline phase changes to uniform amorphous morphology at an N$_2$ fraction of only ~2 mol%. This is well contrasted by the fact, as seen in Fig. 4.2.11, that amorphous phase is only introduced in various inert atom + H$_2$ gas, like

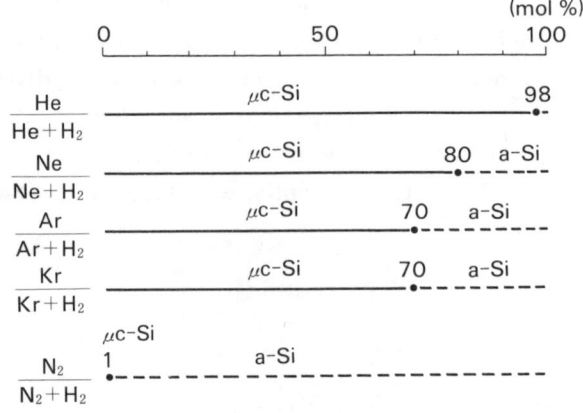

Fig. 4.2.11 Remarkable difference between inert (He, Ne, Ar, Kr) and highly electronegative nitrogen (N) atoms for the fabrication of a- or μc-Si:H films.

Table 4.2.1 Sputtering conditions.

Back pressure	$3\sim4 \times 10^{-7}$ Torr
Sputtering pressure	0.1 Torr
N_2 fraction in $H_2 + N_2$	$0\sim5$ mol%
r.f. power	3.8 W/cm^2
Substrate temperature	250°C

Ar + H_2 gas, with a far higher inert atom fraction (in the vicinity of ~ 80 mol%) under similar sputtering conditions (Table 4.2.1). In addition, corresponding to the transition in IR spectra (Fig. 4.2.12),[6] SiH configuration appears although the vibration frequency shifts due to the presence of nitrogen.

In this section, it has been shown that the strain in the r. f. -sputter deposited Si film due to the rigidness of the Si tetrahedral greatly influences the morpholgy. Except for strain-relieving atoms or impurities (like N)/ or atom groups (like SiH), the film morphology is a columnar one (including microcrystal) which is less dense than the uniform amorphous film. The strain-relieving impurity must be more electronegative than Si and the coordination number should be 2 or 3. Therefore, in addition to nitrogen, oxygen is also expected to play the same role. Requirement for a higer electonegative atom with a lower coordination number is understood as follows.

The rigidness or inflexibility of the Si-tetrahedral is due to the presence of the high

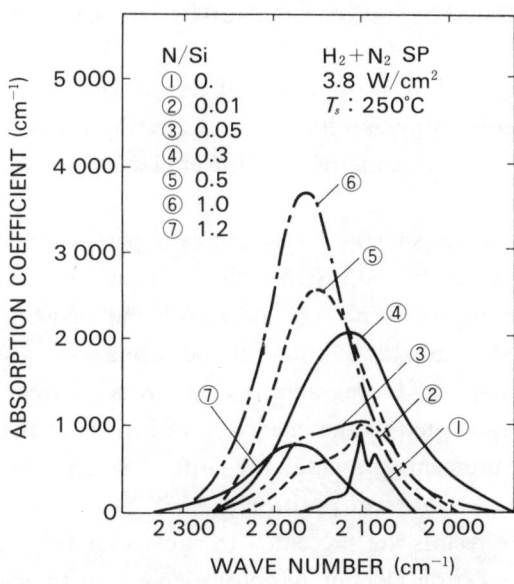

Fig. 4.2.12 Change of IR spectra of μc-Si : H due to addition of nitrogen.

density of the so-called bond charge in the middle of an Si-Si covalent bond. So when more electronegative impurity atoms like N attach to the Si-atoms, the electrons tend to go to the impurity atoms, which results in a reduction of bond charge and makes the Si tetrahedral more flexible. Lower coordination gives more freedom for the impurity atom to interconnect two or three amorphous regions whose size must be ~10 nm in dimension as already discussed.

In other words, a strain-relieving impurity is a kind of connector as shown in Fig. 4.2.13. Of course the bond angle (θ) between the connector atom and connected Si atoms (in the figure N⊲$_\theta$$^{Si}_{Si}$) can vary flexibly due to its higher electronegativity. In this respect H is an imperfect connector.

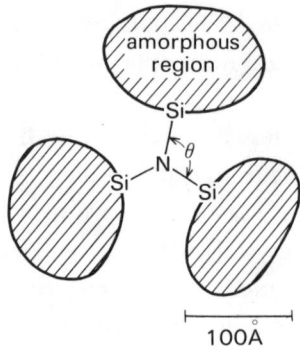

Fig. 4.2.13 Role of nitrogen(N) atom as strain-relieving impurity.

4.2.4 Invasion of Electrode Elements to the Inside of Amorphous Solar Cell

During the investigation of p-i-n solar cell design, cells with a transparent electrode (TE) of SnO_2 have sometimes shown a higher open circuit voltage (V_{oc}) than cells with an In-Sn oxide (ITO) electrode,[7] in spite of such lower electrical resistivity and better optical properties of ITO as TE.

Already surface analyses such as AES have been performed in combination with the ion-sputter etching technique and SIMS in order to measure the depth distribution and redox state of TE elements in the p-layer indicating the invasion from TE.[8] From the correlation between those results and solar cell performances it was concluded that the presence or invasion of reduced In degrades V_{oc}, whereas Sn does not.

Since V_{oc} is closely related to the energy-band profile of the p-i juction, we carried out systematic XPS measurements, in which the shift of Si-2p core electrons was shown to be a good measure of the Fermi-level position.

In this section, the results are presented to show that In acts as a donor and at the same time it compensates Boron (B), an acceptor element in the p-layer of the solar cell. For this p-layer, instead of a-SiH, a-SiC : H is utilized due to its excellent window effect. The p-a-SiC : H films were deposited on glass substrates coated with different TE's in a glow-discharge system and supplied for the present XPS and AES study from Prof. Hamakawa's group of Osaka University. Thicknesses of the films were about 100Å, estimated from the deposition rate.

For XPS a model SHIMADZU ESCA-650 was utilized. Hand-made systems for the pulse counting, accumulating and curve smoothing were used to make the energy resolution as accurate as $\pm 0.05\,eV$. Deposition of a thin Ag layer by evaporation onto the films was always carried out as an energy reference, 368.2 eV of Ag $3d_{5/2}$ peak. The sputter facility was only employed with an ion beam of 2.0 keV Ar in order to measure Fermi-level shift with depth.

Depth distributions of In and Sn and their redox-state were checked with AES equipment (PHI 15-120).

The binding energy of Si-2p core electrons in p-a-SiC : H on the three types TE's for different substrate temperatures is shown in Table 4.2.2. Here, the shift of the binding energy relatively correlates with the Fermi-level position. Evidently the highest binding energy shift to n-type was observed in p-a-SiC : H on ITO (SP), followed by ITO (EB) and then SnO_2 at each substrate temperature.

Table 4.2.2 Si-2p binding energy in p-a-SiC : H on various TE's and substrate temperatures. (eV)

TE	Substrate temperature (°C)		
	210	230	265
SnO_2	99.0	99.1	98.9
ITO (EB)	99.0	99.1	99.2
ITO (SP)	99.3	99.2	99.4

EB : Electron beam deposition
SP : Sputter deposition

Previously it has been shown[8] that reduced In and Sn from the TE were located in the p-layer of these films, and that a larger amount of reduced In in particular existed in a-SiC : H on ITO (SP) than on ITO (EB). The comparison of the binding energies with these depth distributions suggests that reduced In would lift Fermi-level upwards. Since In in the substitutional site is expected to play as an acceptor, this fact might indicate that In is in the interstitial site to act as a donor by the ionization of In into In^+ thereby lifting the Fermi-level upwards. On the contrary, Sn in the interstitial site stays unionized due to its higher ionization potential. Such a case of In acting as a donor was already reported with studying on junctions prepared by alloying In into silicon crystals. The complete physical interpretation for this "pseudotype conversion", however, has not yet been ascertained.

The comparison of V_{oc} with the results in Table 4.2.2 indicates that lower V_{oc} corresponds to higher binding energy. Therefore, the prevention of the diffusion or invasion of these reduced In into the p-layer to act as a donor, has to be achieved to eliminate this problem. But as a TE, ITO is a far superior material to SnO_2 both in terms of transparency for solar radiation and sheet resistivity. From these points of view a SnO_2/ITO bi-layered electrode is expected to prevent the invasion of In into the p-layer without degrading the electrical and optical properties of the whole electrode.

Depth distributions of In and Sn were shown in the previous paper,[8] which indicate that the invasion of In into p-layer was completely blocked by an SnO_2 layer 400 Å thick. But evident degradation of the SnO_2 layer and the invasion of a slight amount of In were observed in the specimen with a 100 Å SnO_2 layer.

The effects of preventing the invasion of In into the p-layer are shown in Fig. 4.2.14. Here, the performance of cells utilizing these TE's is better than that of cells utilizing ITO only.

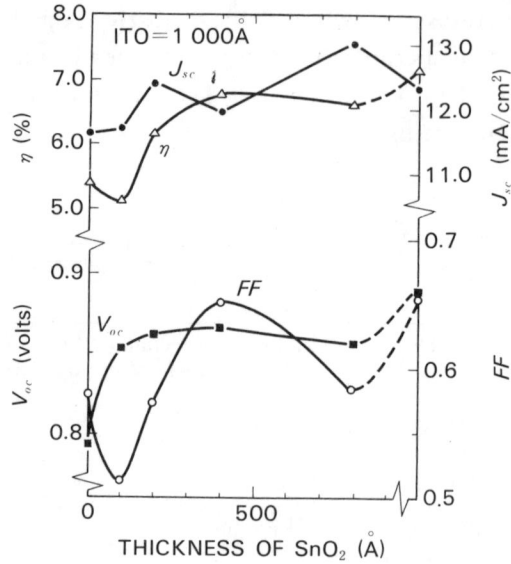

Fig. 4.2.14 Influence of the thickness of coated SnO_2 layer on the cell performance.

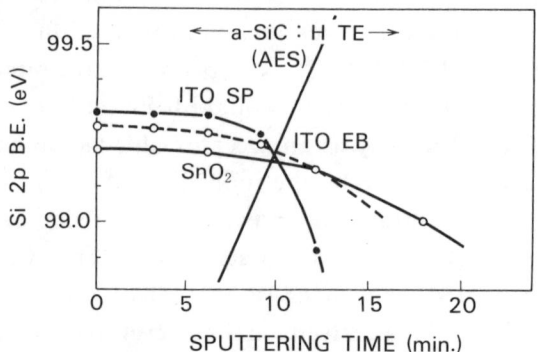

Fig. 4.2.15 Fermi-level shift with depth in p-a-SiC : H.
B_2H_6 : 10^{-4}
sputter rate : 10 Å/ min.

In probably competes with B incorporated as an acceptor if In acts as a donor. In order to investigate the competition between In and B, Fermi-level shift with depth and B-doping concentrations were measured. Fermi-level shift in p-a-SiC : H with depth is shown in Fig. 4.2.15. The result shows that at the surface region Fermi-level is almost constant. However, near TE's (at sputtering time from 10 to 20 min.) it shifts downwards, the reason for which has not yet been clarified. Fermi-level shift with depth does not coincide with the depth profile of In and Sn, but at the surface it exhibits the result of the above-mentioned competition between In and B.

Fermi-level shift in a-SiC : H films on four types of TE's, obtained without sputtering, are plotted against B-doping concentrations in Fig. 4.2.16. At higher B-doping, Fermi-level on each TE is almost the same, indicating that the effect of In is negligible. But at the lower B-doping the influence of In is clearly observed. The order of ITO (SP), ITO (EB), SnO_2/ITO and SnO_2 is determined by the amount of In diffused into a-SiC : H layer. Eventually, at lower B-doping, In affects Fermi-level shift to a considerable extent, so caution has to be taken in the design of p-i-n solar cells.

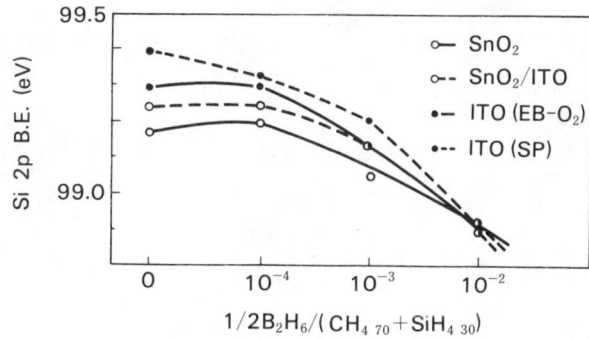

Fig. 4.2.16 Fermi-level shift with the doping concentrations of boron in a-SiC : H.

In this section, one example of impurity invasion into real amorphous solar cells is demonstrated and at the same time a new usage of XPS for the characterization of solar cell devices is described. The principle is based upon the ability of XPS to know the position of the Fermi-level of a specimen. So, without resorting to tedious measurement of electrical resistivity and Hall effect at various temperatures, sufficient information about the Fermi-level can be obtained by the far quicker and simpler XPS measurement thereby allowing doping efficiency and so on to be studied as well.

References

1) A. Hiraki and T. Imura : JARECT, *2* (1982) 52~67.
2) T. Imura and A. Hiraki : JARECT, *8* (1983) 151~171.
3) T. Imura, K. Ushita, and A. Hiraki : Jpn. J. Appl. Phys., *18* (1979) 1923.
4) H. J. Hrostowski and R. H. Kaiser : Phys. Rev., *107* (1957) 966.
5) A. Hiraki, T. Imura, K. Mogi, and M. Tashiro : J. Phys., *42* (1981) C 4-277.

6) A. Hiraki, Y. Fukushima, T. Sato, H. Kiyono, H. Terauchi, and T. Imura: J. Non-Cryst. Solids, *59~60* (1983) 791.
7) Y. Hamakawa: JARECT, *2* (1982) 134 ~155.
8) N. Fukada, T. Imura, A. Hiraki, Y. Hamakawa et al.: Jpn. J. Appl. Phys., *21* (1982) suppl. 21-2, 271.

4.3 Laser Scribing Lithography

Shumpei YAMAZAKI*, Satsuki WATABE* and Kenji ITOH*

Abstract

It has been pointed out that laser scribing lithography would be effective for the development of solar cells.

Up to now, however, no specific reports have been presented on the actual fabrication of solar cells using laser scribing lithography.

As an experiment, we produced an integrated solar cell structure using laser scribing of a TCO-PIN junction type amorphous semiconductor-back electrode structure formed on a glass substrate. That is, a 12-stage, series-connected, integrated solar cell was fabricated by two laser scribing steps and one evaporation step using a metal mask. The conversion efficiency was 6.4% for AM1 (100mW/cm^2) and 7.15% for a 500Lx fluorescent lamp (effective area 69%; substrate area 100cm^2). Furthermore, when a solar cell was produced by the so-called maskless process involving three laser masking steps alone, a 4.11% conversion efficiency was obtained (effective area 83.12%; substrate area 100cm^2). It was found that the effective area can be increased to between 80% to 94% by integration through laser processing.

4.3.1 Introduction

Laser scribing lithography is now being given much attention as its use permits substantial simplification of the integration step involved in the manufacture of photovoltaic cells, and hence is effective for low-cost mass production.[1~7]

Reports issued in the past, however, do not clearly indicate the specific manufacturing step or the manner in which a YAG laser was used, nor do they refer to the resulting conversion efficiencies. In other words, they describe only the merits of laser processing but make no mention of its demerits.

We describe various findings obtained with our actual fabrication of photovoltaic cells using laser scribing lithography.

The laser scribing process possesses the following merits:
(1) No masks are needed.
   Since a metal mask, screen printing mask, or similar mask is not used, manufacturing costs of photovoltaic cells can be reduced as a result, and no additional process is needed for maintaining mask quality.

---

* Semiconductor Energy Laboratory Co. Ltd., 3-11-1 Kamisoshigaya, Setagaya-ku, Tokyo, 157.

(2) Large-area photovoltaic cells can be fabricated.

Masks now in use usually measure 10cm × 10cm. If masks are made with dimensions of 20cm × 30cm or more, cost and the operating expenses will increase exponentially. With the use of the laser scribing technique, however, it is not difficult to produce a 40cm × 120cm panel.

(3) The effective area is increased.

Grooves cut by laser light are 30 to 150 μm wide and, accordingly, electrode coupling portions necessary for integration may be small in area. This makes it possible to increase the entire effective panel area to around 80~94%, even though effective area is 50~75% in the conventional metal mask process.

This paper demonstrates that an 83% effective area (12-stage, series-connected, integrated structure) is obtainable with the laser scribing technique.

(4) The laser scribing process is a computer-aided self-registration process.

The integration of the photovoltaic cell calls for three patterning steps. The registration of respective patterns can be achieved under computer control. Accordingly, it is also possible to adopt an unmanned process for the fabrication of the photovoltaic cell.

Laser scribing lithography has many advantages as described above but, at the same time, it possesses the following demerits:

(1) It is very difficult to control the depth of grooves.

A YAG laser (of a 1.06 μm wavelength) is usually employed for scribing. It is extremely difficult to control the depth of grooves which are cut by the pulsed laser light irradiation. Especially, it is very difficult, from an industrial point of view, to scribe a thin film of 1 μm or less without damaging the underlying layer.

(2) The irradiation by laser light used for scribing gives rise to annealing of the surrounding portion.

As a result of this laser annealing phenomenon, amorphous silicon is transformed into polycrystalline silicon, resulting in OFF current increases as great as 10 to 10^3 times, thereby degrading the characteristics of the photovoltaic cell.

(3) The formation of a given area is difficult in terms of productivity.

For cutting such area, it is necessary to repeatedly perform scanning by laser light having a width of 30 to 150 μm, and this is time-consuming.

Considering these many defects, conventional integrated structures discussed in Fig. 4.3.2 are not yet higher than the level of the conventional integrated structure produced by the metal mask process, and there is strong demand for a practical structure. This paper discusses integrated structures which differ completely from known structures produced by the metal mask process and which are intended to overcome the defects of the laser scribing process and to make use of its strong points.

4.3.2 Outline of Integrated Structures

Figures 4.3.1 (a) and (b) show the outline of the conventional photovoltaic cell manufacturing processes employing laser scribing. The manufacturing processes shown both use only the laser patterning method.

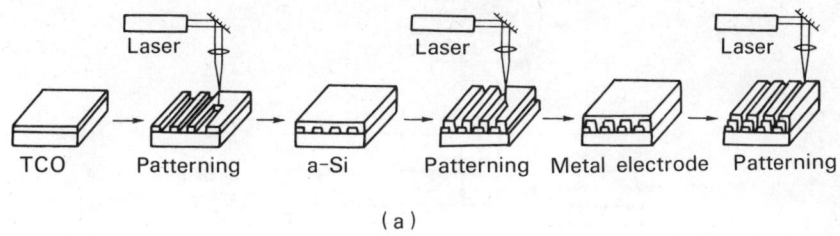

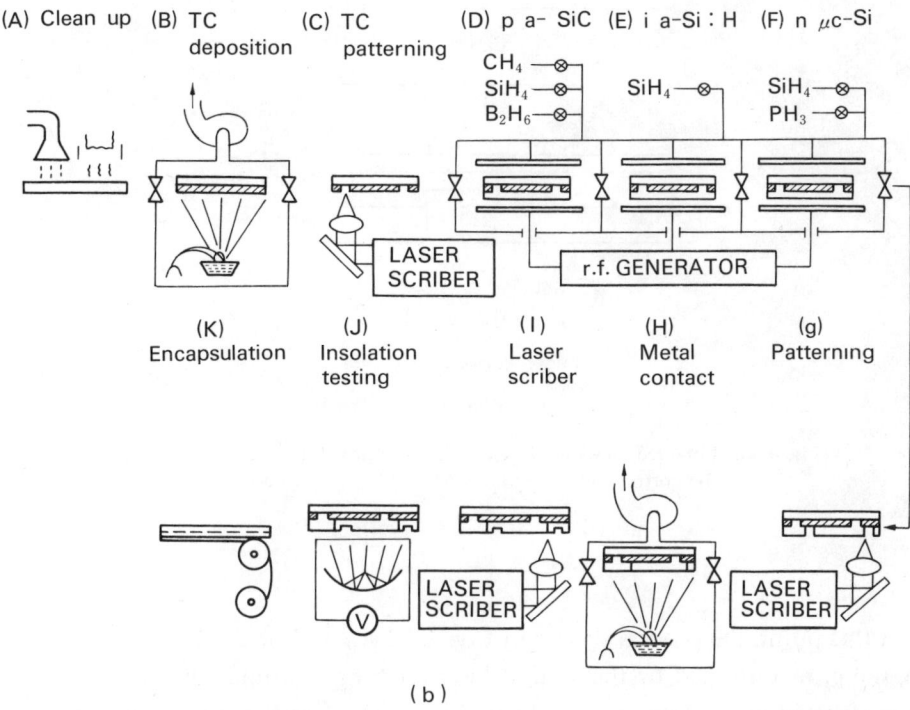

Fig. 4.3.1 Manufacturing processes of integrated a-Si solar cells. (a) [after S. Nakano et al.[4)]], (b) [after Y. Hamakawa et al.[5)]].

Figures 4.3.2 (a) and (b) show an enlargement of the electrode coupling portions necessary for integration in the two manufacturing methods.

In Fig. 4.3.2 (a) respective laser scribing steps overlap. In the second laser scribing of the a-Si layer, much difficulty is encountered in precisely controlling the amount of the cross-hatched portion of the underlying TCO layer that must remain in order to provide a contact for coupling use. Because of the lateral sawing of laser light for the second laser scribing, it becomes increasingly difficult to register, at the same position, contact portions of the back electrode of the left-hand photovoltaic cell and the TCO layer of the right-hand photovoltaic cell.

Furthermore, in the third laser scribing for selectively removing the back electrode, it is impossible to control the laser light irradiation in a manner for removing a portion of the underlying Si layer alone, leaving only the cross-hatched portion of the a-Si layer.

In view of the above, the method shown in Fig. 4.3.2 (a) is not satisfactory for

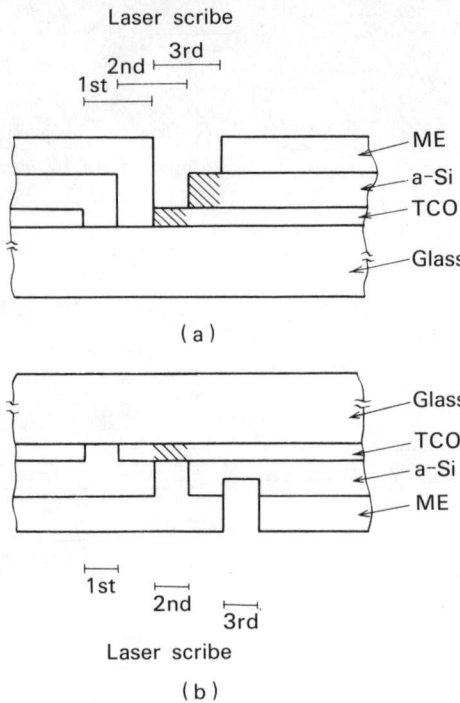

Fig. 4.3.2 Enlarged views of the electrode coupling portion in 4.3.1 (a) and (b) correspond to (a) and (b) of Fig. 4.3.1, respectively.

practical use.

At this point, the process shown in Fig. 4.3.2 (b) will be discussed. According to this process, the groove formed by the second laser scribing is formed at a short distance from the groove formed by the first laser scribing. This is effective but it is not easy to do without effecting the cross-hatched portion of the TCO layer underlying the groove during the second laser scribing. However, this method is far more practical than the method shown in Fig. 4.3.2 (a).

Here we will present new improved structures for those shown in Figs. 4.3.2 (a) and (b). Figures 4.3.3 (a) and (b) and Fig. 4.3.4 show structures produced selectively removing the TCO layer in the first laser scribing and then removing both the TCO layer and the semiconductor layer in the second laser scribing. With this method, it is possible to overcome a serious deficiency in laser scribing lithography that the depths of the grooves are difficult to control.

Further, in the structure shown in Fig. 4.3.3 (a) the back electrode and the a-Si layer were both subjected to a third laser scribing, by which production yield could be raised. In the structure shown in Fig. 4.3.3 (b) the first and second laser scribing steps were carried out in the same manner as in Fig. 4.3.3 (a). In the third laser scribing the intensity of laser light was adjusted so that the grooves were formed to extend down into the a-Si layer as in the case of Fig. 4.3.2 (b).

Figure 4.3.4 is a longitudinal sectional view of an integrated structure in which the

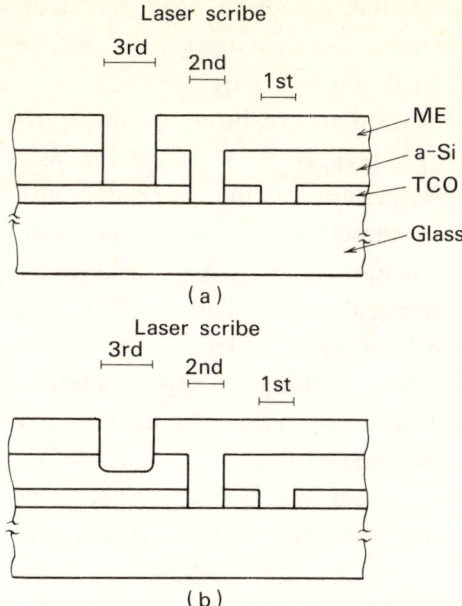

Fig. 4.3.3 Enlarges views of two kinds of electrode coupling portions of integrated a-Si solar cells proposed in this paper. In the both structures the first electrode of a cell is held in so-called side contact with the back electrode of a cell adjacent thereto. For the isolation of the back electrode, (a) both the back electrode and the a-Si layer are simultaneously removed, and (b) portion of the a-Si layer is removed at the same time as the back electrode is removed.

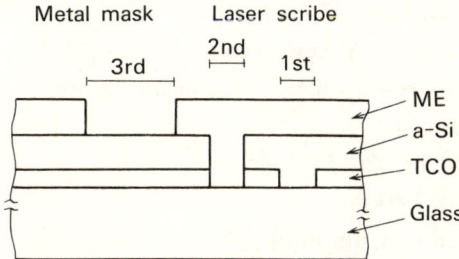

Fig. 4.3.4 Enlarged view of the electrode coupling portion of another integrated a-Si solar cell proposed in this paper. The first electrode of a cell is held in side contact with the back electrode of the adjoining cell. This is a longitudinal sectional view of the integrated a-Si solar cell produced by the hybrid "LS-LS-mask" process in which the back electrode is formed by evaporation using a metal mask.

first and second grooves were cut by the laser scribing but the third grooves were fabricated using the metal mask process and not by laser scribing lithography. Production yield could be improved by using such a hybrid process.

In general, in the metal mask process, precise registration of the first and second patterns is difficult, but the use of a metal mask employed for forming the third pattern presents no particular trouble. On the other hand, the laser scribing process improves productivity with no particular difficulties in the first and second laser scribing steps as shown in Figs. 4.3.3 (a) and (b) and Fig. 4.3.4.

In the formation of the third groove, however, since two structure-sensitive thin films are present immediately below the groove, the use of the laser scribing technique for the formation of the third groove does not lead to high productivity. For this reason, the hybrid "LS-LS-mask" process is still attractive from the viewpoint of production yield.

It is a feature of the structures shown in Figs. 4.3.3 (a) and (b) and Fig. 4.3.4 that not only the semiconductor layer but also the underlying TCO layer are selectively removed at the same time by the second laser scribing. That is, the groove in the TCO layer and the groove made by the second laser scribing are spaced apart and define there between an unremoved portion of the TCO layer. As will be described later, this remaining portion provides a margin for the poly-crystallization of the a-Si semiconductor caused by the annealing accompanying the second laser scribing. That is, the remaining portion of the TCO layer effectively prevents the occurrence of leakage current between the top and bottom electrodes.

In other words, the second previously mentioned drawback of laser scribing lithography could be overcome by leaving a portion of the TCO layer unremoved.

4.3.3 Laser Scriber

We used, as the laser scriber, the YAG laser now in wide use. The general characteristics of this laser are as follows:

Oscillation frequency : $1.064 \mu m$
Laser output : 0.1 to 400 W, variable
Output : 0.01 to 100 J
Laser medium : Nd : YAG crystal
Arrangement : excitation by crypton arc lamp
Voltage : 100 to 440 V
Efficiency : 0.1 to 2%
Weight : 30 to 700 kg
Cooling system : Water cooling
Life time : 1 to 30×10^6 shots
Beam diameter : 5 to 10 mm
Beam divergence : 0.3 to 10 mrad

The YAG laser is capable of oscillating at a wavelengths of $1.34 \sim 1.05 \mu m$ other than the abovementioned $1.064 \mu m$, and is also capable of oscillation at high frequencies of 533 nm (green) and 266/322 nm (ultraviolet).

Some ten other laser light sources are available in addition to the YAG laser. But the YAG laser is now widely employed for the following reasons:

(1) It is widely used as a laser scriber at present.
(2) The laser tube is long-lived.
(3) Oscillation is stable.
(4) An output necessary for scribing is easy to obtain.

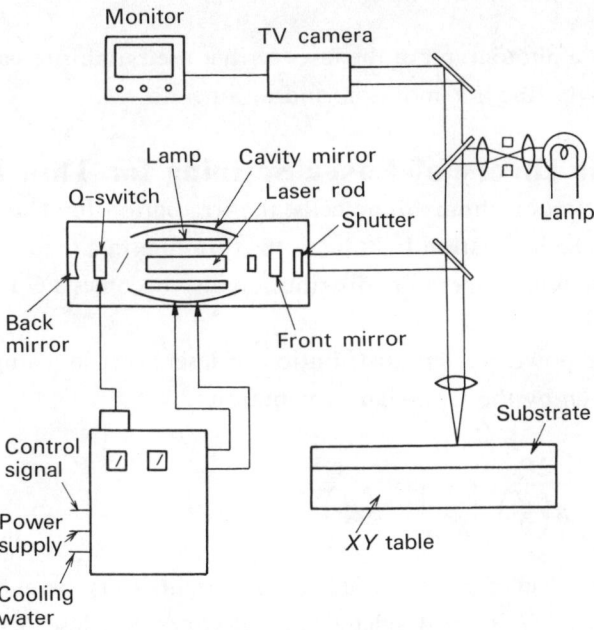

Fig. 4.3.5 General arrangement of the YAG laser scriber.

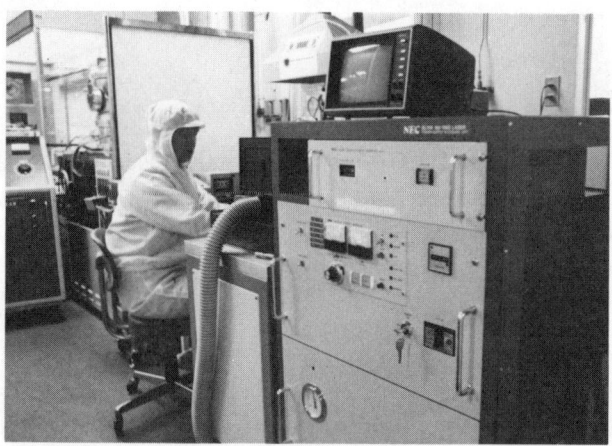

Fig. 4.3.6 Photograph showing the actual operation of a YAG laser scriber.

Figure 4.3.5 shows the outline of the laser system used.

200 V is applied from the power source to a YAG laser oscillator. Energy is provided to the YAG laser from a cavity mirror. The energy is amplified and, by controlling a Q switch cavity, pulsed light is directed to the surface to be processed. The laser light is applied via a dichroic mirror and a condenser lens to the surface to be processed.

In Fig. 4.3.5 light from a visible light lamp is directed, at the same time, to the surface to be scribed, and the light reflected therefrom is monitored through a TV camera. Using the monitor, the entire process can be computerized to control the movement of a specimen table in the X or Y direction, the frequency of laser light, the focal length of the lens and so forth.

Figure 4.3.6 is a photograph of the laser scriber used in this research. The photogaph shows the power source, the TV monitor, and an operator.

4.3.4 General Theory of Laser Scribing for Thin Films

In the fabrication of photovoltaic cells, the irradiation light intensity must be fully taken into account. The irradiation light heats the thin film in a Gaussian distribution and, consequently, the thin film temperature distribution also becomes a Gaussian type expressed by Eq. (4.3.1).[8]

In general, the power density distribution of laser light in a single transverse mode, $P(r)$ (W/cm^2), is given by the Gaussian distribution:

$$P(r) = \frac{2P_0}{\pi \left(\frac{d}{2}\right)^2} \exp\left\{-\frac{8}{\left(\frac{d}{2}\right)^2} r^2\right\} \tag{4.3.1}$$

where r is the distance (cm) from the center of laser light, P_0 is the energy (W) of the laser beam per unit time, and d is the diameter (cm) of condensed laser light.

Here, setting $r = 0$, the power density at the center of the condensed light, P_c (W/cm^2), is obtained as follows:

$$P_c = \frac{2P_0}{\pi \left(\frac{d}{2}\right)^2} \tag{4.3.2}$$

Further, the diameter (L) of a groove made by the laser beam irradiation is equal to $2r$ when $P(r)$ in Eq.(4.3.1) is selected equal to the lower limit of the power density (P_m) for making the groove. Consequently, the diameter L of the groove, the laser light output (P_c), and the diameter (d) of the condensed laser light exhibit the following relation:

$$\frac{L}{d} = \frac{1}{\sqrt{2}} \left(\log \frac{P_c}{P_m}\right)^{1/2} \tag{4.3.3}$$

where d is the diameter of the condensed beam directed to the surface to be processed, P_c is the power density at the center of the condensed light (W/cm^2) (which is effected by the quality of the material to be processed), and P_m is the lower limit of the power density for making grooves (W/cm^2).

The diameter of the condensed beam (d) is given by

$$d = f\theta = 1.27\lambda f C^2/D \tag{4.3.4}$$

where f is the focal length of the condenser lens, θ is the beam spreading angle, λ is the wavelength (1.06 for the YAG laser) and C is a selected parameter, for example, TEM$_{00}$ = 1, TEM$_{10}$ = 1.5, etc.

Considering variations in the output for P_c in this equation, $P_c/P_m > 3$, and it is held substantially constant by the YAG oscillator. This indicates that by increasing the diameter (D) of the beam incident on the condenser lens and by decreasing the focal length

(f) of the condenser lens, the beam diameter can be reduced and, consequently, the width (L) of the groove can be diminished. It is also possible, of course, to decrease the width of the groove by reducing the wavelength λ.

Moreover, the average power (P_a) and the peak power (P_o) are given as follows:

$$P_a = E_p \times R_p \tag{4.3.5}$$

where E_p is the pulse energy (Joule) and R_p is the pulse rate (sec^{-1}), and

$$P_o = E_p/W_p \tag{4.3.6}$$

where W_p is the pulse width (sec^{-1}).

For easy processing, it is important to increase the focal depth of the condenser lens. The focal depth (Z) is given by

$$Z = \frac{\pi}{2\lambda} d^2 = 2\frac{df}{\theta} \tag{4.3.7}$$

where d is the diameter of the beam focused on the surface to be processed, D is the diameter of the beam incident to the condenser lens and f is the focal length of the condenser lens.

In view of variations in scanning laser light between the surface to be processed and the lens, it is necessary to increase the focal depth of the condenser lens. As a result, the beam diameter, and consequently the width of the groove increases. It will be seen that an increased focal depth is important for easy processing.

In addition, the average power is substantially dependent upon the excitation current in the YAG laser. Figure 4.3.7 shows the relation between the average power and the excitation current when the focal length of the lens is 50 mm.

On the other hand, however, increased focal depth prevents variations in the width of the groove resulting from the micro-vibration of the distance between the condenser lens and the specimen table during scanning, but, at the same time, introduces much difficulty in selectively scribing only one thin film of 1 μm or less without damaging the underlying layer.

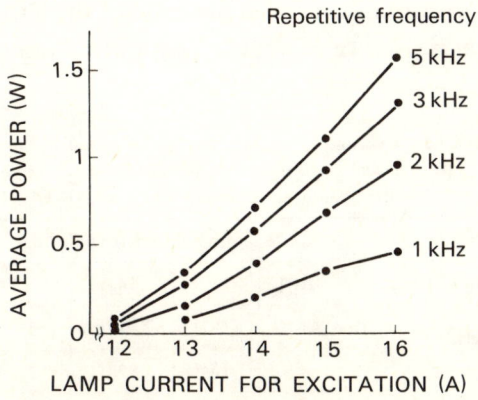

Fig. 4.3.7 Relationship between an excitation current and an average power.

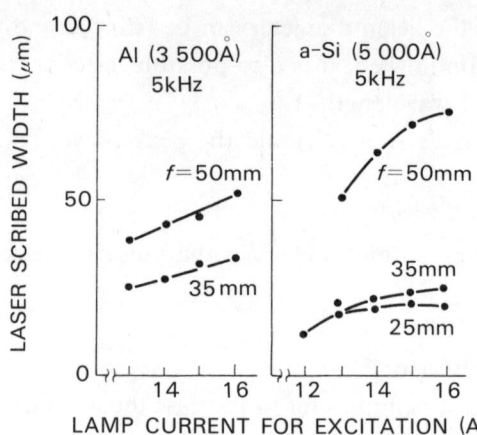

Fig. 4.3.8 Relationship between the lamp current and the scribed width in a-Si and Al when the pulse repetitive frequency was set to 5 kHz under the conditions that lenses of different focal lengths, 25, 35 and 50 mm, were used.

It can be seen that it is necessary to control these factors for optimal processing conditions in the production of photovoltaic cells.

It was thus found that the average power or the excitation current, the repetitive frequency, the focal length, and the scanning speed necessary for groove cutting are basic process parameters. These four parameters will be described hereinafter.

Figure 4.3.8 shows the relation between the width of the groove cut by laser scribing lithography in an aluminum and an a-Si layer, the excitation current, and the focal length.

Form Fig. 4.3.8 it can be seen that the width of the groove increases when the focal length shown in Eq. (4.3.4) is large and the excitation current and the power density at the center of the laser beam are high.

4.3.5 Laser Processing of Transparent Conductive Film

The laser processing of transparent conductive film (TCO) was carried out in the following manner: Table 4.3.1 shows the results obtained when the lamp current and the repetitive frequency were varied and the scanning speed was 6 cm/min. It is understood from Table 4.3.1 that a 14 A lamp current is needed in a case of the TCO layer being a two-layer film (having an initial sheet resistance of $42\,\Omega/\square$) comprised of an ITO layer (1 500 Å) and

Table 4.3.1 Evaluation of the degree of isolation by laser scribing of transparent conductive film when the scanning speed was fixed.

Lamp Current		Scanning Speed (6 cm/min.)			
		16 A	15 A	14 A	13 A
Repetitive Frequency (kHz)	10	4 MΩ	37 kΩ	2.8 kΩ	5 kΩ
	5	∞	5 MΩ	1.8 MΩ	2.8 kΩ
	3	∞	∞	5 MΩ	28 kΩ

Table 4.3.2 Evaluation of the degree of isolation by laser scribing of transparent conductive film when lamp excitation current was fixed.

(a)

		Lamp Current (15A)		
Scanning Speed		60cm/min.	120cm/min.	240cm/min.
Repetitive Frequency (kHz)	10	4 MΩ	∞	∞
	5	∞	∞	∞
	3	∞	∞	∞

(b)

		Lamp Current (14A)		
Scanning Speed		60cm/min.	120cm/min.	240cm/min.
Repetitive Frequency (kHz)	10	∞	∞	∞
	5	20 MΩ	3 MΩ	∞
	3	5 kΩ	0.95 kΩ	∞

an SnO_2 layer (200 Å).

Table 4.3.2 shows the results obtained when the lamp current was held constant (for example, 15 A and 14 A) and the scanning speed was varied in the range of 60 to 240 cm/min.

In the latter case, the electrical resistance between adjacent TCO layers spaced apart by the grooves became greater than 5 MΩ; this clarified the relationship of the conditions for electrical isolation (lamp current 14 to 15 A, repetitive frequency 3 to 10 kHz and a scanning speed of 60 to 240 cm/min.).

4.3.6 Laser Scribing of a-Si Film

Generally, studies are being carried out for the laser scribing of the a-Si layer, on the premise that the underlying TCO layer is left unchanged. It became apparent, however, that

Table 4.3.3 Evaluation of the degree of isolation by laser scribing of a-Si and transparent conductive film when the scanning speed was fixed.
(focusing lens f 50 used)

		Scanning Speed (6cm/min.)				
Lamp Current		17 A	16 A	15 A	14 A	13 A
Repetitive Frequency (kHz)	20	—	2.7 kΩ	1.4 kΩ	120 Ω	35 Ω
	15	—	2 kΩ	7.5 kΩ	8.8 kΩ	37 Ω
	10	—	1.8 MΩ	350 kΩ	—	—
	5	—	30 MΩ	3 MΩ	3 MΩ	—
	3	—	10 MΩ	9 MΩ	5 MΩ	—

there is no appreciable safety margin in removing only the a-Si layer. Therefore, to avoid inflicting damage on the underlying TCO layer, we scribed the TCO and the a-Si layer at the same time and connected the side wall of the TCO layer exposed in the groove to the conductor for the back electrode.

In our experiments, the TCO layer was comprised of an ITO layer (1500Å), a tin oxide layer (200Å) formed thereon and an a-Si layer (5000Å) laminated thereon.

Table 4.3.3 shows the relation between the excitation current and the repetitive frequency in a case where a condenser lens having a 50mm focal length was used and the scanning speed was 6cm/min. With excitation currents below 13A and above 17A, grooves

Table 4.3.4 Evaluation of the degree of isolation by laser scribing of a-Si and transparent conductive film when lamp excitation current was fixed. (focusing lens f 50 used)

(a)

Scanning Speed		60cm/min.	120cm/min.	240cm/min.
		Lamp Current (16A)		
Repetitive Frequency (kHz)	20	20kΩ	10kΩ	2.4kΩ
	15	30kΩ	8.5kΩ	17kΩ
	10	—	—	—
	5	—	—	—
	3	—	—	—

(b)

Scanning Speed		60cm/min.	120cm/min.	240cm/min.
		Lamp Current (15A)		
Repetitive Frequency (kHz)	20	—	—	—
	15	—	—	—
	10	300kΩ	1.5MΩ	3MΩ
	5	9MΩ	7 MΩ	30MΩ
	3	∞	∞	∞

(c)

Scanning Speed		60cm/min.	120cm/min.	240cm/min.
		Lamp Current (14A)		
Repetitive Frequency (kHz)	20	55kΩ	1.35kΩ	300Ω
	15	14kΩ	970Ω	62Ω
	10	—	—	—
	5	—	—	—
	3	—	—	—

observed through a microscope were not neatly formed. The excitation current was then changed to 14 to 16 A and the scanning speed was varied.

Table 4.3.4. shows the scribing conditions when the excitation current was set to 14 A, 15 A and 16 A.

Thus it was found experimentally that there exists an optimum region (excitation current 15 A, repetitive frequency 3 to 5 kHz, and scanning speed 60 to 240 cm/min.) in which laser scribing can be carried out without excessively damaging the underlying substrate.

Table 4.3.5 Evaluation of the degree of isolation by laser scribing of aluminum layer (4000 Å) of TCO/a-Si/Al structure formed on a glass substrate.

(a)

Scanning Speed		6 cm/min.	60 cm/min.	120 cm/min.	240 cm/min.
Lamp Current (16 A)					
Repetitive Frequency (kHz)	20	1.2 kΩ	3.3 kΩ	6.2 kΩ	4.4 kΩ
	15	3.1 kΩ	9.5 kΩ	30 kΩ	160 kΩ
	10	140 kΩ	—	—	—
	5	∞	—	—	—
	3	∞	—	—	—

(b)

Scanning Speed		6 cm/min.	60 cm/min.	120 cm/min.	240 cm/min.
Lamp Current (15 A)					
Repetitive Frequency (kHz)	20	720 Ω	—	—	—
	15	1.9 kΩ	—	—	—
	10	14 kΩ	200 kΩ	1.1 MΩ	660 kΩ
	5	∞	∞	∞	∞
	3	∞	∞	∞	∞

(c)

Scanning Speed		6 cm/min.	60 cm/min.	120 cm/min.	240 cm/min.
Lamp Current (14 A)					
Repetitive Frequency (kHz)	20	850 Ω	1.3 kΩ	1.4 kΩ	1.95 kΩ
	15	1 kΩ	3 MΩ	2.8 kΩ	3.1 kΩ
	10	×	10.5 kΩ	15.5 kΩ	60 kΩ
	5	100 kΩ	5 MΩ	10 MΩ	8 MΩ
	3	∞	30 MΩ	∞	∞

4.3.7 Laser Scribing of Back Electrode

For the laser scribing of the back electrode, we sought the conditions under which the regions defined by laser scribing would be electrically isolated from each other and the underlying TCO layer would not be damaged.

As can be seen from Table 4.3.5, it was found that with a scanning speed of 6 to 240 cm/min., a repetitive frequency of 3 to 5 kHz, and an excitation current of 14 to 15 A, electrical isolation is ensured and the underlying TCO layer is not damaged.

4.3.8 Trial Fabrication of Laminated Photovoltaic Cell

Figure 4.3.9 shows, in section, a sequence of steps involved in the trial production of the photovoltaic cell.

In Fig. 4.3.9 (a), a TCO layer comprised of an ITO layer (about 1 500 Å) and an SnO_2 layer (200 Å) was formed on the top surface of a glass substrate. A laser beam from a YAG

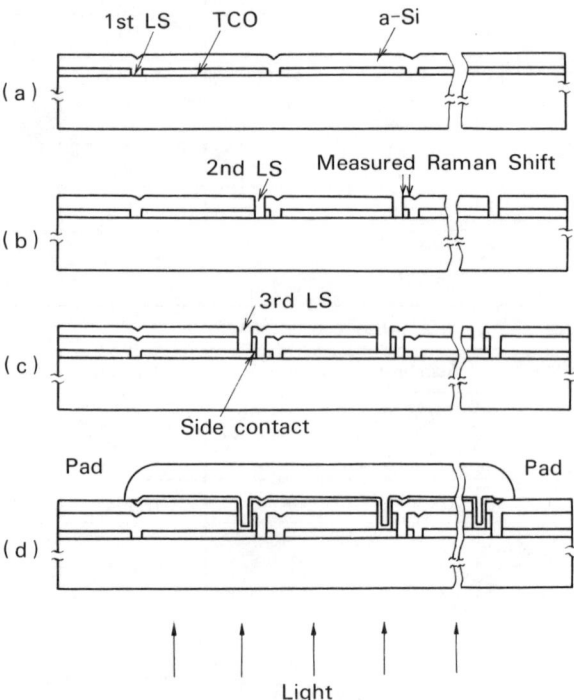

Fig. 4.3.9 Manufacturing process of integrated solar cell.
(a) A TCO layer on a glass substrate was removed by first laser scribing, and an a-Si semiconductor layer was formed.
(b) The a-Si layer and the underlying TCO layer were simultaneously removed by second laser scribing.
(c) A metal layer for the back electrode was formed in side contact with the first electrode, after which the back electrode and the underlying a-Si layer were selectively removed by third laser scribing.
(d) A passivation film was formed and then an organic resin coating was given for mechanical enforcement.

laser was directed to the TCO layer with an average power of 0.5 to 3 W and with a spot diameter of 30 to 70 μm^ϕ, first cutting grooves to isolate individual cell regions and electrode regions for external connection. In this way, a first electrode was produced.

Next, a semiconductor layer was formed by a plasma CVD method to a thickness of about 0.5 μm to form a PIN junction. A typical example of this semiconductor layer is one that is comprised of a P-type semiconductor layer (Si_xC_{1-x}, where $x = 0.8$, 50 to 150 Å thick), an I-type amorphous silicon semiconductor layer (0.4 to 0.6 μm), and an N-type microcrystalline layer (300 to 500 Å).

Then, as shown in Fig. 4.3.9 (b), a second series of grooves was cut in the semiconductor layer and the underlying TCO layer and, in this case, the second grooves were spaced 50 μm apart from the first grooves, leaving the portion of the TCO layer between the first and second grooves unremoved.

Thus, the side walls of the first electrodes were exposed by the formation of the second grooves.

In the absence of the unremoved portions of the TCO layer, the first electrodes are isolated from adjacent ones by filling in the grooves cut by the first laser scribing with a-Si, but the a-Si becomes polycrystalline due to the laser annealing phenomenon. It was found that this increased the generation of leakage current.

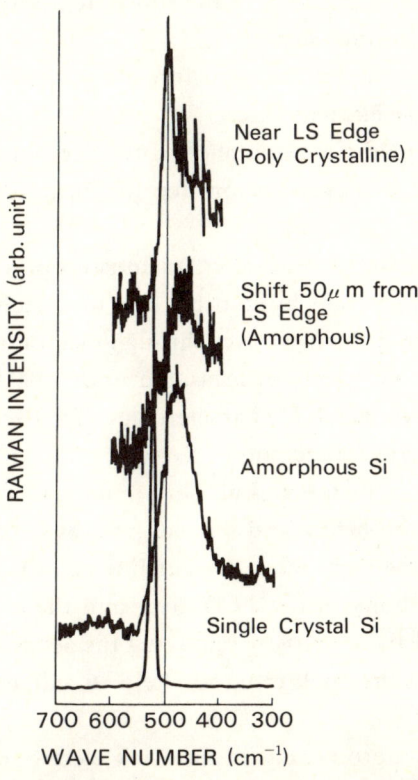

Fig. 4.3.10 The order of crystallization of a-Si in a region surrounding the laser scribed portion of the a-Si layer, investigated by the laser Raman scattering method. Standard a-Si and singlecrystal Si are also shown.

Figure 4.3.10 shows a shift of Raman scattering at the portion indicated by arrows in Fig. 4.3.9 (b). This reveals that a-Si in the vicinity of the laser-scribed region was transformed into a polycrystalline structure. Further, it became clear that the a-Si in regions more than $50\,\mu$m apart from the scribed edge remained amorphous. From this it was clarified that no leakage would occur between electrodes if the electrodes were spaced more than $100\,\mu$m apart which is the sum of the above distance of more than $50\,\mu$m from each groove and the length of the amorphous semiconductor layer necessary for isolation. In other words, the remaining portions of the TCO layer are necessary for isolation.

The cell structure shown in Fig. 4.3.9 (b) was designed with a view to preventing the annealing which inevitably accompanied laser scribing of the a-Si layer, and was very effective in enhancing production yield.

Next, a metal layer was formed for the back electrode as shown in Fig. 4.3.9 (c). In this case, the metal layer was also formed to extend into the second groove in contact with the first electrode.

Further, the metal layer for the back electrode was selectively removed by laser light irradiation to form a series of third grooves, resulting in the structure shown in Fig. 4.3.9 (c).

In this way, a plurality of cells could be connected in series through coupling portions and, at the same time, electrodes for external connection (pads) could be formed on the right- and left-hand end portions of the substrate assembly in the drawings without the involvement of any additional steps.

Figure 4.3.9 (d) shows the finished structure with a passivation film of silicon nitride and an organic resin film for encapsulation.

It is also possible to manufacture a photovoltaic cell measuring 10cm × 10cm using two laser scribing steps and one evaporation step for forming the back metal electrode by using a metal mask.

Figure 4.3.11 (a) and (b) show I-V characteristics of a photovoltaic cell fabricated on a 10cm × 10cm glass substrate by cutting first and second grooves using the laser scribing technique and cutting third grooves through evaporation technique using a metal mask ((a) measured under AM1 and (b) measured under 500 Lx with a fluorescent lamp).

Figure 4.3.11 (c) shows the I-V characteristics of a 10cm × 10cm photovoltaic cell fabricated using the laser scribing technique alone.

Figure 4.3.12 (a) is a metallurgical photomicrograph of the electrode coupling portion of a structure with both first and second grooves cut by laser scribing.

In this photograph, the right-most straight line is a first groove (about $50\,\mu$m wide) cut by laser scribing lithography in the TCO layer and the center line is a second groove (about $50\,\mu$m wide) formed by selectively removing the semiconductor layer and the TCO layer at the same time, with the back electrode held in side contact with the TCO layer in the second groove.

The left-most line is a groove cut in the back electrode.

Figure 4.3.12 (b) is a metallurgical photomicrograph of the electrode coupling portion of a structure having the first and second grooves formed by laser scribing lithography as in the case in Fig. 4.3.12 (a) and the third groove formed by using a metal mask.

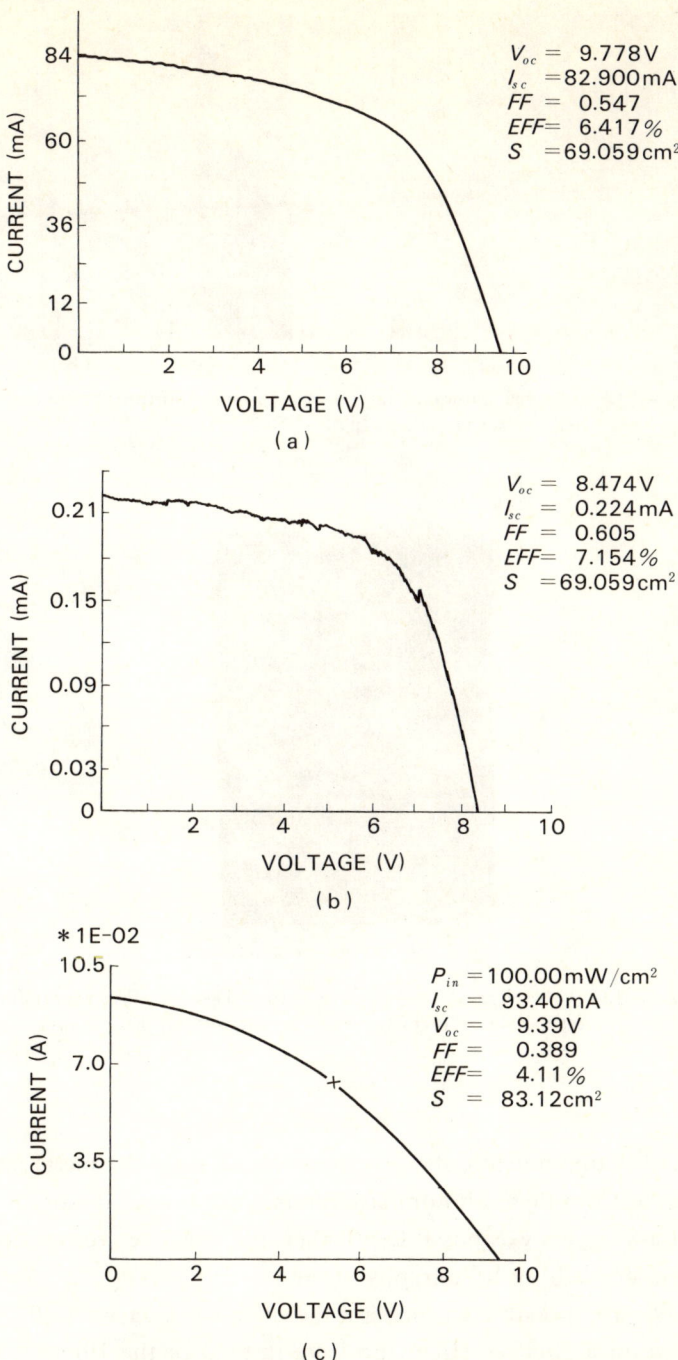

Fig. 4.3.11 Integrated solar cells produced using 10cm×10cm glass substrates.
(a) Characteristic of the solar cell produced by using the laser scribing for the first and second steps and evaporation for the third step using a metal mask. (AM1 100mW/cm^2)
(b) The characteristic (a) measured under a fluorescent lamp (500 Lx).
(c) Characteristic of the solar cell produced by using the laser scribing for the first to third steps, i. e. by the so-called maskless process. (AM1 100mW/cm^2)

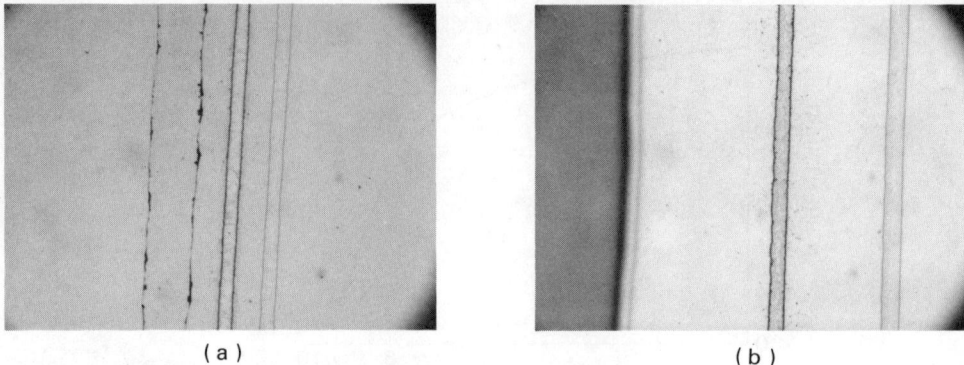

Fig. 4.3.12 Enlarged photomicrograph of a coupling portion corresponding to Figs. 4.3.11 (c) and (a)

Fig. 4.3.13 Solar cell panel comprised of 24 (6 × 4) solar cells, each produced on a 10cm × 10cm substrate, by the hybrid "LS-LS-mask" process.

In Fig. 4.3.12 (b), only one portion of a groove is shown since the metal mask used is wide.

As is apparent from these photomicrographs, laser processing of all the thin films or the two TCO and a-Si layers was possible. Further, the effective area could be increased up to 83% by using laser scribing lithography alone.

It is theoretically possible to increase the effective area up to 92% by forming the integrated structure on a 20cm × 60cm substrate instead of the 10cm × 10cm substrate.

Figure 4.3.13 is a photograph of a 40cm × 60cm solar cell panel comprised of 24 glass substrates, each measuring 10cm × 10cm, produced by the hybrid process.

4.3.9 Conclusion

A 10cm × 10cm a-Si solar cell was fabricated using the laser scribing process. As it was clarified that it is possible in a integrated structure to couple electrodes by the side contact method while at the same time avoiding difficulty in controlling the depth of the groove in laser scribing. Moreover, a portion of the TCO layer was left unremoved between the grooves cut by the first laser scribing of the TCO layer and those cut by the second laser scribing of the a-Si and the TCO layer. This was effective for preventing the occurrence of leakage caused by the polycrystallization of the a-Si.

As a result of this, a 4.11% conversion efficiency (83.12cm^2 effective area) was obtained, under the AM1 condition, with a 12-stage, series-connected, integrated solar cell produced by three laser scribing steps. Further, with two laser scribing steps and one evaporation step using a metal mask, a 6.4% conversion efficiency was obtained under the AM1 condition, and a 7.15% conversion efficiency was obtained under a fluorescent lamp (500 Lx).

References

1) J. J. Hanak : U. S. Pat. 4, 292, 092 (Sep. 29, 1981).
2) J. J. Hanak : Patent application laid open Pat. Pub. Disc. No. 12568/82 (1982. 1. 22).
3) G. R. Swartz : Patent application laid open Pat. Pub. Disc. No. 176778/82 (1982. 10. 30).
4) S. Nakano, M. Ohonishi, K. Kawada, H. Nishiwaki, S. Tsuda, and Y. Kuwano : 5th EC Photovoltaic Solar Energy Conference, Kavouri (Athens) Oct. 17~21, (1983).
5) Y. Hamakawa : 10th International Conf. on Amorphous and Liquid Semiconductor, Tokyo, Aug. 22~26 (1983).
6) S. Yamazaki : Saishin Taiyohko Hatsuden (Latest Photovoltaic Solar Generation) (1984. 4), Maki Shoten, (edited by Y. Hamakawa).
7) S. Yamazaki, K. Itoh, S. Watabe, A. Mase, K. Urata, H. Shinohara, and K. Shibata : 17th IEEE Photovoltaic Specialists Conf. May 1~4, (1984).
8) A. Siegman : An Introduction to Laser and Maser McGraw-Hill Co. N. Y. (1968).

4.4 Dry Etching Processing

Akira YOSHIKAWA* and Yasushi UTSUGI*

Abstract

This paper clarifies the dry etching techniques required for inorganic resist processing. Dry etching techniques play an important role in the lithographic processing of Se-Ge inorganic resists. A highly accurate pattern without undercutting can be developed using a reactive ion etching (RIE) technique through proper selection of the etching gas. Strong resistance to oxygen plasma of inorganic resist films is a desirable property for the top resist in two-layer systems. These dry-processed properties make the inorganic resist and its two-layer application technology the most effective one for realizing a submicron photolithographic process for VLSI fabrication.

4.4.1 Introduction

Dry etching as an alternative to the conventional, wet-chemical methods has become an essential technique for large-scale integrated-circuit (LSI) fabrication. This is because the dry process provides both high-fidelity pattern transfer from the resist mask to the underlying substrate layer and better control of the etching process. The use of dry etching is also increasing in amorphous semiconductor technology, particularly in lithographic application, which is better known as inorganic resist technology. In this paper, clarification is given to the dry etching techniques required in inorganic resist processing.

An inorganic resist utilizing Ag-photodoping effect in amorphous chalcogenide films was first developed in 1976.[1,2] Continued research efforts have been made to exploit the potential of this inorganic resist and to determine its many advantages over conventional organic polymer resists.[3,4] The inorganic resist is also currently of considerable interest in LSI microlithography. This is mainly due to its high resolution capability, dry-developable nature,[5~8] and the excellent performance of inorganic/organic two-layer resists using an inorganic resist as the top layer.[9~11] Dry development and two-layer resist processing are directly related to dry etching. They are described in detail in 4.4.2 and 4.4.3, respectively.

4.4.2 Dry Development Using Reactive Ion Etching (RIE)

For the past several years, one of the major aims of resist research has been to perfect dry development of the resist. Dry development was first achieved in inorganic resist

---

* Atsugi Electrical Communication Laboratory, Nippon Telegraph and Telephone Public Corporation, 1839, Ono, Atsugi-shi, Kanagawa, 243-01.

systemes, using the high etch selectivity between Ag-photodoped and undoped chalcogenide films in CF_4[5~7] or SF_6[7,8] gas plasma. Since previous experiments were predominantly performed under isotropic etching conditions using a barrel-type plasma reactor, undercutting was more or less unavoidable. This was because the Ag-photodoped layer was relatively thin, typically several hundred angstroms in the Ag/Se-Ge system, and the underlying undoped layer was etched isotropically, as shown in Fig. 4.4.1. Such undercutting damages the dimensional accuracy of delineated patterns. To overcome this problem, an anisotropic RIE technique has been developed. Recently, an experiment was conducted to determine the effectiveness of this technique.

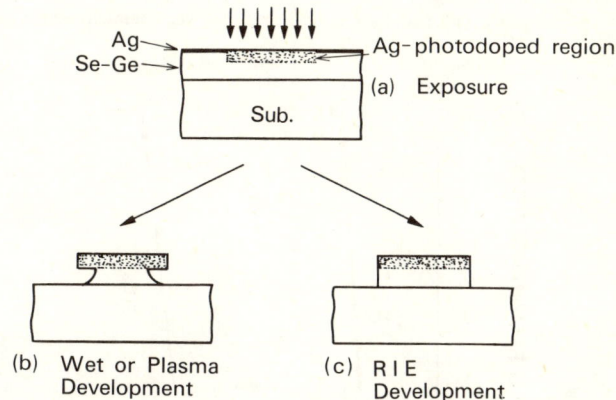

Fig. 4.4.1 Schematic representation of developed resist edge profiles viewed in cross section,
 (a) Before development
 (b) Resulting profile for isotropic wet or plasma development
 (c) Resulting profile for anisotropic RIE development.

Preparation of Ag/$Se_{80}Ge_{20}$ (atomic %) inorganic resists was performed in the same manner as described in previous studies. Photoexposure was made in a 10:1 g-line reduction projection aligner with a numerical aperture of NA = 0.28. After exposure, excess Ag remaining on unexposed regions was removed through treatment in a diluted solution of aqua regia. RIE was carried out with a conventional apparatus under variation of such etching conditions as chamber pressure and rf power density. Fluorocarbons employed in the experiments were CF_4, C_2F_6, C_3F_8, and their mixtures with O_2. It was found that Se-Ge films were etched slightly with pure C_3F_8 gas. On the other hand, although fast etching was observed with both CF_4 and $CF_4 + O_2$, undercutting did occur. Lowering the gas pressure to reduce undercutting resulted in resist surface deterioration due to ion bombardment. Satisfactory results were obtained using a mixture of $C_2F_6 + O_2$, in which the O_2 was added to increase the etch rate.

For comparison, a SEM micrograph of the RIE-developed inorganic resist pattern is shown in Fig. 4.4.2 (a), together with that of a wet-chemical-developed pattern (Fig. 4.4.2 (b)). As shown, a steep profile without undercutting is easily realized in RIE development.

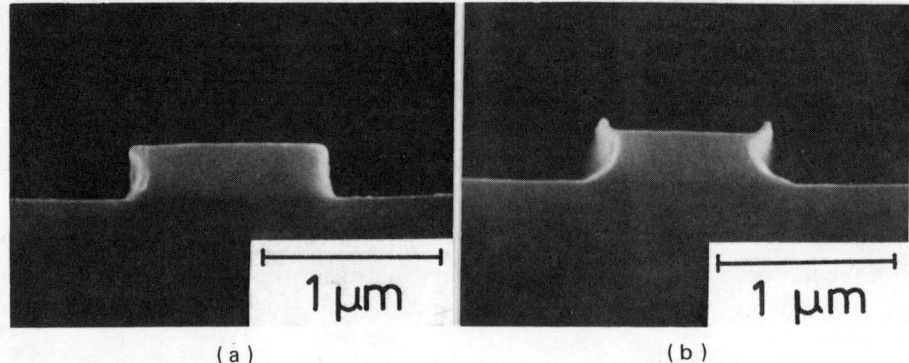

Fig. 4.4.2 SEM photographs showing Ag/Se-Ge inorganic resist pattern edge profiles developed by (a) RIE and (b) wet chemical techniques.

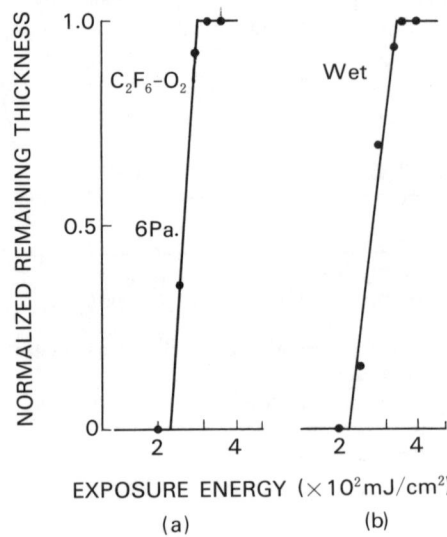

Fig. 4.4.3 Exposure characteristics of Ag/Se-Ge inorganic resist. Remaining film thicknesses after exposure and (a) RIE and (b) wet development are normalized in terms of the initial 2000 Å thickness.

Pattern edge bend-up observed in the wet-developed pattern is probably due to stress relief following development. This RIE develpment technique was applied to determine its exposure characteristics. The normalized remaining thickness after development was plotted as a function of the exposure dose for both RIE and wet development, as shown in Fig. 4.4.3. Sensitivity and contrast are almost equal. These results clearly show that the RIE technique is a very useful means of inorganic resist development.

4.4.3 Dry Processing in Two-Layer Resist Applications

Multi-layer resist technology[12] has attracted strong interest due to its ability to overcome many of the difficulties that plague conventional single-layer resist pattern generation. It is expected to become essential for fabrication of VLSI's.

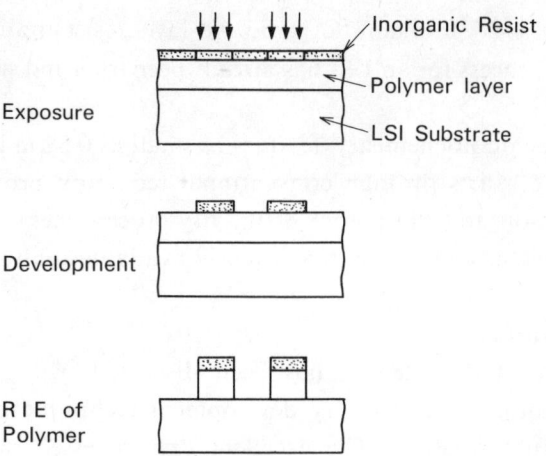

Fig. 4.4.4 Inorganic/organic two-layer resist process steps.

Pattern delineation in a multi-layer resist is basically performed according to the following process steps. First, an organic polymer layer coating is applied to the substrate to planarize the surface topology. A thin top-layer resist is then coated on the planarizing layer (two-layer system) or on an intermediate layer through its deposition process (three-layer system). After patterning the top resist, pattern transfer to the bottom planarizing layer is accomplished using dry etching techniques.

The process steps for an inorganic/organic two-layer resist are shown in Fig. 4.4.4. Requirements for the top resist are
(1) strong resistance to O_2 RIE and
(2) easy coating and removal.

The O_2 RIE etch-rate ratio between the Se-Ge film and Shipley AZ-1370 resist, which is usually employed as the bottom layer, is as small as 1/100. Therefore, inorganic resists satisfy requirement (1). Se-Ge films can easily be deposited on the bottom layer without any deformation or crack generation in the bottom layer by rf sputtering (dry) method. Thus requirement (2) is also satisfied. An SEM micrograph of this two-layer resist

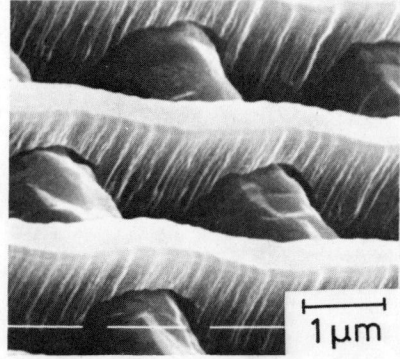

Fig. 4.4.5 SEM photograph showing inorganic/organic two-layer resist pattern only 0.6μm wide, formed on Al substrate with deep steps.

pattern is shown in Fig. 4.4.5. Inorganic/organic two-layer resist application to the Al interconnection patterning process for an LSI has already been tried and satisfactory results have been obtained.[11]

Due to its unique photochemistry, features as small as 0.5 μm lines and spaces can be delineated in inorganic resists through conventional reduction projection printing. Such unusually high resolution and the nature of the dry-process mentioned above make this two-layer resist the most attractive among currently available multi-layer systems.

4.4.4 Conclusion

The importance of dry etching has been discussed, with elaboration of Se-Ge inorganic resist processing. The RIE dry development technique presented here ensures highly accurate pattern formation. The excellent dry-processed properties of inorganic resists for use as the top layer of two-layer resist systems offer the finest resolution optical resist system now available. The authors believe that this two-layer resist technology coupled with RIE development will become the essential submicron photolithography technique for VLSI fabrication.

References

1) A. Yoshikawa, O. Ochi, H. Nagai, and Y. Mizushima: Appl. Phys. Lett., *29* (1976) 677.
2) A. Yoshikawa, O. Ochi, H. Nagai, and Y. Mizushima: Appl. Phys. Lett., *31* (1977) 161.
3) Y. Mizushima and A. Yoshikawa: Japan Annual Reviews in Electronics, Computers & Telecommunications, Amorphous Semiconductor Technologies & Devices 1982, ed. Y. Hamakawa (OHM * North-Holland, Tokyo, Amsterdam, 1982) 277-295.
4) See for example, Extended Abstract of Symposium on Inorganic Resist Systems, 161st Meeting of the Electrochemical Society, Montreal, Canada, (1982) 152-180.
5) M. S. Chang and J. T. Chen: Appl. Phys. Lert., 33 (1978) 892.
6) A. Yoshikawa, O. Ochi, and Y. Mizushima: Appl. Phys Lett., *36* (1980) 107.
7) S. A. Lis, J. M. Lavine, and J. I. Masters: Proc. Microcircuit Engineering 82, 275-284.
8) P. G. Huggett, K. Frick, and H. W. Lehmann: Appl. Phys. Lett., *42* (1983) 592.
9) K. L. Tai, W. R. Sinclair, R. G. Vadimsky, J. M. Moran, and M. J. Rand: J. Vac. Sci. Technol., *16* (1979) 1977.
10) K. L. Tai, R. G. Vadimsky, C. T. Kemmerer, J. W. Wagner, V. E. Lamberti, and A. G. Timko: J. Vac. Sci. Technol., *17* (1980) 1169.
11) Y. Utsugi, A. Yoshikawa, and T. Kitayama: to be published in Microelrctron. Eng.
12) J. M. Moran and D. Maydan: J. Vac. Sci. Technol., 16 (1979) 1620.

4.5 Quantum Wells in Amorphous Semiconductors

Toshio OGINO* and Yoshihiko MIZUSHIMA**

Abstract

Optical properties in multi-layered amorphous films are reviewed. A one-dimensional quantum size effect similar to that seen in crystalline superlattice is observed. Quantum well structure does not influence localized phenomena such as photodarkening and photoluminescence. Long-rang interactions in atomic vibration are, for the first time, discovered by using multi-layered structure. It is proposed that multi-layered structure can be a new method which demonstrates long-range interaction in amorphous materials.

4.5.1 Introduction

The supperlattice concept[1] has brought about the recent development of a new field in physics and the fabrication of crystalline material devices. In the crystalline supperlattice, the periodicity of atomic arrangement is modulated spatially. As a result, the overall energy band structure is perturbed by Brillouin zone folding. On the other hand, the constituent atoms of amorphous materials have no periodic order. In this sense, amorphous multi-layer is not a so-called supperlattice. However, most crystalline supperlattice devices utilize only the quantum well effect formed by a multi-layered structure. Since the potential periodicity is also formed in amorphous materials, quantum size effects should be observed more or less according to the degree of electron wavefunction localization. Therefore, the construction of quantum well devices utilizing amorphous materials are possible.

Moreover, the multi-layered periodicity is the only existing long-range ordering in amorphous materials due to the lack of a definite crystalline ordering. Therefore, long-range interaction effects in amorphous materials should be clearly demonstrable. Hitherto, such effects were regarded as minor. For example, a molecular model has been widely accepted[2] in which the infrared vibrational spectra in multi-component materials can be broken down into independent vibrational modes of local structures. Nevertheless, it is recognized that such a simple model is not sufficient for a complete understanding of amorphous properties.[3-5] The multi-layered structure is a new experimental method for investigating long-range interactions.[6,7]

---

* Atsugi Electrical Communication Laboratory, Nippon Telegraph and Telephone Public Corporation, Ono, Atsugi-shi, Kanagawa 243-01.
** Hamamatsu Photonics Co., Ltd., Ichino-cho, Hamamatsu-shi, Shizuoka 435.

4.5.2 Experimental Technique

Formation of a crystalline supperlattice requires lattice matching between two kinds of layer materials. Moreover, a tedious formation process is needed for single crystal growth. In amorphous multi-layered films, such restrictions do not exist. Therefore, freedom for material selection is large, and various fabrication techniques can be employed. For example, vacuum evaporation[8] and RF sputtering[7] have been used for chalcogenides, and glow discharge deposition[9] for hydrogenated silicon (a-Si : H) have been recently used.

When the individual layer thickness ratio is kept constant, the average composition of the whole film is also kept constant. Therefore, any change due to layer pitch can be attributed only to the long-range interaction between layers.

4.5.3 One-Dimensional Quantum Size Effect

One-dimensional quantum size effect in amorphous semiconductors has been observed in an $As_{40}Se_{60}/Ge_{25}Se_{75}$ multi-layer[6,7] and in an a-Si : H thin layer sandwiched between $Si_{0.2}C_{0.8}$ thick layers.[9] This effect is most apparent in the optical absorption edge shift.

Figure 4.5.1 shows a typical absorption spectra for the $As_{40}Se_{60}/Ge_{25}Se_{75}$ multi-

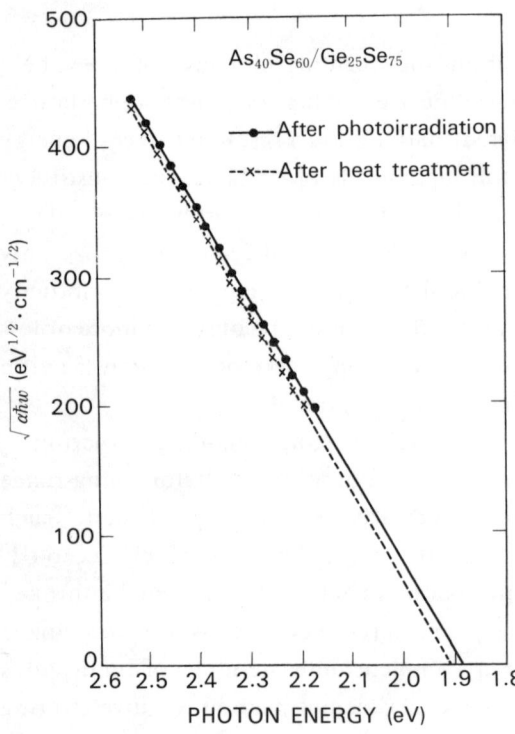
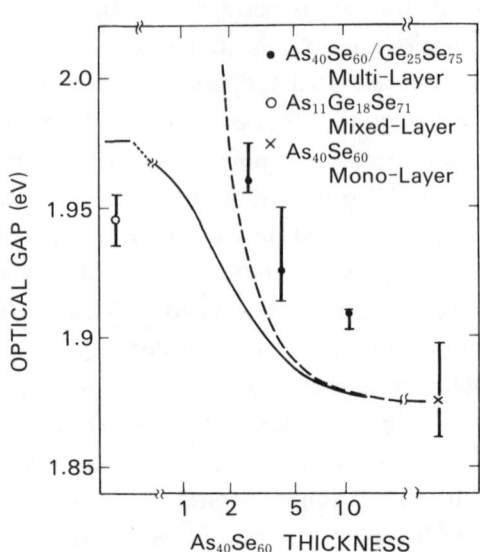

Fig. 4.5.1 Dependence of $\alpha\hbar\omega$ on photon energy in the $As_{40}Se_{60}/Ge_{25}Se_{75}$ multi-layered film. Solid and broken lines indicate αs after photoirradiation and after heat treatment just below $As_{40}Se_{60}$ glass transition temperature, respectively.

Fig. 4.5.2 Variation in measured optical gap ($E_{g,opt}$) with thickness of the $As_{40}Se_{60}$ well layer. The broken line is a calculated curve for a single quantum-well model. The solid line is a curve for periodic potential model.

layered film. In an amorphous chalcogenide, the absorption coefficient, α, obeys the relation $(\alpha\hbar\omega)^{1/2} \propto (\hbar\omega - E_{g,opt})$ in the high absorption region, where $\hbar\omega$ and $E_{g,opt}$ are photon energy and the optical gap, respectively. Since $E_{g,opt}$ defined by this relation can be precisely determined, the quantum well effect can be evaluated by measuring $E_{g,opt}$ changes resulting from layer pitch changes.

Figure 4.5.2 shows the measured $E_{g,opt}$ in $As_{40}Se_{60}/Ge_{25}Se_{75}$ multi-layered film. Since the absorption coefficient of $As_{40}Se_{60}$ is much larger than that of $Ge_{25}Se_{75}$, the observed $E_{g,opt}$ is determined almost completely by absorption in the $As_{40}Se_{60}$ layer. The $E_{g,opt}$ shift shown in Fig. 4.5.2 is similar to that for crystalline supperlattice. The electronic states in the multi-layered structure are represented by the following two models. One is based on a single quantum well, where the interaction between neighboring wells is negligible. This model is valid when the layer pitch is relatively large as compared with the spatial extent of electron wavefunction. The other model is a periodic potential model, i. e. a Kronig-Penny model. In this model, it is assumed that electron wavefunction extends over the film and is not localized. The broken and solid lines in Fig. 4.5.2 are the transition energies associated with the lowest quantized level calculated for the single well model and for the periodic potential model, respectively. For calculation, effective mass is assumed to be equal to a free electron mass, m_e. Since the thickness of the $Ge_{25}Se_{75}$ barrier layer is 2.5 times that of the $As_{40}Se_{60}$ well layer, $E_{g,opt}$ for multi-layered film is believed to obey the single well model. The calculated curve explains the experimental results qualitatively. Better agreement is obtained by assuming an effective mass less than m_e. The mixed ternary $As_{11}Ge_{18}Se_{71}$ film is regarded as a limit when the layer pitch approaches zero. $E_{g,opt}$ for this film is less than that for a multi-layered film with small layer pitch. It is suggested that the electronic state in the multi-layered film changes from the state described by the single well model to that described by the periodic potential model as the barrier layer decreases. This transition occurs when the barrier layer thickness decreases to less than 5 nm corresponding to a well layer thickness of 2 nm.

In the a-Si : H thin layer, a similar $E_{g,opt}$ shift is observed when the well layer thickness is less than 10 nm.

4.5.4 Localized Phenomena

Since a short-range order in a multi-layered structure is regarded as being identical with that in a thick film, a localized phenomenon is not influenced by the quantum well structure. Figure 4.5.1 shows the difference between the absorption coefficient of an $As_{40}Se_{60}/Ge_{25}Se_{75}$ multi-layer after photoirradiation and after heat treatment just below the glass transition temperature. This change, called photodarkening,[10] is reversible and interpreted as an increase in atomic configuration randomness. The magnitude of the change is the same as that in homogeneous materials. The difference between $E_{g,opt}$ after photoirradiation and after heat treatment is independent of the layer pitch.[7] This evidence supports the idea that the photodarkening shift is determined only by the short-range order and is not influenced by the long-range structure.

Another example of localized phenomena is photoluminescence. In amorphous semiconductors, a large Stoke's shift is observed ; the transition occurs within the localized states formed by lattice relaxation. The photoluminescence spectrum in a-Si : H thin film has been found to be identical with that in thick film.[8]

4.5.5 Long-Range Interaction in Atomic Vibrations[6,7]

The infrared (ir) vibrational spectrum of multi-component amorphous chalcogenide has been regarded as being satisfactorily explained by the molecular model. The multi-layered structure is a suitable means to examine the applicability of such a model, because the long-range interaction, if it exists, is clearly demonstrated in this structure.

Figure 4.5.3 shows the ir atomic vibration absorption spectra in $As_{40}Se_{60}/Ge_{25}Se_{75}$ multi-layered films with various layer pitches, as well as in $As_{40}Se_{60}$ and $Ge_{25}Se_{75}$ homogenious mono-layers and in an $As_{11}Ge_{18}Se_{71}$ mixed ternary film. The peaks at $221 \sim 238$ cm^{-1} and at $256 \sim 262$ cm^{-1} correspond to the vibrational modes associated with a pyramidal As-Se$_3$ cluster and a tetrahedral Ge-Se$_4$ cluster.[11,12] Both peaks are separate even in the mixed ternary film; the molecular model is roughly valid. However, the spectrum evidently changes with decreases in the layer pitch, though the total number of the local clusters is conserved in every film.

Figure 4.5.4 shows the absorption peak shift for both vibrational modes. Both peaks monotonically shift to higher wavenumbers as the layer pitch decreases. This shift is not

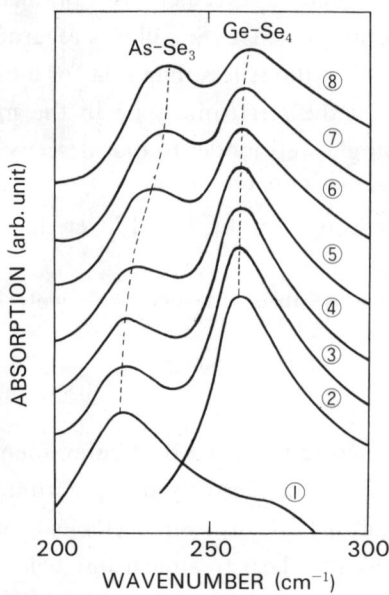
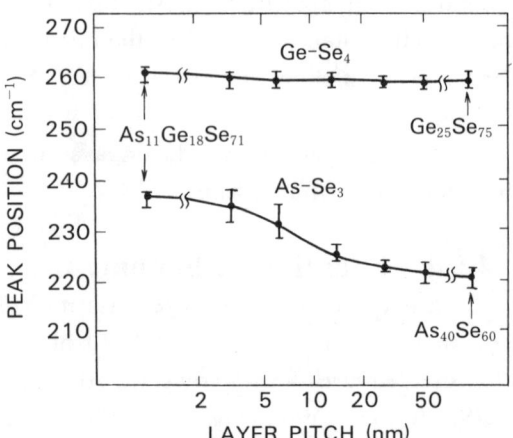

Fig. 4.5.3 Variation in infrared absorption spectra for atomic vibration with layer pitch change. ① $As_{40}Se_{60}$ mono-layer. ② $Ge_{25}Se_{75}$ mono-layer. ③ $d_1 = 14$ nm, $d_2 = 36$ nm. ④ $d_1 = 8$ nm, $d_2 = 21$ nm. ⑤ $d_1 = 4$ nm, $d_2 = 10$ nm. ⑥ $d_1 = 1.8$ nm, $d_2 = 4.6$ nm. ⑦ $d_1 = 0.9$ nm, $d_2 = 2.4$ nm. ⑧ $As_{11}Ge_{18}Se_{71}$ mixed ternary film. d_1 and d_2 are thickness of the $As_{40}Se_{60}$ and the $Ge_{25}Se_{75}$ layers, respectively.

Fig. 4.5.4 Absorption peak shift for the As-Se$_3$ 221 cm^{-1} band and the Ge-Se$_4$ 256 cm^{-1} band. Points for $As_{11}Ge_{18}Se_{71}$ correspond to the limits when the layer pitch approaches zero. Points for $As_{40}Se_{60}$ and $Ge_{25}Se_{75}$ correspond to the infinite layer pitch.

attributable to a simple coupling between two modes, because the two peaks would approach each other in such a case. Therefore, some other interaction between As-Se$_3$ and Ge-Se$_4$ clusters, or between two layers, is newly induced by the formation of multi-layered structure. Since the shift begins at a layer pitch of more than 10 nm, this interaction is a long-range one. Usually, a coulomb force is the only long-range force. The origin of this long-range force is believed to be a charge transfer between As atoms and Ge atoms due to an electronegativity difference between both atoms. This charge transfer is also suggested from oscillator strength values resulting from layer pitch changes. From an analysis based on a classical osillator model, the oscillator strength for the As-Se$_3$ vibration increases with decreases in layer pitch, whereas that for the Ge-Se$_4$ vibration decreases. Since the oscillator strength is proportional to the square of the effective charge in the oscillator dipole, this change suggests that the charge transfer occurs between the As-Se$_3$ and the Ge-Se$_4$ clusters.

There are two possible interpretations for the peak shift based on the charge transfer. One is the non-linear effect due to the built-in DC field formed between two layers. In glassy materials, the Kerr effect is predominant, so that the built-in field effectively increases the dielectric function for LO-like vibrations. The LO phonon component is also excited by ir absorption in the case of randomly distributed oscillator dipoles. The LO-like vibrational frequency increases with an increase in dielectric function. Therefore, both absorption peaks shift to higher wavenumbers when the built-in field is formed. This model is applicable to a large layer pitch case where the built-in DC field is not averaged out. The other model is based on a microscopic coulomb force between As and Ge atoms formed by the charge transfer. This force, which is not averaged out even in the mixed ternary film, acts as an additional constraining force because it does not exist in the $As_{40}Se_{60}$ or $Ge_{25}Se_{75}$ binary mono-layers. Therefore, this force shifts both resonance frequencies to higher wavenumbers.

4.5.6 Conclusion

(1) Multi-layered structure is a new vehicle for long-range interaction in amorphous materials. Since there is no tedious fabrication process, the range of materials and formation techniques is large.

(2) One-dimensional quantum size effect has been observed in $As_{40}Se_{60}/Ge_{25}Se_{75}$ multi-layered films and in a-Si : H thin films sandwiched between $Si_{0.2}C_{0.8}$ thick layers.

(3) Quantum well structure does not influence localized phenomena such as photo-darkening and photoluminescence.

(4) A new mode of long-range interaction in atomic vibration has been demonstrated by using $As_{40}Se_{60}/Ge_{25}Se_{75}$ multi-layered structure.

References

1) L. Esaki and R. Tsu : IBM J. Res. & Dev., *14* (1970) 61.
2) G. Lucovsky and R. M. Martin : J. Non-Cryst. Solids, *8-10* (1972) 185.
3) R. J. Nemanich, S. A. Solin, and G. Lucovsky : Solids, State Commun., *21* (1977) 273.
4) G. Lucovsky and F. L. Galeener : J. Non-Cryst. Solids, *35 & 36* (1980) 1209.
5) S. Onari, O. Sugino, M. Kato, and T. Arai : Jpn. J. Appl. Phys., *21* (1982) 418.
6) T. Ogino, A. Takeda and Y. Mizushima :

Collected Papers 2nd Int. Symp. Molecular Beam Epitaxy & Related Clean Surface Techniques, Tokyo, (1982) 65.
7) T. Ogino, and Y. Mizushima: Jpn. J. Appl. Phys., *22* (1983) 1674.
8) E. Maruyama: Jpn. J. Appl. Phys., *21* (1982) 213.
9) H. Munekata and H. Kukimoto: Jpn. J. Appl. Phys., *22* (1983) L 544.
10) T. Igo and Y. Toyoshima: Proc. 3rd Conf. Solid State Devices, Tokyo (1971) 61.
11) G. Lucovsky: Phys. Rev. *B 6* (1972) 1480.
12) G. Lucovsky, R. J. Nemanish, S. A. Solin, and R. C. Keezer: Solid State Commun., *17* (1975) 1567.

CHAPTER 5

AMORPHOUS SILICON SOLAR CELLS

5.1 p-i-n & n-i-p Basis Solar Cells

Yoshiyuki UCHIDA*

Abstract

Recent advances in p-i-n- and n-i-p-based a-Si : H solar cells are reviewed based on the extensive work being done in Japan. It is shown by experimental studies that the boron profile in the i-layer of such solar cells definitely plays a very important role in determining photovoltaic performance. Optical enhancement by adoption of a textured substrate and a highly-reflective back-electrode improves conversion efficiency in both glass-substrate and stainless steel-substrate solar cells. An a-Si : H solar cell formed on a 30 × 40cm^2 glass-substrate is demonstrated, as well as technologies to produce a homogeneous a-Si : H film over a large area. The stability of a-Si : H solar cells to light exposure is also discussed, with emphasis on the effects of i-layer thickness and compensation of the i-layer with boron on performance stabilization.

5.1.1 Introduction

Hydrogenated amorphous silicon (a-Si : H) is a promising and rapidly advancing material for thin film solar cells. Although both Schottky barrier and MIS types of a-Si : H solar cells were produced and demonstrated at an early stage, most a-Si : H solar cells peresently produced in both mass production and laboratory scales are based on the p-i-n (or n-i-p) junction structure. In fact, a-Si : H solar cells used in hand-held calculators, the industrialization of which was initiated by Japanese industries (Fuji Electric, Sanyo) in 1980 and whose monthly production reached several million units in 1983, have this type of junction structure. The p-i-n (and n-i-p) type of solar cell structure has been recognized as having several advantages over the other types, not only in performance and stability but also in reproducibility and automation of their manufacturing processes.

The basic structure of the solar cells discussed in this section are schematically shown in Fig. 5.1.1. The primary substrates on which a-Si : H solar cells are formed are stainless steel (SS) and glass . To maximize the amount of light reaching the intrinsic (i) layer, where photocurrent is generated, the front layer is kept as thin as possible and is typically on the order of 100Å. Another technique to maximize the light reaching the i-layer is to employ a material with high optical transmission as the front layer. In SS-substrate solar cells (Fig. 5.1.1 (a) and (b)), microcrystalline Si : H with high optical transparency is sometimes used in place of conventional a-Si : H.[1] In glass-substrate solar cells (Fig. 5.1.1 (c) and (d)), light

---

* Fuji Electric Corporate Research and Development Ltd., 2-2-1 Nagasaka, Yokosuka-shi, 240-01.

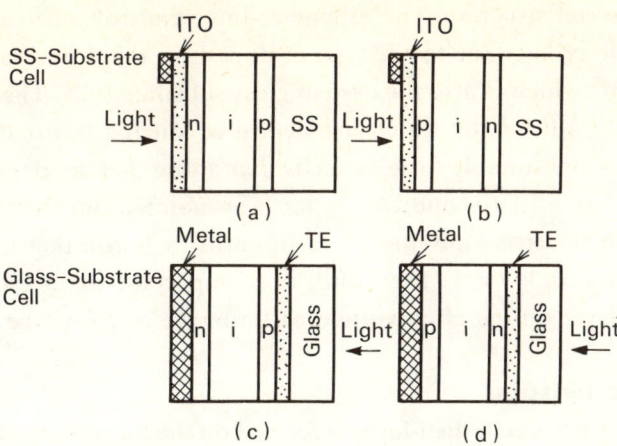

Fig. 5.1.1 Schematic structures of a-Si : H solar cells discussed in this section ; (a) ITO/n-i-p/SS, (b) ITO/p-i-n/SS, (c) glass/Transparent Electrode (TE) /p-i-n/metal and (d) glass/TE/n-i-p/metal types.

comes through both the substrate and the transparent electrode (TE). Hydrogenated amorphous silicon carbide (a-SiC : H) with a wide optical-gap is currently utilized as the front layer material in glass-substrate solar cells.[2]

In the last few years, the photovoltaic performance of the a-Si : H solar cell has been improved remarkably. In addition to the above improvement in the front layer, several techniques, including optimization of the boron-profile in the i-layer and optical enhancement utilzing a textured substrate in conjunction with a highly reflective back-electrode, have been proposed. Sections 5.1.2 and 5.1.3 are a review of recent progress in these techniquse . The efforts to develop fabrication processes for an a-Si : H solar cell on a large-area substrate are presented in 5.1.4.

Since Staebler and Wronski[3] reported light-induced changes in the conductivity and photoconductivity of a-Si : H films, the stability of photovoltaic performance to sunlight exposure has been an important technical problem. The degradation mechanism in a-Si : H solar cells and techniques to supress light-induced degradation have been intensively investigated. This is discussed in 5.1.5. Finally, significant results recently obtained for an a-Si : H solar cell with a single p-i-n (or n-i-p) junction are summarized in 5.1.6.

5.1.2 Effect of Boron Doping and Its Profile on Photovoltaic Performance

The photovoltaic performance of p-i-n- and n-i-p-based solar cells is strongly dependent on the photovoltaic properties of the a-Si : H film and cell construction parameters, such as the thickness of each a-Si : H layer and the doping fraction of the p- and n-layers. There is also a dependence on the order of deposition of the p-,i- and n-layers. It has been shown that the photovoltaic performance of ITO/n-i-p/SS cells is superior to that of ITO/p-i-n/SS cells.[4,5] However, for glass-substrate cells, a glass/ITO/p-i-n/metal structure has been commonly adopted because of its better cell performance compared with that of a glass/ITO/n-i-p/metal structure. The difference in conversion efficiency between the two types of SS-substrate cells is due to the type of dopant in the doped layer deposited prior to i-layer deposition; which is also the case in glass-substrate cells. The cells shown in Fig. 5.1.1 (a) and (c), the i-layers of which are deposited on the boron-doped p-layer, have junction properties more suitable for solar cells than those that are deposited on the n-type layer, as shown in Fig. 5.1.1 (b) and (d). It has been pointed out that the i-layer of the p-i-n structure contains a certain amount of phosphorus or boron that is incorporated from the pre-deposited doped layer.[6] Therefore, it is important to investigate the effects of impurities in the i-layer on the photovoltaic performance of p-i-n type a-Si:H solar cells.

Photovoltaic characteristics

In an ITO/n-i-p/SS cell, the i-layer is formed on the boron-doped p-layer. The boron concentration, as measured by SIMS, decreases rapidly in the p-i interface region and reaches a relatively constant concentration.[6] The amount of boron in this constant concentration in the i-layer can be controlled by varying the B_2H_6/SiH_4 gas ratio. Therefore, ITO/n-i-p/SS cells with various boron concentrations, ranging from 10^{16} to 10^{18} atoms/cm^3, in the i-layer were fabricated and their photovoltaic performance under AM1 illumination was measured.[7] Figure 5.1.2 shows conversion efficiency plotted as a function of boron concentration in the i-layer. The highest conversion efficiency is obtained in the 1 to 3×10^{17} atoms/cm^3 range.

As described earlier, photovoltaic performance of an ITO/p-i-n/SS cell is inferior to that of an ITO/n-i-p/SS solar cell. This difference in conversion efficiency between the two

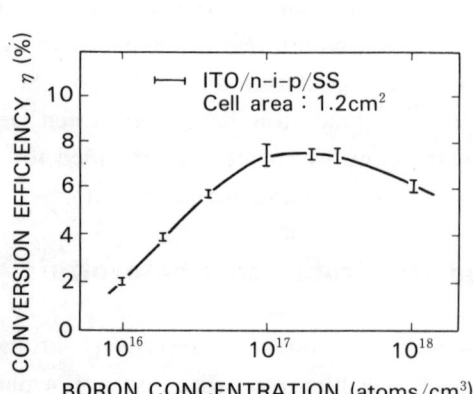

Fig. 5.1.2 The conversion efficiency as a function of boron concentration in the i-layer for ITO/n-i-p/SS solar cells. [after H. Haruki et al.[7]]

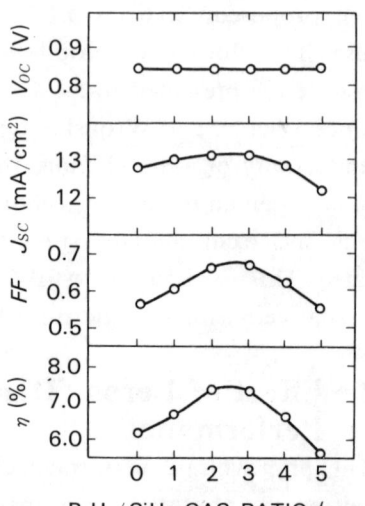

Fig. 5.1.3 Photovoltaic performances of the ITO/p-i-n/SS solar cell as a function of B_2H_6/SiH_4 gas mixture ratio for the i-layer deposition. [after H. Sakai et al.[8]]

types of SS-substrate solar cells may be due to a difference in the concentrations of dopants in the i-layer. From this point of view, ITO/p-i-n/SS solar cells with various boron concentration were fabricated by doping B_2H_6 of less than 5 ppm in SiH_4 plasma during i-layer deposition.[8] The photovoltaic performance of an ITO/p-i-n/SS solar cell is shown in Fig. 5.1.3 as a function of B_2H_6/SiH_4 gas mixture ratio during i-layer deposition. The open-circuit voltage, V_{oc}, of the solar cell remains constant over the examined gas mixture ratio, although the short-circuit current density, J_{sc}, and the fill factor, FF, change with increased B_2H_6. It was found that J_{sc} and FF reach their maximum values at a B_2H_6 gas mixture of 2 to 3 ppm. The gas mixture ratio that prodvuced the highest conversion efficiency, η, also lies in the 2 to 3 ppm range.

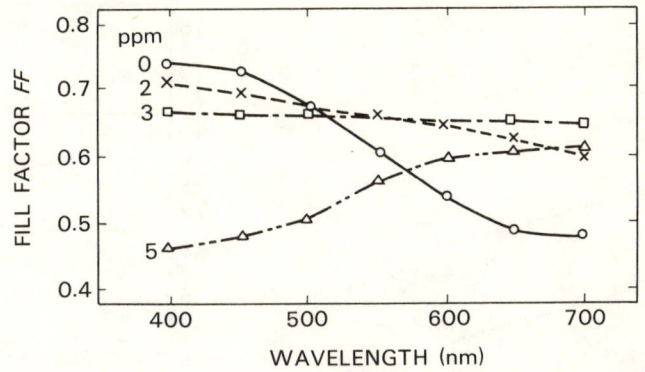

Fig. 5.1.4 The wavelength dependence of FF for the ITO/p-i-n/SS solar cell i-layers of which were produced from SiH_4 mixed with $0\sim5$ ppm B_2H_6. [after H. Sakai et al.[8]]

Spectral photocurrent was then measured for ITO/p-i-n/SS solar cells, the i-layers of which were produced from SiH_4 containing $0\sim5$ ppm B_2H_6. The response in longer wavelength light is enhanced by doping boron into the i-layer; however, at 5 ppm B_2H_6 photocurrent decreases in short wavelength light, which causes a decrease in J_{sc}, as shown in Fig. 5.1.3. The wavelength dependence of FF was also measured, as shown in Fig. 5.1.4. In a solar cell with an undoped i-layer, FF shows maximum at 400 nm wavelength for incident light and monotonically decreases as the wavelength increases. Increasing the amount of B_2H_6 during i-layer deposition caused FF in the solar cell to decrease in the shorter wavelength region and increase in the longer wavelength region. FF in a solar cell prepared in a doping gas containing $2\sim3$ ppm B_2H_6 had the highest value, as shown in Fig. 5.1.3, but showed little wavelength dependence.

The effect of boron doping in the i-layer was also investigated for glass-substrate solar cells with a glass/ITO/SnO_2/p (SiC)-i-n/metal structure.[9] In solar cells with uniform boron-doped i-layer, a maximum conversion efficiency of about 6% was obtained at a doping level of $0.35\sim0.40$ ppm, proving that boron doping in the i-layer is a very important factor for achieving high photovoltaic performance. However, with uniform profile for boron doping, an FF of more than 0.6 could not be obtained in cells with an i-layer as thick as 0.5 μm. In order to achieve a stronger electric field throughout the i-layer and a correspondingly larger FF, the i-layer was formed with a graded boron profile, as shown in Fig. 5.1.5. Figure 5.1.6 shows the photovoltaic performance as a function of B_2H_6 doping level at the p/i interface ($B_{p/i}$) with doping at the i/n interface held constant at 0.15 ppm. It was found that photovoltaic performances depended on $B_{p/i}$, and that the best conversion efficiency could be achieved when $B_{p/i}$ was around 0.42 ppm. As expected, FF in a solar cell with a graded boron-doped i-layer exceeds 0.6, as shown in Fig. 5.1.6.

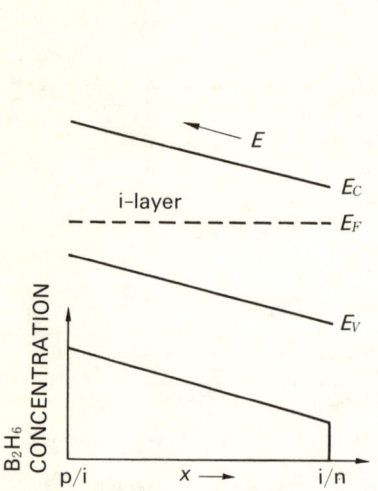

Fig. 5.1.5 Band diagram model of a graded boron-doped i-layer. [after P. Sichanugrist et al.[9)]]

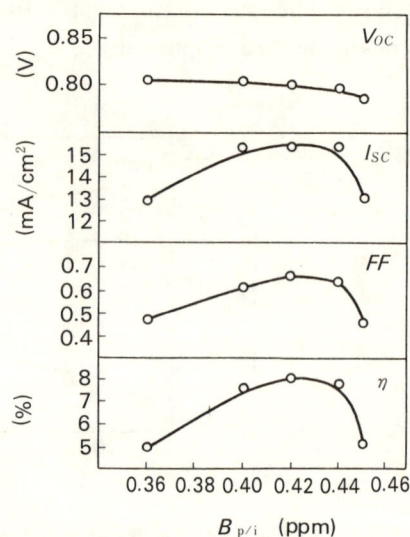

Fig. 5.1.6 Photovoltaic performances for the glass/ ITO/SnO$_2$/p(SiC)-i-n/metal solar cell as a function of B_2H_6 doping level at p/i interface ($B_{p/i}$). [after P. Sichanugrist et al.[9)]]

As described above, improved conversion efficiency is obtained in a solar cell with an i-layer having an optimized boron profile. In other words, boron doping in the i-layer plays a very important role in improving photovoltaic performance of a p-i-n- or n-i-p-based a-Si : H solar cell.

Carrier transport properties

Mobility-lifetime ($\mu\tau$) products of photo-generated carriers in the i-layer have been estimated from the bias voltage dependences of photocurrents in a-Si : H solar cells.[7,8,10)] The $\mu\tau$ products for p-i-n- and n-i-p-based solar cells with i-layers in which boron atoms are doped at a concentration level of $10^{16} \sim 10^{17}$ atoms/cm^3 lie on the order of 10^{-8} cm^2/V.

To obtain more information about the doping effect of boron atoms on photo-generated carrier transport, the spectral photocurrent, $J(\lambda, V)$, was measured for solar cells under three bias voltage conditions : $-0.4, 0, 0.4$ V. Measurement was done for both ITO/n-i-p/SS[7)] and ITO/p-i-n/SS[8)] solar cells. Figures 5.1.7 and 5.1.8 show normalized $J(\lambda, V)/J(\lambda, V = 0)$ as a function of wavelength for the former and latter types of solar cells,

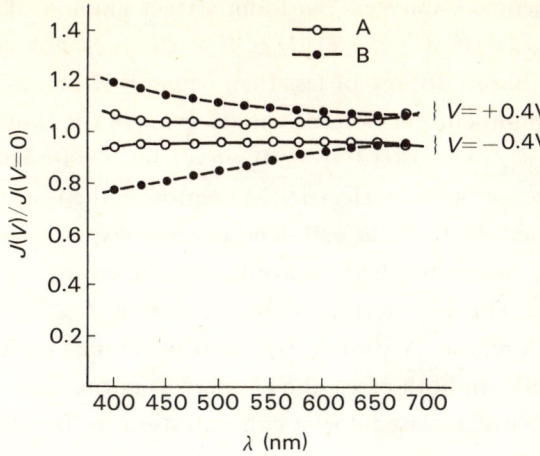

Fig. 5.1.7 Ratio of photocurrent $J(\lambda, V)/J(V=0)$ versus wavelength for the ITO/n-i-p/SS cells with the appropriately boron-doped ($\sim 10^{17}$ atoms/cm^3) i-layer (curve A) and insufficiently boron-doped ($\sim 10^{16}$ atoms/cm^3) i-layer (curve B). [after H. Haruki et al.[7]]

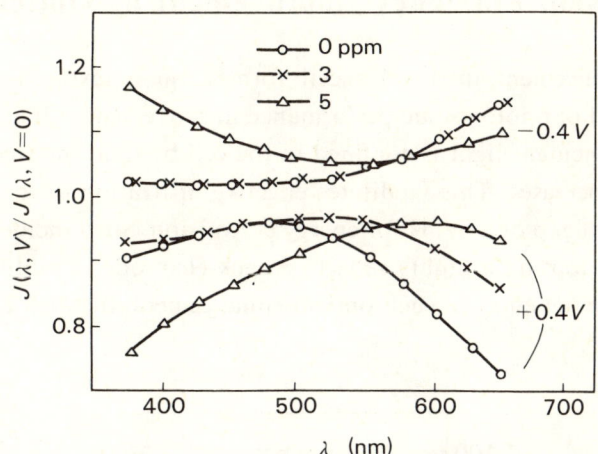

Fig. 5.1.8 Ratio of photocurrent $J(\lambda, V)/J(V=0)$ versus wavelength for the ITO/p-i-n/ SS cells with i-layers of various B$_2$H$_6$ doping level (0, 3, 5 ppm). [after H. Sakai et al.[8]]

respectively. In the shorter wavelength region, the bias dependence of $J(\lambda, V)$ for a n-i-p cell with an insufficiently boron-doped i-layer is much greater than that for a cell with an appropriately boron-doped i-layer, as shown in Fig. 5.1.7. $J(\lambda, V)/J(\lambda, V=0)$ for the latter is nearly independent of wavelength and is very close to unity. In contrast, the bias dependence of $J(\lambda, V)$ for a p-i-n cell with an undoped i-layer is larger in the longer wavelength region, as shown in Fig. 5.1.8. $J(\lambda, V)/J(\lambda, V=0)$ increases toward unity in the longer wavelength region with increased B$_2$H$_6$/SiH$_4$ during i-layer deposition, resulting in improved *FF* in the solar cell. It should be noted that $J(\lambda, V=0.4\text{V})/J(\lambda, V=0)$ and $J(\lambda, V=-0.4\text{V})/J(\lambda, V=0)$ are asymmetric with respect to unity for a solar cell with an undoped i-layer, but become symmetric by doing a trace amount of boron atoms into the i-layer. The value of $J(\lambda, V = \pm 0.4\,\text{V})/J(\lambda, V = 0)$ does not change in the shorter wavelength region with boron doping of less than 3 ppm B_2H_6. However, at 5 ppm, the bias voltage dependence of photocurrent becomes pretty great in the shorter wavelength region.

The asymmetry of $J(\lambda, V)/J(\lambda, V = 0)$ curves for a p-i-n solar cell with undoped i-layer suggests the existence of a low electric field region in the i-layer when a forward bias voltage of 0.4 V is applied to the solar cell. One possible explanation for this is that light doping of boron makes the electric field uniform over the entire i-layer and eliminates the low electric field region. This also results in an increase in $J(\lambda, V = 0.4\,\text{V})/J(\lambda, V = 0)$ in the longer wavelength region. A similar explanation can be applied to the experimental results for n-i-p solar cells. In both cases, the changes in normalized spectral photocurrent, $J(\lambda, V)/J(\lambda, V = 0)$, with increased boron concentration in the i-layer, shown in Figs. 5.1.7 and 5.1.8, indicate that the important role of doped boron atoms is to assist photo-generated holes traversing the i-layer. Computer modeling of a-Si : H solar cells[11] also shows that the boron profile in the i-layer plays a very important role in determining the electric field distribution throughout the i-layer.

5.1.3 Conversion Efficiency Improvement by Optical Enhancement

Optical enhancement in a solar cell formed on a textured substrate has been suggested as a method of photovoltaic performance improvement.[12] If at least one of the cell surfaces is textured, incident light is confined in the cell by total internal reflection and the optical pass-length increases. This facilitates effective utilization of incident photons with energy near the optical-gap of a-Si : H for energy conversion. Such incident light can be used more effectively by adopting a highly reflective back-electrode in addition to the textured substrate. To estimate the effect of such optical enhancement, incident angle dependence of

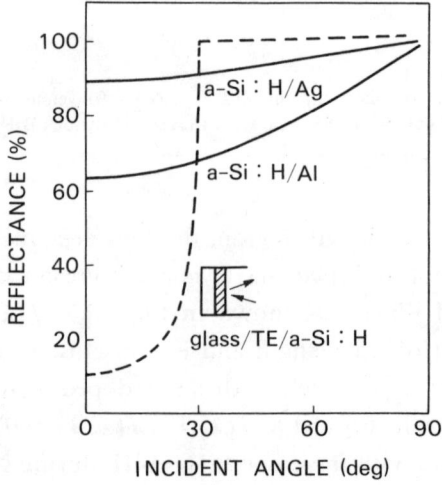

Fig. 5.1.9 Reflectance caluculated as a function of the incident angle. [after H. Sakai et al.[13]]

reflectance at the interface between a-Si : H and the back-electrode was evaluated.[13] Figure 5.1.9 shows reflectance at the a-Si : H/Ag interface as a function of incident angle, together with the data for the most widely used a-Si : H/Al structure. In the calculation, $1.44 + 3.63i$ and $0.2 + 17.2i$ were used as the respective complex reflective indices. We can adopt 0.76 and 0.93 as the average reflectance at the a-Si : H/Al and the a-Si : H/Ag interfaces on the basis of the assumption that the incident light is fully scattered at the cell surface and the angle of scattering at the back surface is distributed uniformly from 0° and 90°. As shown in Fig. 5.1.9, reflectance at the Al back-electrode is 0.63 when the substrate surface is completely smooth (the incident angle is 0°). Adopting a textured substrate together with an Ag back-electrode increases the average reflectance to 0.93. The internal reflectance at the transparent eletrode (TE)/a-Si : H interface is also shown in Fig. 5.1.9 as a function of incident angle. As the incident angle increases, the reflectance increases steeply, until it reaches 100% at 30°.

The glass-substrate solar cell reported as being the first 10% efficiency a-Si : H solar cell had an optical enhancement structure[14] : a transparent electrode with a rough surface and an Ag back-electrode. Various other solar cell structures have been proposed and demonstrated, with emphasis on optical enhancement. They include ITO/n-i-p/ITO/Ag/textured glass,[15] ITO/n-i-p/ITO/textured glass/Ag,[15] glass/TE/p (SiC) in/Ag,[14,16,17] glass TE/p (SiC) in/ITO/Ag,[13,18] ITO/n-i-p/TiO$_2$/Ag/SS,[19] and ITO/p-i-n/Ag/glass or SS.[20] In most of them, conversion efficiencies of more than 9% have been observed. The effect of optical enhancement appears mainly as distinct improvement in wavelength response above

Table 5.1.1 Grain size and height of five kinds of SnO$_2$ films in Å. [after H. Sakai et al.[13]]

Sample	A	B	C	D	E
Size	200~1200	←	500~2000	→	
Height	~200	~300	~500	~800	~1200

A

E

Fig. 5.1.10 Scanning electron micrograghs of SnO$_2$ surfaces. The surface roughnesses of SnO$_2$ A and E are shown in Table 5.1.1. [after H. Sakai et al.[13]]

600 nm, which also results in short-circuit current improvement.

A systematic experiment to determine the effect of transparent electrode (TE) surface roughness on short-circuit current was performed using glass/TE/p(SiC)in/metal solar cell.[13] A 2000 Å thick SnO_2 layer was used as the TE and either Al or Ag was used as the back-electrode metal. The p-, i- and n-layers were made 100, 5000 and 500 Å thick, respectively. Surface roughness was identified as A, B, C, D or E, as shown in Table 5.1.1. Figure 5.1.10 shows scanning micrographs for samples 'A' and 'E'. Figure 5.1.11 shows the relative short-circuit current, J_{sc}, observed in galss/SnO_2/p (SiC) in/Ag solar cells with the TE surface roughnesses listed in Table 5.1.1. As shown, J_{sc} increased monotonically with increasing TE surface roughness. Additionally, J_{sc} in cells with an Al back-electrode is lower than that shown in Fig. 5.1.11. Table 5.1.2 lists the relative short-circuit current of solar cells with different TE surface roughnesses and different back-electrode materials. The effect of back-electrode reflectance and TE surface roughness on short-circuit current is clearly observed in Table 5.1.2 and Fig. 5.1.11.

With regard to SS-substrate solar cells, a metal layer with a higher reflectance should be formed on top of the SS-substrate to more effectively utilize blue light for energy

Table 5.1.2 The short-circuit current observed in glass/SnO_2/p(SiC)-i-n/metal type a-Si: H solar cells with different SnO_2 films and back-electrode metals in relative unit. [after H. Sakai et al.[13]]

Back-electrode	Al	Ag
Flat SnO_2 (A)	1	1.12
Rough SnO_2 (E)	1.03	1.21

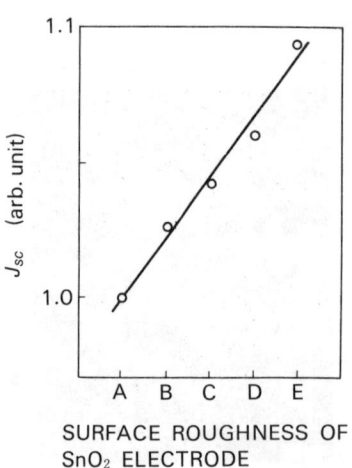

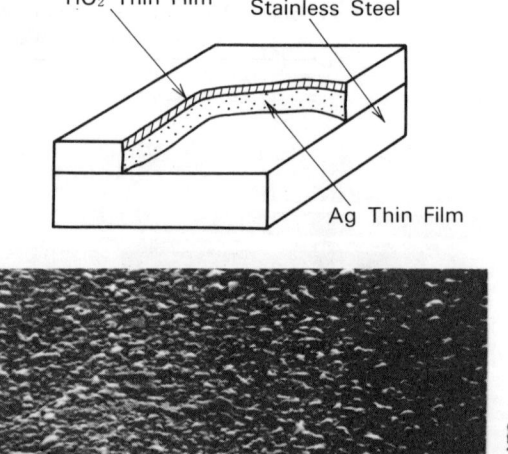

Fig. 5.1.11 Relative short-circuit currents observed in glass/SnO_2/p(SiC)-i-n/Ag solar cells with various SnO_2 surface roughnesses shown in Table 5.1.1. [after H. Sakai et al.[13]]

Fig. 5.1.12 Schematic configuration of the TiO_2/Ag/ SS substrate and the scanning microgragh of its surface morphology. [after K. Fujimoto et al.[19]]

conversion. A double-coated TiO$_2$/Ag/SS semi-textured substrate has been developed for this purpose. Substrate surface irregularity has been obtained by deposition of TiO$_2$ on an Ag/SS substrate.[19] Figure 5.1.12 shows the configuration of this TiO$_2$/Ag/SS substrate and a scanning electron micrograph of its surface morphology. Chemically mirror-etched stainless steel was used as the base substrate. TiO$_2$ and Ag thin films with thicknesses of about 100 Å and 1000 Å, respectively, were prepared by vacuum evaporation. Numerous grains, ranging from several hundred angstrom units to submicrom size, can be seen on the surface of this substrate.

Spectral response of these cells was also measured. Figure 5.1.13 shows the results. In the wavelength region below 550 nm, no definite difference in collection efficiency was noted in these cells. In contrast, a remarkable increase in carrier collection efficiency in the TiO$_2$/Ag/SS cell, compared with that in Ag/SS and SS cells, has been observed in the wavelength region above 600 nm.

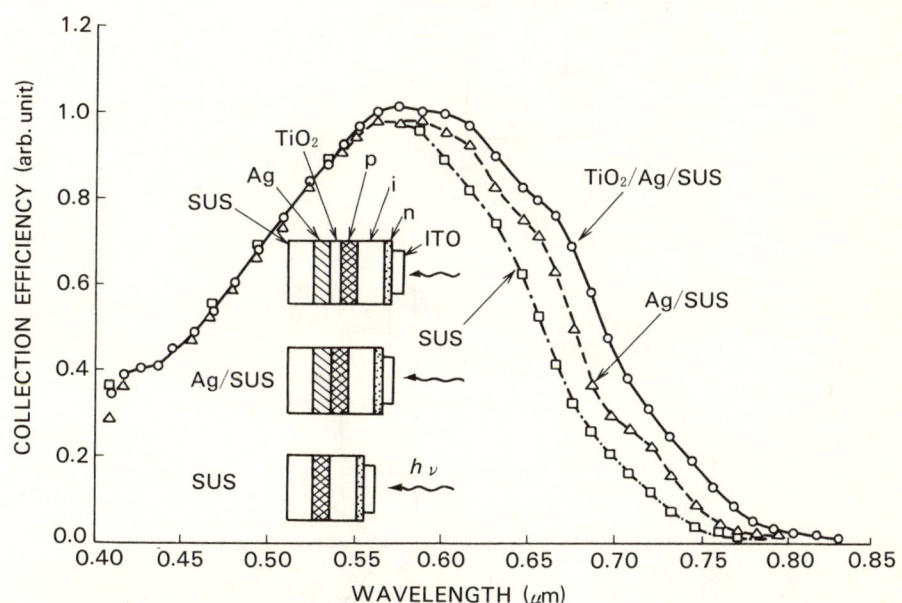

Fig. 5.1.13 Carrier collection efficiency of a-Si:H solar cells formed on SS, Ag/SS and TiO$_2$/Ag/SS substrates. [after K. Fujimoto et al.[19]]

5.1.4 Development of Large-Area Solar Cells

An a-Si:H solar cell can be formed on a foreign substrate with a large area, such as glass, polymer or ceramic coated with a conductive film, as well as on sheet metals. From an economical point of view, large-area producibility is a definite advantage of a-Si:H solar cells over conventional crystalline solar cells. This facilitates reducing the cost for assembling solar cells into a module, as well as for fabricating the cell itself. Figure 5.1.14 shows a "single-substrate module" as an example of a low-cost module structure using a

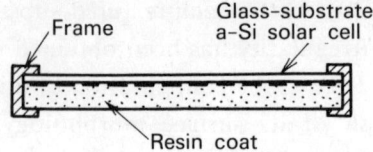

Fig. 5.1.14 Schematic structure of the "single substrate module" using a large-area glass-substrare cell. [after Y. Uchida et al.[21]]

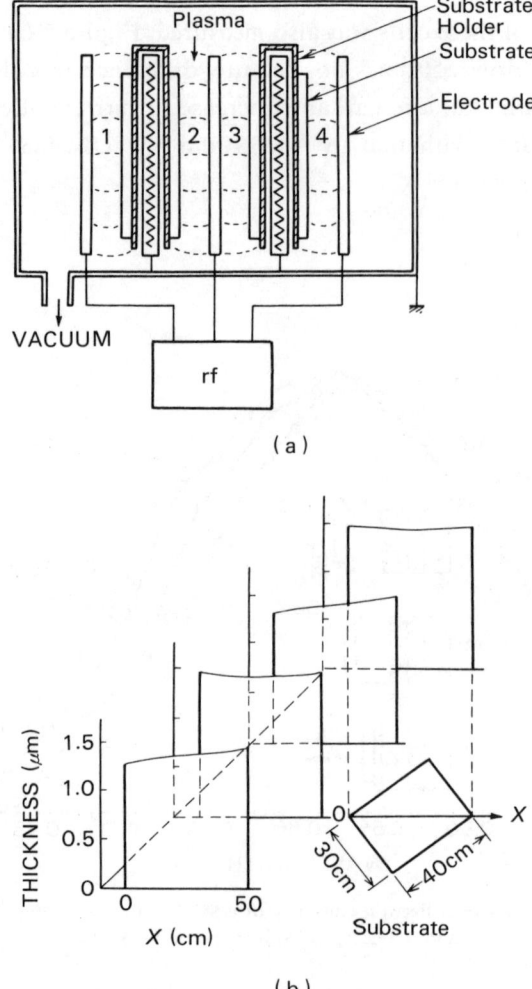

Fig. 5.1.15 (a) The a-Si : H deposition apparatus named IVE (Interdigital-Vertical-Electrode) system. (b) The thickness distribution of a-Si : H films formed on $30 \times 40 cm^2$ glass-substrates in the IVE system. [after H. Sakai et al.[22] © 1984 IEEE]

large-area glass-substrate cell.[21]

Developments in fabrication of large-area a-Si : H solar cells with high performance have been intensively pursued. One of the most important techniques is to produce a homogeneous a-Si : H film over a large area. In the development of an a-Si : H deposition technique, anything that causes pinholes to form in the a-Si : H film should be carefully avoided. To produce large-area solar cells at low cost and at a high throughput, a promising a-Si : H deposition apparatus has been proposed.[22] Figure 5.1.15 (a) shows an a-Si : H depostion apparatus called IVE (Interdigital-Vertical-Electrode) system that consists of three rf electrodes arranged interdigitally around two grounded electrodes. These electrodes are placed vertically to prevent dust and flakes that may cause pinholes in the a-Si : H film from falling on the substrate or rf electrode surfaces. Since this configuration provides four discharge stations, one can deposit a-Si : H simultaneously on four substrates held on both sides of the grounded electrodes. Figure 5.1.15 (b) shows the film thickness distribution of a-Si : H formed in the IVE system. The thickness measured along the diagonal of 30×40 cm^2 substrates is distributed within $\pm 10\%$ of the mean value. Figure 5.1.16 shows an a-Si : H solar cell formed on a 30×40 cm^2 glass-substrate in the IVE system.

Another proposed method to produce a-Si : H on a large-area substrate at high throughput is the roll-to-roll system.[23,24] Figure 5.1.17 shows an a-Si : H solar cell formed

Fig. 5.1.16 The a-Si : H solar cell formed on a 30×40 cm^2 glass-substrate. [after H. Sakai et al.[22] © 1984 IEEE]

Fig.5.1.17 The a-Si : H solar cell formed on a flexible polymer substrate. [presented by Teijin Ltd.]

on a flexible polymer substrate in such a deposition apparatus.[24] Details of the roll-to-roll system are presented in Section 5.3.

5.1.5 Stability under Light Exposure

Since Staebler and Wronski[25] reported on light-induced changes in the conductivity and photoconductivity of a-Si:H films, degradation of cell performance under sunlight exposure has been a very important problem. It was reported that some solar cells degraded rapidly while others were relatively stable, even though they had the same structure,[26] but the reason was not clarified. Recently, intensive efforts have been concentrated on improving the stability of a-Si:H solar cells, similar to those on conversion efficiency improvement. Systematic studies have shown that the stability of a-Si:H solar cells to light exposure strongly depends on the junction structure, and that light-induced changes in cell performance can be controlled.[27]

Samples of ITO/p-i-n/SS and ITO/n-i-p/SS solar cells with various i-layer thicknesses were prepared to test the effect of light exposure on cell performance.[27] The i-layer of some ITO/p-i-n/SS cells was doped with a small number of boron atoms using a B_2H_6/SiH_4 mixture (2 ppm B_2H_6 concentration). The area of all sample cells was 1.2 cm^2. Light passes through the ITO for both types of cell. Microcrystalline Si:H was used as the front layer of the sample. In all samples, the thicknesses of the front and back layers (p-layer and n-layer, respectively, in the case of an ITO/p-i-n/SS cell) were about 100 Å and 500 Å. The initial conversion efficiencies of the cells lay between about 6% and 7.5%.

The solar cells were continuously exposed to a simulated AM1 (100 mW/cm^2) light. Since there is a tendency for the performances of unstable cells to degrade in the first part of the exposure and then stabilize, change for light exposure in a relatively short period of 240 min was investigated. Solar cells were simultaneously exposed to light under short-circuit, optimum load and open-circuit conditions in order to examine the effect on performance stability of the electric field in the i-layer under light exposure.

Figure 5.1.18 shows the exposure time dependence of conversion efficiency normalized to its initial value for ITO/p-i-n/SS cells. The change in conversion efficiency after 240 min exposure was less than 2% for cells with a 0.5 μm i-layer and less than 4% for cells with a 1.2 μm i-layer. Cells exposed under short-circuit conditions were the most stable and those under open-circuit conditions showed the greatest change in conversion efficiency. The fill factor, FF, and open-circuit voltage V_{oc} scarcely changed during 240 min light exposure. Most of the changes were due to a decrease in the short-circuit current, J_{sc}.

Since J_{sc}, FF, and thus the conversion efficiency of the ITO/p-i-n/SS cell, were improved by doping the i-layer with a small amount of boron[8] as described in 5.1.2, the stability of an ITO/p-i-n/SS cell with a 0.55 μm i-layer doped with boron was tested. This type of cell is more suitable than the ITO/n-i-p/SS cell for accurate evaluation of the effects of boron doping in the i-layer, because, in this cell, the boron-doped layer is formed after deposition of the i-layer and it is not necessary to consider autodoping of boron in the i-layer.[6] Figure 5.1.19 shows the stability of short-circuit current and conversion efficiency of ITO/p-i-n/SS cells with a boron-doped i-layer under light exposure. This figure shows that these solar cells were extremely stable under light exposure and the change in conversion

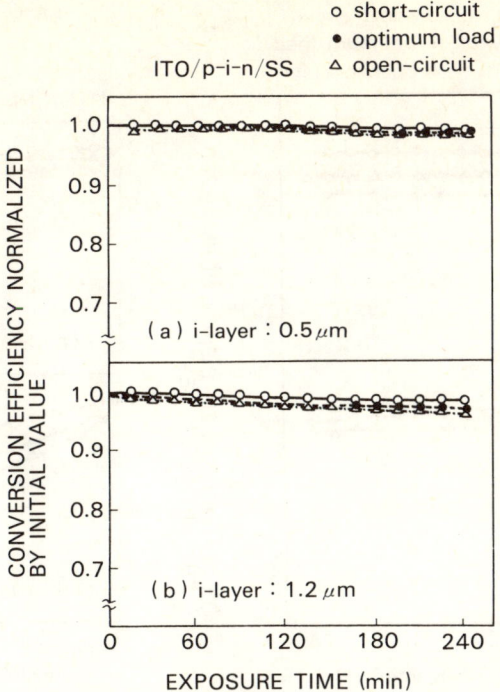

Fig. 5.1.18 Light exposure time dependence of the conversion efficiency for various load conditions in ITO/p-i-n/SS solar cells with i-layers of thickness (a) 0.5 μm and (b) 1.2 μm: ○, short-circuit; ●, optimum load; △, open-circuit. [after Y. Uchida et al.27)]

efficiency after the 240 min exposure was within ± 1% of the initial value.

Figure 5.1.20 shows the exposure time dependence of conversion efficiency of ITO/n-i-p/SS solar cells with a (a) 0.35 μm i-layer and (b) 0.57 μm i-layer. Degradation of the conversion efficiency strongly depended on the i-layer thickness and on the load condition during exposure. Conversion efficiency in cells of type (a) underwent little change, while that in cells of type (b) underwent a relatively large change. Such changes are mainly caused by decreases in J_{sc} and FF. The stability of an ITO/n-i-p/SS cell depends on the load condition of the cells when exposed to light, in the same manner as that of ITO/p-i-n/SS cells does. However, ITO/n-i-p/SS cells are less stable than ITO/p-i-n/SS cells with similar i-layer thicknesses.

Figure 5.1.21 summarizes the performance dependence of ITO/n-i-p/SS cells on i-layer thickness up to 240 min exposure, with the quantities normalized to the initial values. The i-layer thickness dependence and the load condition dependence of stability on light exposure are clearly observed in J_{sc}, FF, and thus the conversion efficiency. In the case of cells exposed under short-circuit conditions, the conversion efficiency of cells with an i-layer thickness below 0.52 μm remained at more than 95% of the initial value after 240 min light exposure, while that of cells with a thicker i-layer showed more than 10% degradation. Cells exposed under optimum load and open-circuit condition exhibited greater changes in conversion efficiency. The short-circuited cell has a built-in electric field in the i-layer, while

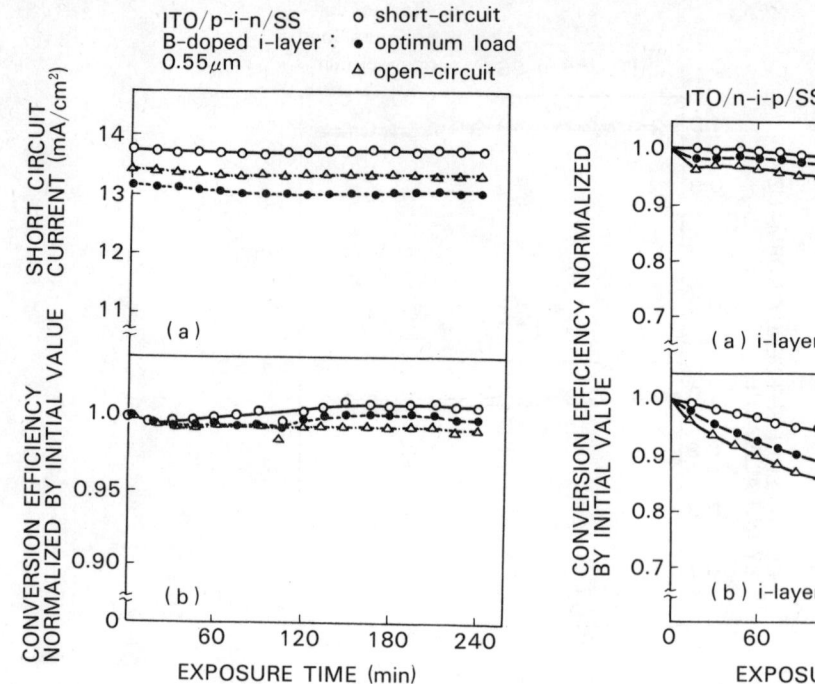

Fig. 5.1.19 Stability of (a) the short-circuit current and (b) the conversion efficiency under light exposure of an ITO/p-i-n/SS cell with a 0.55 μm i-layer doped with 2 ppm B_2H_6 (symbols as in Fig. 5.1.18). [after Y. Uchida et al.[27]]

Fig. 5.1.20 Light exposure time dependence of the conversion efficiency for various load conditions in ITO/n-i-p/SS cells with i-layers of thickness (a) 0.35 μm and (b) 0.57 μm (symbols as in Fig. 5.1.18). [after Y. Uchida et al.[27]]

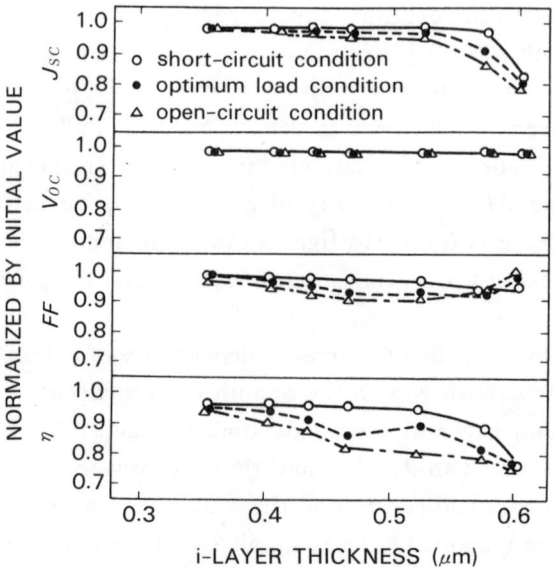

Fig. 5.1.21 The i-layer thickness dependence of the photovoltaic performances in ITO/n-i-p/SS cells after 240 min light exposure (symbols as in Fig. 5.1.18). [after Y. Uchida et al.[27]]

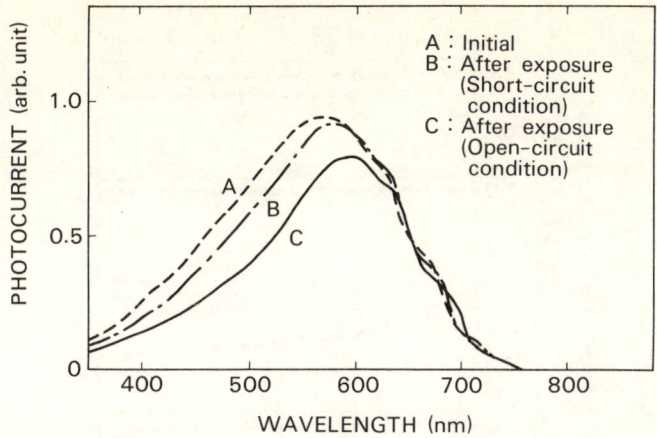

Fig. 5.1.22 Wavelength dependence of the photocurrent in an ITO/n-i-p/SS cell with an i-layer thickness of 0.57 μm : curve A, initial values : curve B, values for the cell exposed under short-circuit condition ; curve C, values for exposure in open-circuit conditions. [after Y. Uchida et al.[27]]

cells under optimum load or open-circuit conditions are forward biased. These results show that degradation in cell performance is suppressed as the electric field in the i-layer during light exposure increases.

Figure 5.1.22 shows the wavelength dependence of photocurrent in an ITO/n-i-p/SS cell with an i-layer thickness of 0.57 μm (measured under incident light at a constant power density of 150 μW/cm^2). Curve A gives the initial photocurrent, while curve B shows the values after exposure under short-circuit conditions and curve C gives the photocurrent under open-circuit conditions. In all cases the short-wavelength response decreased after light exposure, but the change depends on the load condition.

To examine effects of injecting carriers into the i-layer separately from illumination, a forward bias current density of 10 mA/cm^2 was applied to the cell in the dark for 240 min. Figure 5.1.23 shows the changes in cell performance of ITO/n-i-p/SS solar cells with 0.35 and 0.60 μm i-layers. Cell performance of the sample with a 0.35 μm i-layer was stable, while that of the cell with a 0.60 μm i-layer showed dagradation. The i-layer thickness dependence of stability was also observed in this forward bias current test and was quite similar to that obtained in the light exposure test. The change in wavelength dependence of photocurrent was also similar to that in the light exposure test, but not exactly the same.

Table 5.1.3 shows light-induced effects on glass/SnO$_2$/p (SiC) -i-n/Al type a-Si : H solar cells.[28] Light exposure was with a He-Ne laser (6 328 Å) and an Ar laser (4 880 Å). During light exposure, $\mu\tau$ products decreased and carrier recombination velocity at the n/i interface, S_n, increased, as shown in Table 5.1.3. These changes caused degradation of *FF* and were reversible by heat treatment, as shown in Table 5.1.3. This reversible change is in agreement with reversible conductivity change in a-Si : H films (Staebler-Wronski effect).

As described above, light-induced change in cell performance are much more marked in ITO/n-i-p/SS cells than in ITO/p-i-n/SS cells. In the former, the short-wavelength

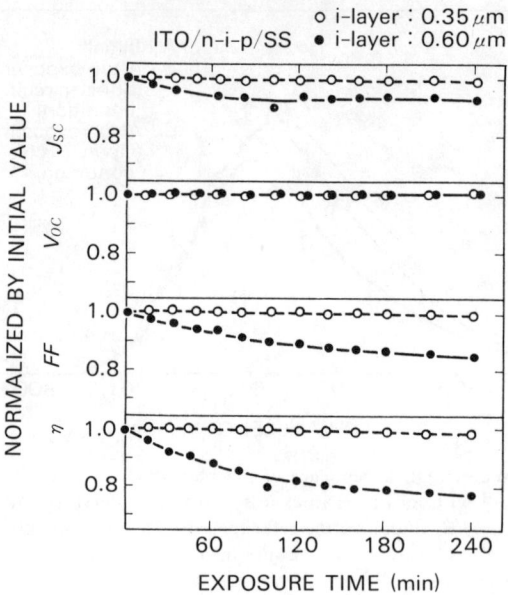

Fig. 5.1.23 Changes in the phtovoltaic performances of ITO/n-i-p/SS cells with i-layers of thickness 0.35 μm (○) and 0.60 μm (●) to which a forward bias current of 10mA/cm² was applied in the dark. [after Y. Uchida et al.[27]]

Table 5.1.3 Light-induced effects on glass/p(SiC)-i-n/metal a-Si:H solar cells. The light exposure was done with a He-Ne laser and an Ar laser. [after Y. Tawada et al.[28]]

		V_b	$\mu_n\tau_n$	$\mu_p\tau_p$	S_n	S_p	r_{sp}	V_{oc}	J_{sc}	FF
Light Induced Effect	He Ne	→	⬈	↘	↗	→		→	↘	⬈
	Ar	→	↘	↘	⬈	→		→	→	⬈
Heat Treatment Effect	Initial Heat Treatment	→	↗	↗	⬈	→	↘	→	↗	⬈
	After Heat Treatment	→	↗	↗	↘	→		→	↗	⬈
		Potential Profile	Bulk Effect		Interface Effect		Contact			

response decreases after light exposure. This suggests that collection efficiency of photogenerated holes is adversely affected by a decrease in $\mu\tau$ product or by collapse of the built-in field in the i-layer due to light-induced gap states. Doping of the i-layer in an ITO/p-i-n/SS solar cell with a trace amount of boron atoms, which changes the i-layer from slightly n-type to intrinsic,[7] improves cell performance and makes it quite stable to light exposure. This may correlate with a feature of light-induced change in a-Si:H: that the Fermi level moves toward the mid-gap under light exposure.

Light-induced changes in photovoltaic performance depend on the i-layer thickness of the solar cells and the changes are smaller in cells with a thinner i-layer. Furthermore, the changes are suppressed as the electric filed in the i-layer becomes higher during light exposure. Very similar performance changes are observed in solar cells after application of a forward bias current in the dark. These results suggest a degradation mechanism in which recombination of injected carriers in the i-layer plays an important role in creating additional gap-states and degrades carrier collection efficiency.

5.1.6 Summary

Recent developments in research on the a-Si : H solar cells with a single p-i-n (or n-i-p) junction have been reviewed with emphasis on conversion efficiency improvement, enlargement of cell area and performance stabilization to light exposure. The significant results from the research described in the foregoing sections are :

(1) The boron profile in the i-layer of both p-i-n- and n-i-p-based solar cells plays a very important role in determining photovoltaic performance. In other words, a solar cell with an i-layer having an optimized boron profile exhibits higher conversion efficiency than one with an undoped i-layer.

(2) Optical enhancement by adopting a textured substrate together with a highly reflective back-electrode improves wavelength response above 600 nm, which results in conversion efficiency improvement. For this purpose, an SnO_2 transparent electrode with a rough surface and an Ag back-electrode have been effectively applied to glass-substrate solar cells and a double coated TiO_2/Ag/SS semi-textured substrate has been developed for SS-substrate solar cells.

(3) To produce a homogeneous a-Si : H film over a large area at high throughput, an a-Si : H deposition apparatus called the IVE (Interdigital-Vertical-Electrode) system has been proposed. Since this apparatus is provided with three rf electrodes interdigitally arranged around two grounded electrodes, one can deposit a-Si : H simultaneously on four substrates held on both sides of the grounded electrodes. The rf and grounded electrodes are placed vertically to prevent dust and flakes that may cause pinholes in the a-Si : H film, falling on substrate or the rf electrode surface.

(4) a-Si : H solar cells formed on a 30 × 40 cm² glass-substrate and a wide, flexible substrate have been demonstrated.

(5) There are fewer light-induced changes in photovoltaic performance in ITO/p-i-n/SS cells than in ITO/n-i-p/SS cells and such changes are smaller in cells with a thinner i-layer. Furthermore, these changes are suppressed as the electric field in the i-layer becomes higher during exposure.

(6) Compensation of the i-layer with boron results in a more stable cell to light exposure.

(7) Performance changes similar to the light-induced effects are observed in solar cells after forward current biasing in the dark.

(8) The above facts about stability suggest a degradation mechanism in which recombination of injected carriers in the i-layer plays an important role in creating additional gap-states and degrades carrier collection efficiency.

Acknowledgements

The author is grateful to Prof. Yoshihiro Hamakawa of Osaka University and to Dr. Hiromu Haruki and Dr. Masaya Yabe of Fuji Electric for their helpful comments. A large part of the work described in this section was supported by the Agency of Industrial Science and Technology and the New Energy Development Organization under contract to Sunshine Project.

References

1) Y. Uchida, T. Ichimura, M. Ueno, and H. Haruki: Jpn. J. Appl. Phys., *21* (1982) L 586.
2) Y. Tawada, H. Okamoto, and Y. Hamakawa: Appl. Phys. Lett., *39* (1981) 237.
3) D. L. Staebler and C. R. Wronski: Appl. Phys. Lett., *31* (1977) 292.
4) D. E. Carlson: Solar Energy Mater., *3* (1980) 503.
5) Y. Uchida, H. Sasaki, M. Nishiura, and H. Haruki: Proc. 15th IEEE Photovoltaic Specialists Conf. (Florida, 1981) 922.
6) H. Haruki, Y. Uchida, H. Sakai, and M. Nishiura: Proc. 13th Conf. on Solid State Devices (Tokyo, 1981) and Jpn. J. Appl. Phys., *21* (1982) Suppl. 21-1, 283.
7) H. Haruki, H. Sakai, M. Kamiyama, and Y. Uchida: Solar Energy Mater., *8* (1983) 441.
8) H. Sakai, M. Kamiyama, Y. Uchida, and H. Haruki: J. Non-Cryst. Solids, *59 & 60* (1983) 1151.
9) P. Sichanugrist, M. Kumada, M. Konagai, K. Takahashi, and K. Komori: J. Appl. Phys., *54* (1983) 6705.
10) Y. Higaki, M. Aiga, S. Terazono, Y. Yukimoto, H. Okamoto, and Y. Hamakawa: Proc. Fifth EC Photovoltaic Solar Energy Conf. (Athens, 1983) 156.
11) M. Hack, M. Shur, W. Czubatyj, J. Yang, and J. McGill: J. Non-Cryst. Solids, *59 & 60* (1983) 1115.
12) E. Yablonovitch and G. Cody: IEEE Trans. Electron Devices, *ED-29* (1982) 300.
13) H. Sakai, M. Kamiyama, M. Ueno, Y. Uchida, and H. Haruki: Proc. Fifth EC Photovoltaic Conf. (Athens, 1983) 808.
14) A. Catalano, R. V. D'Alello, J. Dresner, B. Faughnan, A. Firester, J. Kane, H. Shade, Z. E. Smith, G. Swarts, and A. Triano: Proc. 16th IEEE Photovoltaic Specialists Conf. (San Diego, 1982) 1421.
15) H. W. Deckman, C. R. Wronski, and H. Witzke: J. Vac. Sci. Technol., *A 1-2* (1983) 578.
16) Y. Hamakawa, K. Fujimoto, K. Okuda, Y. Kashima, S. Nonomura, and H. Okamoto: Appl. Phys. Lett., *43* (1983) 644.
17) J. Tajika, H. Mizukami, T. Miyake, S. Sano, and O. Kuboi: Abst. 44th Autumn Meeting of the Jpn. Soc. of Appl. Phys. (1983) 350 [in Japanese].
18) H. Iida, T. Mishuku, and Y. Hayashi: Suppl. to the Extended Abst. of 15th Conf. on Solid State Devices and Mater. (Tokyo, 1983) 38.
19) K. Fujimoto, H. Kawai, S. Nonomura, H. Okamoto, and Y. Hamakawa: Proc. Fifth EC Photovoltaic Solar Energy Conf. (Athens, 1983) 813.
20) A. Asano, M. Kamiyama, H. Sakai, and Y. Uchida: Abst. 31st Spring Meeting of the Jpn. Soc. of Appl. Phys. and of the Related Soc. (1984) 2a-R-4 [in Japanese].
21) Y. Uchida, H. Sakai, N. Furusho, M. Nishiura, and H. Haruki: J. Electrochem. Soc., *130* (1983) 712.
22) H. Sakai, K. Maruyama, T. Yoshida, M. Kamiyama, T. Ichimura, and Y. Uchida: Proc. 17th IEEE Photovoltaic Specialists Conf. (Florida, 1984) (to be published).
23) see Section 5. 3.
24) H. Okaniwa, K. Nakatani, M. Asano, M. Yano, and Y. Hamakawa: Proc. 16th IEEE Photovoltaic Specialists Conf. (San Diego, 1982) 1111.
25) D. L. Staebler and C. R. Wronski: Appl. Phys. Lett., *31* (1977) 292.

26) D. L. Staebler, R. S. Crandall, and R. Williams: Appl. Phys. Lett., *39* (1981) 733.
27) Y. Uchida, M. Nishiura, H. Sakai, and H. Haruki: Solar Cells, *9* (1983) 3.
28) Y. Tawada, K. Nishimura, S. Nonomura, H. Okamoto, and Y. Hamakawa: Solar Cells, *9* (1983) 53.

5.2 Heterojunction Stacked Solar Cells

Yoshihiro HAMAKAWA* and Hiroaki OKAMOTO*

Abstract

The concept of high-efficiency solar cell employing a multi-band gap stacked structure is described. Design rule and optimizations of cell parameters on two- and three-stacked heterojunction solar cells are discussed, and the theoretical limit of conversion efficiencies is clarified for both two types of solar cells. The current state-of-the-art in experimental trials on these multi-band gap stacked solar cells with various material combinations are reviewed.

5.2.1 Concept of a Multi-Layered Stacked Cell

Among the wide variety of solar photovoltaic R & D projects, improvement of solar cell performance is one of the important items. In amorphous silicon solar cell projects, tremendous efforts have been mode to increase cell efficiency using various approaches, such as film-quality improvement,[1] new junction structures with a-SiC,[2] c-Si,[3] a-SiGe[4] and the use of new electrode materials.[5] For example, an a-SiC/a-Si heterojunction solar cell, which first broke through the 8% efficiency barrier in 1980,[6] in a typically successful one, and has now become a routine technology for fabrication of high-efficiency solar cells. In fact, the recent records of more than 10% efficiency reported by such companies of RCA,[7] Sanyo,[8] Komatsu[9] and TDK-SEL[10] were all attained by using this a-SiC/a-Si heterojunction structure.

Another possibility for further improvement of amorphous solar cell efficiency is heterostructure stacked junctions utilizing internal carrier exchange effect through localized states at the heterojunction interface. This effect was first discovered by an Osaka University group in 1979,[11] and was applied to the high voltage solar cell called "HOMLAC (Horizontally Multi-Layered Photovoltaic Cell)".[12] Figure 5.2.1 shows the energy band diagram (a), schematic representation of hole-electron exchange effect at the junction (b) and an equivalent circuit explanation of the HOMLAC device.[13] combining this effect with more efficient "Optical Confinement Effect" and "Carrier Confinement Effect" would produce a double heterojunction stacked cell. Figure 5.2.2 shows a schematic band diagram of this type of cell. It should be noted that, in this multi-layered structure, not only wide band gap window effect but also short wavelength photon collections are possible at the front junction. Moreover, longer wavelength photon energy might be completely absorbed

---

* Faculty of Engineering Science, Osaka University, Toyonaka, Osaka 560.

by the back-surface junction. Table 5.2.1 shows a list of candidate materials and their electrical, optical and opto-electronic properties in tetrahedrally-bonded amorphous semiconductors.[14]

5.2.2 Theory of Heterojunction Stacked Cells

This section describes the theoretical procedure for analyzing photovoltaic operation of heterojunction stacked solar cells, and presents theoretical conversion efficiencies expected for various combinations of photovoltaic materials. To begin, consider a heterojunction stacked cell consisting of m unit cells connected in series by internal carrier exchange contact (p/n) junctions. The equivalent circuit for three-stacked ($m = 3$) cell is given in Fig. 5.2.1 (d), where each p/n junction connecting neighboring p-i-n cells acts as a shunt resistance. Thus, the common current through the stacked cell can be written as

$$I = I_{Li} - I_{si}\left[\exp\left\{\frac{q(V_i + IR_{si})}{n_i kT}\right\} - 1\right] - \frac{V_i + IR_{si}}{R_{shi}} \quad \text{for } i = 1, m \quad (5.2.1)$$

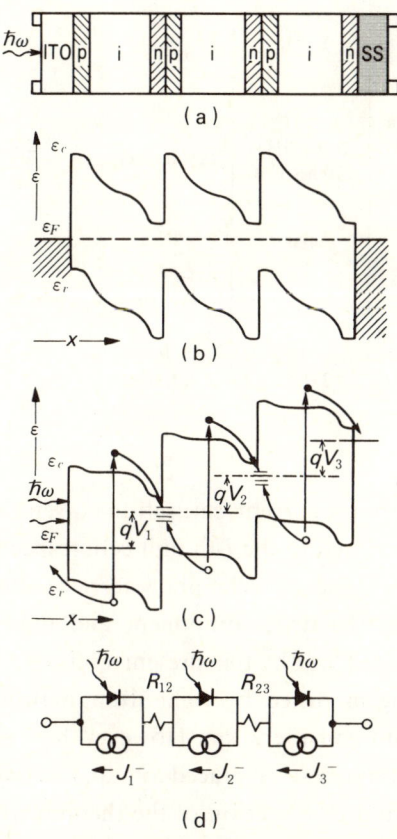

Fig. 5.2.1 Schematic representation for the principle of the mutilayered device. Device construction (a), equivalent circuit (b), band diagram in the dark (c) and band diagram under illumination (d). [after Y. Hamakawa et al.[11]]

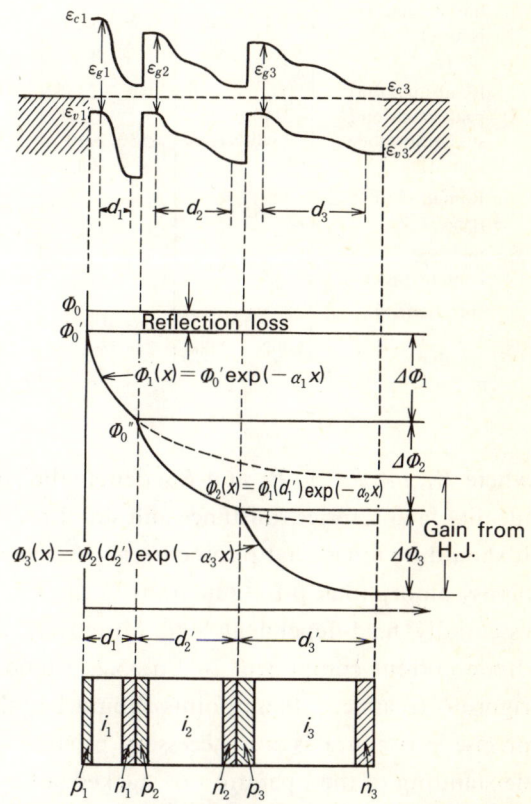

Fig. 5.2.2 Band diagram explanation of the multi-band gap stacked Solar cell (a) and its photon energy collection (b).

Table 5.2.1 Tetrahedrally bonded amorphous semiconductors and their physical properties.

	a-SiN	a-SiC	μc-Si	a-Si	a-SiGe	a-SiSn
source gases	$SiH_4 + NH_3$	$SiH_4 + CH_4, C_2H_4$ or $Si(CH_3)_4$	$SiH_4 + H_2$	SiH_4	$SiH_4 + GeH_4$	$SiH_4 + Sn(CH_3)_4$
optical energy gap (eV)	$1.8 \sim 5.5$	$1.8 \sim 2.8$	$\alpha < 10^4 cm^{-1}$ ($h\nu = 2$ eV)	$1.7 \sim 2.0$	$1.0 \sim 1.7$	$1.0 \sim 1.7$
dark conductivity ($1/\Omega \cdot cm$)	$10^{-10} \sim 10^{-8}$	$< 10^{-10}$	$10^{-5} \sim 10^{-2}$	$10^{-12} \sim 10^{-8}$	$10^{-8} \sim 10^{-5}$	$10^{-10} \sim 10^{-6}$
activation energy (eV)	$0.7 \sim 1.0$	< 0.9	—	$0.7 \sim 0.9$	$0.5 \sim 0.9$	$0.3 \sim 0.6$
photo-conductivity: AM1,100mW/cm² ($1/\Omega \cdot cm$)	$< 10^{-4}$	$< 10^{-7}$	$< 10^{-4}$	$< 10^{-3}$	$< 10^{-3}$	$< 10^{-6}$
mobility (cm²/Vs)	—	—	< 1 (ambipolar)	< 1 (e) $< 10^{-2}$ (h)	< 0.1 (e) $< 10^{-2}$ (h)	—
lifetime (s)	—	—	$< 10^{-5}$ (ambipolar)	$< 10^{-6}$ (e) $< 10^{-5}$ (h)	—	—
photo-luminescence peak energy (eV)	$1.4 \sim 1.6$	$1.5 \sim 2.3$	$1.2 \sim 1.4$ 0.8	$1.2 \sim 1.4$	$0.6 \sim 1.2$	—
IR absorption peak position: streching mode (cm^{-1})	2140(SiH) 2340(NH)	$2013 \sim 2135$ (SiH) 2850, 2910 (CH_2)	$2089 \sim 2108$ (SiH_2) 2140, 2158 (SiH_3)	2000(SiH) 2100(SiH_2)	1850(GeH)	—
Raman shift (cm^{-1})	450	750	480, 520	480	370	—
hydrogen content (at.%)	$10 \sim 50$	$10 \sim 50$	< 8	$5 \sim 50$	$4 \sim 40$	—
spin density (cm^{-3}) g-value	$> 10^{17}$ $2.002 \sim 2.006$	$> 10^{16}$ 2.0028(C)	$> 10^{17}$ 2.0049	$> 10^{15}$ 2.0055	$> 10^{16}$ 2.021 (Ge)	—

where V_i, I_{Li}, I_{si}, n_i, R_{si} and R_{shi} denote the voltage, photocurrent, saturation current, diode ideality factor, series resistance and shunt resistance related to the i-th unit cell, respectively. It should be noted that photocurrent I_{Li} depends on V_i because the photocarrier collection across amorphous p-i-n junctions is mainly due to the drift component and thereby is essentially field-dependent.[15,16] Moreover, as pointed out by the present authors,[17] the diode current component in Eq. (5.2.1) tends to be modified by light illumination. For rigorous treatment, these points should be taken into consideration. However, here such a precise procedure is not necessary because central interest is placed on qualitative understanding of the operation of stacked solar cells and clarification of the theoretical limit of their conversion efficiencies.

Equation (5.2.1) can be rewritten with additional simplification of $n_i = n$ and $R_{shi} = $ infinity for $i = 1, m$.

$$V + IR_s = \frac{nkT}{q} \ln \prod_{i=1}^{m} \left(\frac{I_{Li} + I_{si} - I}{I_{si}} \right) \tag{5.2.2}$$

with

$$V = \sum_{i=1}^{m} V_i \text{ and } R_s = \sum_{i=1}^{m} R_{si}.$$

It can be easily clarified by simple mathematical treatment of Eq. (5.2.2) that both the short circuit current, I_{sc}, and open circuit voltage, V_{oc}, of the stacked cell attain their maxima when each photocurrent, I_{Li}, is equal to $1/m$ of photocurrent $\bar{I}_L$, defined by $\sum_{i=1}^{m} I_{Li}$. Under this optimal condition, Eq. (5.2.2) can be reduced to[13]

$$\bar{V} + \bar{I}\bar{R}_s \simeq \frac{nkT}{q} \ln \left(\frac{\bar{I}_L + \bar{I}_s - \bar{I}}{\bar{I}_s} \right) \tag{5.2.3}$$

with definitions

$$\bar{V} = \frac{V}{m}, \quad \bar{I} = mI, \quad \bar{I}_s = m \left(\prod_{i=1}^{m} I_{si} \right)^{1/m} \text{ and } \bar{R}_s = \frac{R_s}{m^2}$$

Equation (5.2.3) indicates that the heterojunction stacked cell is equivalent to a single-junction cell characterized by photocurrent $\bar{I}_L$, saturation current $\bar{I}_s$, and series resistance $\bar{R}_s$, as far as the curve-fill factor and conversion efficiency are concerned. Thus, the maximum obtainable conversion efficiency can be readily evaluated by utilizing Eq. (5.2.3):

$$\eta_{max} \simeq \frac{\bar{V}_m \bar{I}_L}{P_{in}} \left\{ 1 - \frac{\exp(q\bar{V}_m/nkT) - 1}{\exp(q\bar{V}_{oc}/nkT) - 1} \right\} \tag{5.2.4}$$

where P_{in} is the incident solar power, $\bar{V}_{oc}$ denotes

$$\bar{V}_{oc} = \frac{nkT}{q} \ln \left(\frac{\bar{I}_L}{\bar{I}_s} \right) \tag{5.2.5}$$

and $\bar{V}_m$ is given by the solution of

$$\exp \left(\frac{q\bar{V}_m}{nkT} \right) \cdot \left(1 + \frac{q\bar{V}_m}{nkT} \right) = \frac{\bar{I}_L}{\bar{I}_s} + 1. \tag{5.2.6}$$

Photocurrent I_{Li} can be approximated as

$$I_{Li} \cong q \int \Phi_j(\lambda) \eta_c(\lambda, V_i) d\lambda \tag{5.2.7}$$

where $\Phi_i(\lambda)$ denotes incident photon flux upon the i-th cell and η_c the collection efficiency. In the simplest treatment, in which both the carrier collection process and optical interference effect are neglected, η_c and Φ_i may be approximated by utilizing absorption coefficient $\alpha_i(\lambda)$ and the thickness of the i-th photovoltaic layer, and by defining the incident photon flux $\Phi_0(\lambda)$ as

$$\eta_c(\lambda) \sim 1 - \exp\{-\alpha_i(\lambda) d_i\} \tag{5.2.8}$$

and

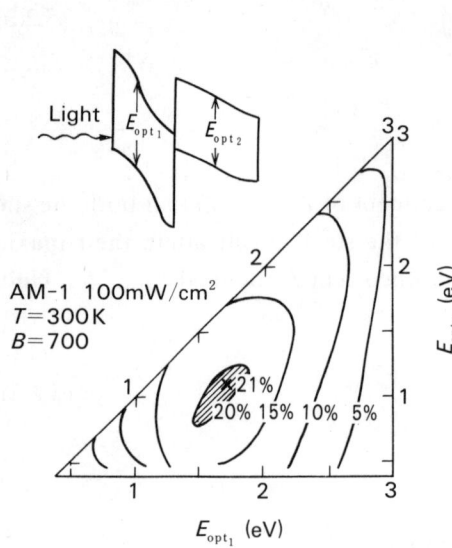
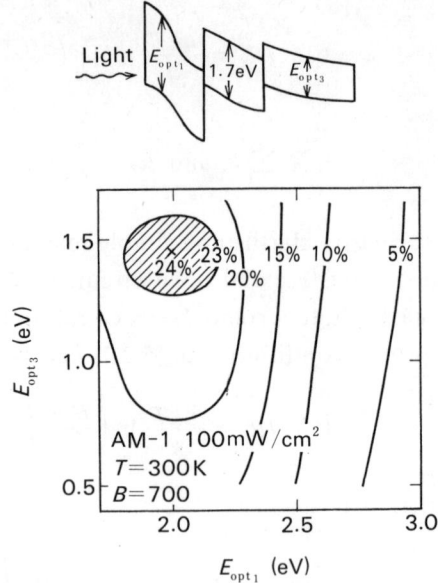

Fig. 5.2.3 Contour map of calculated conversion efficiency as a function of E_{opt1} and E_{opt2} for the double stacked solar cell. [after S. Tsuda et al.[18]]

Fig. 5.2.4 Contour map of calculated conversion efficiency as a function of E_{opt1} and E_{opt2}; $E_{opt3} = 1.7\,\text{eV}$ for the triple stacked solar cell. [after S. Tsuda et al.[18]]

$$\Phi_i(\lambda) \sim \Phi_0(\lambda) \prod_{j=1}^{i-1} \exp\{-\alpha_j(\lambda)d_j\} \tag{5.2.9}$$

These approximations give the maximum limit of I_{Li}. Optimization of the stacked cell is performed by choosing a set of d_i that satisfies the condition $I_{Li} = \bar{I}_L/m$ ($i = 1, m$).[13] This procedure is schematically illustrated in Fig. 5.2.2. When I_{Li} or $\bar{I}_L$ is thus calculated, the maximum limit of conversion efficiency can be obtained from Eq. (5.2.4).

On the basis of the procedure outlined above, Tsuda et al.,[18] Nakamura et al.[19] and Fan et al.[20] have calculated the maximum conversion dfficiencies for various material combinations. Figures 5.2.3 and 5.2.4 show contour maps of calculated conversion efficiencies for the two- and three-stacked solar cells presented by Tsuda et al.[18] In making the calculations, they took into account shunt resistance effect and adopted a somewhat improved formula for the collection efficiency spectrum. The material quality of amorphous materials of all energy gaps is assumed to be equal to that of the best a-Si : H films. In the case of a two-stacked solar cell, maximum efficiency of 21% is obtained with $E_{opt1} = 1.75$ eV and $E_{opt2} = 1.15\,\text{eV}$. For a three-stacked solar cell where the second cell material has an optical energy gap of $1.7\,\text{eV}$ (corresponding to a-Si : H), the maximum efficiency is estimated to be 24% with $E_{opt1} = 2.0\,\text{eV}$ and $E_{opt3} = 1.45\,\text{eV}$. The maximum conversion efficiencies and corresponding energy gap choices for two- and three-stacked cells evaluated by other authors are also listed in Table 5.2.2.

Table 5.2.2 Summary of the theoretical efficiencies of multi-band gap stacked amorphous solar cell.

	two stacked cell			three stacked cell			
	E_{opt1} (eV)	E_{opt2} (eV)	η (%)	E_{opt1} (eV)	E_{opt2} (eV)	E_{opt3} (eV)	η (%)
G. Nakamura et al.*	1.95	1.45	14	1.95	1.45	1.0	19
S. Tsuda et al.**	1.75	1.15	21	2.0	(1.70)	1.45	24
J. C. C. Fan**	1.85	1.35	19.6	2.10	(1.65)	1.35	21.4

* based upon i-type a-Si$_{0.6}$Ge$_{0.4}$ ($E_{opt}=1.45$ eV) as the center cell
** based upon i-type a-Si : H ($E_{opt}=1.65\sim1.7$ eV) as the center cell

5.2.3 Review of Experimental Approaches

Homojunction stacked cell

The most important key for improving a-Si solar cell efficiency is to decrease the localized state density in the i-layer, as has been shown elsewhere, and tremendous attention has been paid to various deposition techniques for a-Si : H, a-Si : F : H and others. The second strategy for such reduction might be optimization of cell structures to absorb higher optical energy and surpress the loss component. Another approach is saving of the geminate recombination loss component by employing a multi-layered stacked cell structure. The concept behind this approach is represented schematically in Fig. 5.2.5 for the three-stacked case ($m=3$). To form a multi-layered cell, the Osaka University group utilized a-Si : H p/n junctions as an internal shunt electrode to connect each p-i-n unit cell in series with a multi-layered p-i-n/p-i-n/···/p-i-n cell structure.[11] A schematic illustration of this cell structure is shown in Fig. 5.2.1 (a). As can be seen from Fig. 5.2.1 (c), photogenerated carriers recombine at the p/n junctions, and hole-electron exchange occurs. Therefore, summing up of all the unit cell voltages in series would be the output voltage of the cell. The output current of the cell, on the other hand, is restricted by the minimum photocurrent

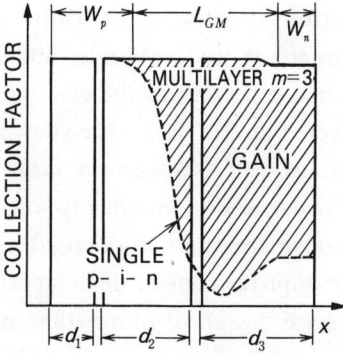

Fig. 5.2.5 Schematic illustration of the cell efficiency improvement in the multilayered device against single p-i-n basis a-Si : H solar cell. [after Y. Hamakawa et al.[12)]

in the unit cells in series connection.

In order to optimize cell efficiency, the thickness of each cell should be chosen so as to equalize the photocurrents of all the unit cells, as described previously. Multiple layers of p-i-n a-Si : H cells were deposited continuously by plasma deposition from SiH_4 with appropriate control of the dopant gas (PH_3 or B_2H_6). Photovoltaic performance measured under $80 mW/cm^2$ sunlight on the fabricated devices are shown in Fig. 5.2.6. The almost constant conversion efficiencies of about 4% obtained for all $m = 1 \sim 5$ show good evidence of the validity of the optimum design theory for this device. As expected from theory, V_{oc} increases in direct proportion to m, and I_{sc} varies inversely with m. An important example is the case of $m = 2$ for $V_{oc} = 1.35 V$, which could be substituted for a commercial mercury cell with a non-load output voltage of $1.35 V$.

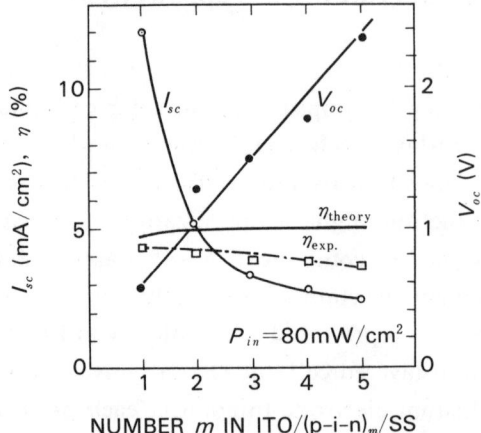

Fig. 5.2.6 Photovoltaic performance V_{oc}, I_{sc} and η as the function of the number of p-i-n repetitions m in the multilayered photovoltaic cell. The divinations from the theoretical expected efficiency might be mostly due to the high absorption in the p-layers. [after Y. Hamakawa et al.[12]]

It should be noted that one can optimize the cell structure by adjusting the ith-layer thickness for the spectra of various light sources. Multi-layered devices designed for a fluorescent lamp yield of $3 \sim 4$ micro Watts/cm^2 under 400 Lux white light with an output voltage of more than 1 V, and thus can serve as the electric power sources for pocke table calculators driven by a multi-layered device. From the view point of fabrication technology, this type of thin-film high-voltage cell has certain other advantages, such as ease of integration to line sensors. Such real sensors might provide a wide variety of applications for optically integrated circuit elements and high-speed color sensors having a desired spectral photoresponse. Further improvement of device performance can be expected with improved film quality, more precise design of i-layer thicknesses and a proper chosen grid pattern for the front electrode.

a-Si : H/poly Si : H stacked solar cell

One of the major limiting factors in a-Si solar cell is low photosensitivity in the longer wavelength region of the solar radiation spectrum, which basically comes from the energy band gap of 1.8eV in a-Si. On the other hand, there are many materials having narrower energy gaps in which the preparation technology for wide-area thin films is well established. Table 5.2.3 shows possible candidate materials, their energy gaps and fundamental edge types.[21]

Figure 5.2.7 shows schematic representations of the collection efficiency spectra in a-Si and c-Si solar cells. Comparing these data with the solar radiation spectrum (shown as the dashed line), one big deficiency in the s-Si solar cell is a lack of photosensitivity for longer wavelength photons from the sun. One way to overcome this problem is to stack it on a solar cell that has good spectral response in the infrared region of sunlight. It has been

Table 5.2.3 Possible bottom cell materials, their energy gaps and edge types stacked with amorphous silicon.

material	E_g(eV)	type
Ge	0.66	ind.
Si	1.11	ind.
GaAs	1.42	dir.
GaSb	0.72	dir.
InP	1.35	dir.
Zn_3P_2	1.50	dir.
CdTe	1.44	dir.
Cu_2S	1.20	dir. ?
Cu_2Se	1.20	dir. ?
$CuInS_2$	1.55	dir.

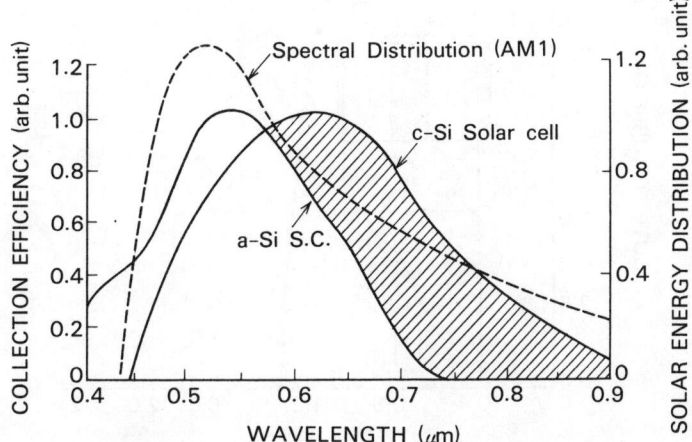

Fig. 5.2.7 Collection efficiency spectra of a-Si and c-Si solar cells, and solar energy radiation spectrum. [after Y. Hamakawa et al.[21] © 1984 IEEE]

shown that the interface of an a-Si junction acts as a good internal contact (e-h carrier exchange contact).[11] Utilizing this effect, one could make good stacked multi-layer solar cells with lower energy band-gap materials. Recently, Okuda et al.[22] developed a-Si solar cell stacked with a polycrystalline Si substrate. The concrete junction structure is ITO//n-i-n a-Si//n a-Si/poly c-Si//Al, as shown in Fig. 5.2.8. The polycrystalline silicons used were polyc-Si wafers (SILSO) prepared from casting of Wacker-Chemitronics (p type, 0.5 ~10 ohm·cm). The a-Si layers were deposited successively by the GD method. This cell consists of two unit cells, a top n-i-p a-Si solar cell and a bottom μc-Si/poly c-Si heterojunction solar cell, which are connected in optical and electrical series. The photovoltaic active layer of the top cell (i a-Si layer) has a band gap of about 1.75 eV, while that of the bottom cell (poly c-Si) has a band gap of about 1.1 eV. Therefore, the light in the long wavelength region, which cannot be absorbed by the top cell, can be utilized in the bottom cell.

In a solar cell using a polycrystalline semiconductor substrate, passivation of the grain boundary is an important key technology. In order to berify this point, spatial distribution of the LBIC in solar cell was measured. The result shows that hydrogenation passivation of the grain boundary occurs during a-Si deposition. In addition to this, a low temperature process for junction formation also preserves this passivation condition. Optimization of both this layer thickness and the a-Si top cell is now in progress. At the present stage of investigation, a conversion efficiency of 12.5% with $V_{oc} = 1.42$ V, $J_{sc} = 13.4$ mA/cm² and $FF = 65\%$ has been obtained under AM1 illumination, as shown in Fig. 5.2.8.

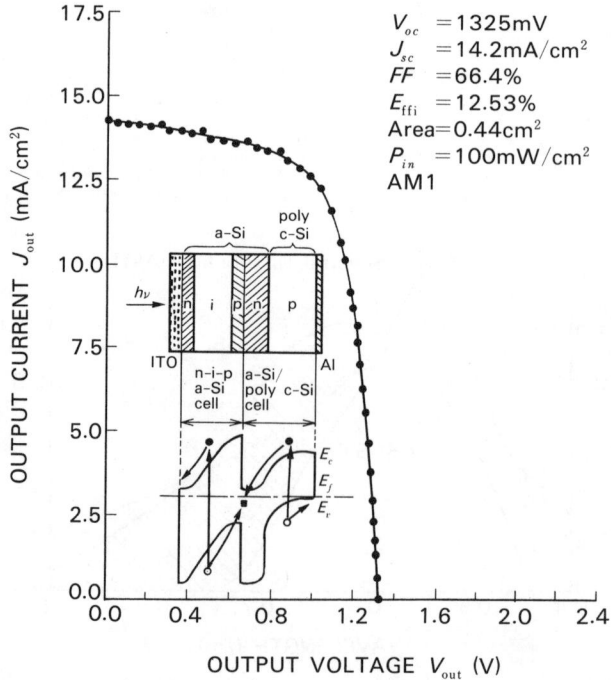

Fig. 5.2.8 J-V characteristic of the a-Si/poly Si stacked solar cell and its junction structure. [after Y. Hamakawa et al.[21] © 1984 IEEE]

The efficiency reported here does not seem to be the obtainable maximum one, since it is still limited by the photocurrent generated in the a-Si top cell. Furthermore, other crystalline materials (Ge, GaAs etc.) can be used instead of poly c-Si. Using crystalline semiconductor thin films prepared by such thin film formation techniques as CVD, MOCVD and MBE instead of a wafer should futher reduce the cost of solar cell production.

a-Si:H/a-SiGe:H stacked solar cells

A series of intensive experiments on a-SiGe:H has been done by the Mitsubishi group since 1980.[23] The electrical and optical properties of hydrogenated a-SiGe films have been clarified for a wide variety of germanium contents.[24] On the basis of these parametric data, they have developed a-Si:H/a-SiGe:H heterostructure tandem-type stacked cells. Figure. 5.2.9 shows the structures of such two-stacked and three-stacked solar cells.[24]

The thicknesses of the n-type and p-type a-Si:H layers are approximately 100Å at the front layer and the interface layers between the 1st and 2nd (2nd and 3rd) junctions, and 200Å at the bottom p-type layer. These layers can be deposited sequentially in a glow-discharge reaction chamber by controlling the gas composition and RF power for plasma excitation. After depositing the amorphous semiconductor layers, ITO film was deposited by an electron-beam evaporation apparatus to a thickness of about 680Å and Ti/Ag films were also deposited by evaporation. The two-stacked cell has an a-Si:H n-i-p cell stacked with an n-i-p cell having an i-layer of a-$Si_{0.6}Ge_{0.4}$:H film. A three-stacked cell is made of a two-stacked tandem-type cell with a-Si:H n-i-p cells and an n-i-p cell with an i-layer of a-$Si_{0.6}Ge_{0.4}$:H film. All cells have sensitive areas of 3 × 3mm². The i-layer thickness for the upper two junctions includes the short-circuit current of the bottom cell. This short-circuit current was about half the value for a single layer n-i-p junction cell. The sum of the i-layers in a-Si:H n-i-p cells for a three-stacked cell is approximately equal to the thickness of an i-layer in the upper a-Si:H n-i-p cell for a two-stacked cell.

Fig. 5.2.10 shows the J-V characteristics of single-junction, two-junction and three-junction tandem-type solar cells under illumination of simulated AM1 sunlight. As can be

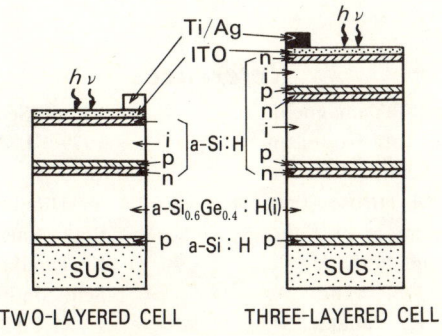

Fig. 5.2.9 Schmatic diagram of two kind of tandem type solar cells used of the a-Si:H and a-SiGe:H films. [after Y. Yukimoto.[23]]

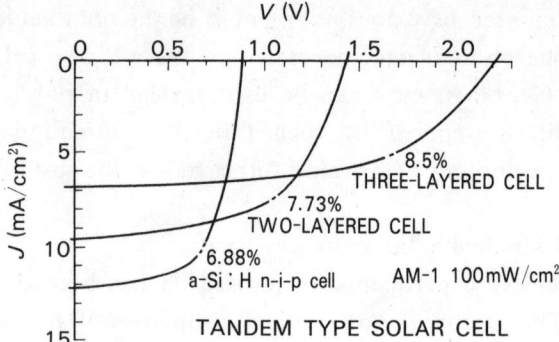

Fig. 5.2.10 Photovoltaic performance measured under simulated AM-1 sunlight of 100mW/cm² on the two type tandem type solar cells compared with a characteristic of a normal single junction inverted p-i-n type a-Si:H cell. [after G. Nakamura et al.[19]]

seen from this figure, open-circuit voltage increases with the number of stacked layers, m, while the short-circuit current decreases with m. An efficiency of 8.5% has been obtained with a three-stacked tandem solar cell.

The Sharp ECD group has also reported high efficiency solar cell with a two-junction tandem-type structure using a-Si:H and a-SiGe:H films with an efficiency of 8.3%.[25] The size of the cell was 0.03 cm². An open-circuit voltage of 1.37 V, a short-circuit current of 10.84 mA/cm², and a fill factor of 56% have been reported.

5.2.4 Conclusion

A brief review on the multi-band gap stacked solar cell has been given. Theoretical background on high-efficiency photon-energy collection in a heterostructure stacked cell was discussed, together with the efficiency limits for possible combinations of candidate materials. Recent progress in experimental approaches were also overviewed. Expetimental trial of a a-SiC/a-Si/a-SiGe double hetero-material solar cell now in progress was also described. Apart from this, basic reseach on other candidate meterial, such as a-SiSn:H and a-SiN:H, are also being made for triple-stacked solar cells.[26]

References

1) For example, H. Okamoto, T.Yamaguchi, and Y. Hamakawa: 8th Int. Conf. on Amorphous and Liquid Semiconductors, Cambridge,(1979) XE-7, M. Hirose: Proceedings of the 13th Conference on Solid State Devices (Tokyo), Aug. 1981, JJAP., *21*, suppl. 21-1 (1982) 275. And A. Matsuda, K. Nakagawa, K. Tanaka, M. Matsumura, S. Yamasaki, H. Okushi, and S. Izumi: Proceedings of the 8th International Conference on Amorphous and Liquid Semiconductors (Cambridge) Aug. 1979; J. Non-Cryst. Solid. *35-36* (1980) 183.

2) Y. Hamakawa and Y. Tawada: Int. J. Solar Energy, *1*, 1 (1982) 125.

3) K. Tanaka and A. Matsuda: Amorphous Semiconductor Technologies & Devies, JARECT, *6*,Chapt. 4. 3 ed. Y. Hamakawa, OHM-SHA * North Holland (1983) 161.

4) Y. Yukimoto: ibid, Chapt. 4. 1,(1983) 135.

5) K. Fujimoto, H. Kawai, H. Okamoto, and

Y. Hamakawa : Solar Cells, *11* (1984) 357.
6) Y. Tawada, H. Okamoto, and Y. Hamakawa : PVSEC-2 (1980) 52.
7) T. Catalano, A. Firostar, and B. Fanghaman : Proc. 16th IEEE Photovoltaic Specialists Conf., San Diego (1982).
8) M. Ohnishi, H. Nishiwaki, K. Enomoto, Y. Nakashita, S. Tsuda, S. Takahama, H. Tarui, M. Tanaka, H. Dojo, and Y. Kuwano : Proc. 10th ICALS, Tokyo, (Aug. 1983) 23-a-3.
9) A. Tanabe, A. Tajika, H. Minakami, T. Miyako, S. Sano, and O. Kuboi : No. 44 Annaul JSAA Meeting,(1983) 25a-L-8.
10) S.Yamazaki, K. Itoh, S. Watabe, A. Mase, K. Urata, K. Shibata, and H. Shirohara : Proc. 17th IEEE PV. Specialists Conf. Florida,(1984) 1-1 B-1.
11) Y. Hamakawa, H. Okamoto, and Y. Nitta : Appl. Phys. Lett., 35, July (1979) 15.
12) Y. Hamakawa, H. Okamoto, and Y. Nitta : Proc. 14th PV. Specialists Conf. San Diego,(1980) 1074.
13) H. Okamoto, Y. Nitta, and Y. Hamakawa : Jpn. J. Appl. Phys.,*19* (1980) 545.
14) Y. Hamakawa : Oyobutsuri,*53*, 8 (1984) in press.
15) Y. Hamakawa : Device Physics and Optimum Design of the Amorphous Silicon Photovoltaic Devices, Chapt. 4. 1 of Amorphous Semiconductor Technology and Devices, OHM-SHA ∗ North-Holland.
16) H. Okamoto, T. Yamaguchi, S. Nonomura, and Y. Hamakawa : Drift type Photovoltaic Effect in a-Si p-i-n Junction, J. de Physique, *42*, Suppl. 10 (1981). C4-507
17) H. Okamoto, H. Kida, and Y. Hamakawa, Solar Cells, *8* (1983).
18) S. Tsuda, N. Nakamura, Y. Nakashima, H. Tarui, H. Nishiwaki, M. Ohnishi, and Y. Kuwano : Jpn. J. Appl. Phys. *21* suppl *21-2* (1982) 251.
19) G. Nakamura, K. Sato, H. Kondo, Y. Yukimoto, and K. Shirahata : Photovolataic Solar Energy Conf. (PVSEC), (Stresa, 1982) 616.
20) J. C. C. Fan and B. J. Palm : Solar Cells, *10* (1983) 81.
21) Y. Hamakawa, K. Okuda, H. Takakura, and H. Okamoto : Proc. 17th IEEE PV. Specialists Conf. Orando, Florida, (1984) Late News.
22) K. Okuda, H. Okamoto, and Y. Hamakawa : JJAP. *22*,9 (Sep. 1983) L605.
23) For example, general review is : Y. Yukimoto ; Hydrogenated a-SiGe alloy and its Optoelectronic Properties, JARECT *6*, Chapt. 4 ed. Y. Hamakawa, (1983) 135.
24) G. Nakamura, K. Sato, and Y. Yukimoto : Proc. 3rd Photovoltaic Science and Engineering (PVSEC) (Kyoto, 1982) Jpn. J. Appl Phys. *21* suppl., 21-2 (1982) 297.
25) T. Morimoto : Private communication, also reported by Y. Yukimoto : in JARECT *6* ed. Y. Hamakawa, (1983) 237.
26) N. Tsuda, N. Nakamura, Y. Nakajima, G. Takehara, H. Tarui, H. Nishiwaki, M. Ohnishi, and Y. Kuwano : The 29th Annu. Meeting Appl. Phys. SOC. Japan, (1982) 4p-Z-1.

5.3 Mass Production Technology in a Roll-to-Roll Process

Hiroshi MORIMOTO* and Masatsugu IZU**

Abstract

A roll-to-roll mass production process for fabricating multilayered amorphous silicon alloy solar cells has been developed. A glow discharge deposition machine continuously fabricates a thin film photovoltaic structure. The structure includes six amorphous silicon alloy layers which are successively deposited onto a roll of 400 mm wide, 300 m long substrate. A transparent conductive oxide layer is also continuously deposited atop the amorphous silicon alloy layers in a roll-to-roll process. Other downstream processes for the fabrication of solar cell modules also utilize a roll-to-roll process. A commercial plant utilizing this automated process for mass producing tandem amorphous silicon alloy solar cells is now in operation. A model of a future roll-to-roll process is proposed.

5.3.1 Introduction

Since 1980, when amorphous solar cell modules were first commercially utilized to provide power for pocket calculators, the annual production of solar cells in Japan has increased exponentially. At present, amorphous silicon alloy solar cells have replaced crystalline silicon solar cells in pocket calculators and may soon replace conventional button batteries.

The research and development of technologies for the low cost amorphous solar cells has been sponsored by the Japanese government funded Sun Shine Project as well as by funding from private companies. Presently, several promising mass production technologies are being enthusiastically developed.

Sharp-ECD is using low-cost, flexible, stainless steel substrates to produce solar cells for powering pocket calculators.[1,2] Flexible polyimide and polyester film substrates have been investigated by Teijin,[3] and lowcost, light-weight aluminum substrates have been tested by Konagai et al,[4] Conventional glass subtrates are being widely developed by many researchers; Kyocera is developing a low-cost ceramic substrate.[5] The mass production of

---

* Solar Systems Group, Sharp Corporation, 282-1 Hajikami Shinjo-cho, Kitakatsuragi-gun, Nara 639-21.
** Energy Conversion Devices, Inc., 1675 West Maple Road Troy Michigan 48084, U. S. A.

one type of glow discharge deposition process has been improved with the development of an interdigital-vertical-electrode type machine which is adapted to simultaneously process four 30 × 40cm² substrates.[6] High conversion efficiencies of 6.5% at deposition rates of 11 Å/s, , have been achieved with an a-Si : H solar cell fabricated from disilane;[7] and a dry patterning, laser scribing process for TCO, a-Si : H and back contact lithography has been reported by RCA[8] and Sanyo.[9]

The fabrication process of amorphous silicon alloy solar cells, compared to that of crystalline silicon solar cells, lends itself easily to mass production, due to the following features :
(1) Low-cost substrates may be utilized (flexible stainless steel, polymer film, glass ets.) ;
(2) Large area solar cells may be produced (30 × 34cm² cells by Sharp-ECD[1,2] and 30 × 40cm² by Fuji[6]) ;
(3) Production processes are simple and easily automated.

In order to utilize the above-mentioned technologies, two automated glow discharge deposition processes were proposed. One process is a consecutive, seperated reaction chamber process utilized by Sanyo.[10] The other process utilizes a roll-to-roll multilayer deposition machine which is being operated in the Sharp-ECD Solar Manufacturing plant.[1,2] The proprietary continuous roll-to-roll multilayer glow discharge deposition machine[2] has been developed and constructed by ECD. The former is a batch process utilizing glass substrate, which is in some respects analogous to an IC wafer process. The latter is a continuous, roll-to-roll deposition process utilizing thin flexible stainless steel substrate. Obviously, various other types of rolled substrate material can also be used.

The success of the Sharp-ECD plant represents the first commercial use of roll-to-roll mass production technology for manufacturing solar cells and may prove to be of historical significance as the beginning of a new photovoltaic industry.

This paper presents an example of an automated mass production process utilizing roll-to-roll technology. Also, a model of a future roll-to-roll process is proposed.

5.3.2 Roll-to-Roll Mass Production Process for an Amorphous Silicon Solar Cell

Figure 5.3.1 depicts diagrammatically a roll-to-roll process for the continuous fabrication of solar cells adapted for use in powering consumer products. A stainless steel substrate having a width of 400mm and a thickness of 0.2mm is first subjected to a substrate cleaning process. Then the substrate travels through isolated reaction chambers wherein a tandem solar cell structure of amorphous silicon alloy layers is deposited. Next, a transparent electrode (ITO) layer is deposited by a reactive evaporation process. Resist ink is then printed in a desired pattern in a mask printing process. The ITO layer is chemically etched and the resist ink pattern is removed in an electrode printing process. In a grid electrode printing process, silver paste is applied for collecting current, and a transparent polymeric film is placed atop the ITO pattern in order to protect the solar cell from moisture and mechanical damage. Lastly, every cell is tested under illumination and removed from the roll in an inspection and punching process. An amorphous silicon alloy solar cell fabricated according to the steps outlined above is shown in Fig. 5.3.2.

MASS PRODUCTION TECHNOLOGY IN A ROLL-TO-ROLL PROCESS

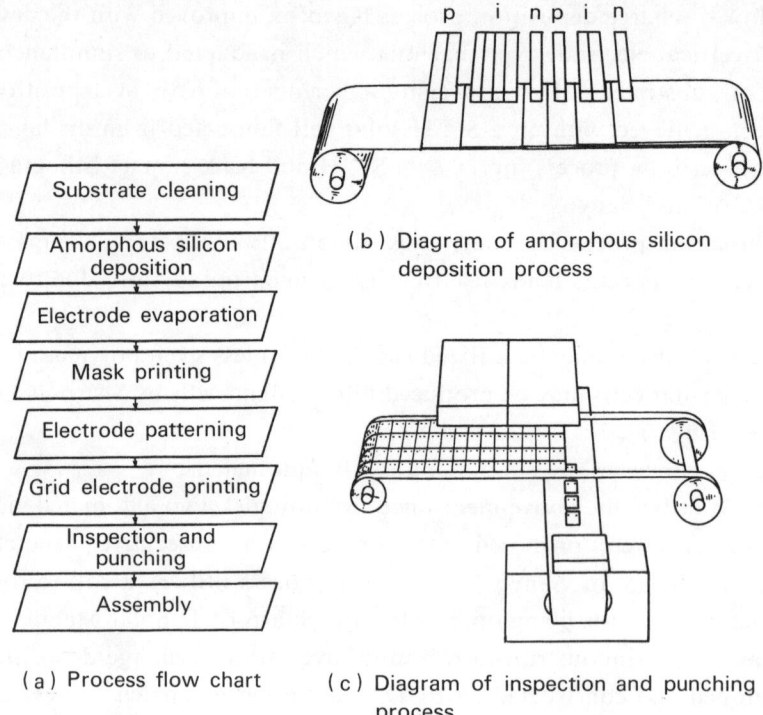

(a) Process flow chart

(b) Diagram of amorphous silicon deposition process

(c) Diagram of inspection and punching process

Fig. 5.3.1 Roll-to-roll mass production line.

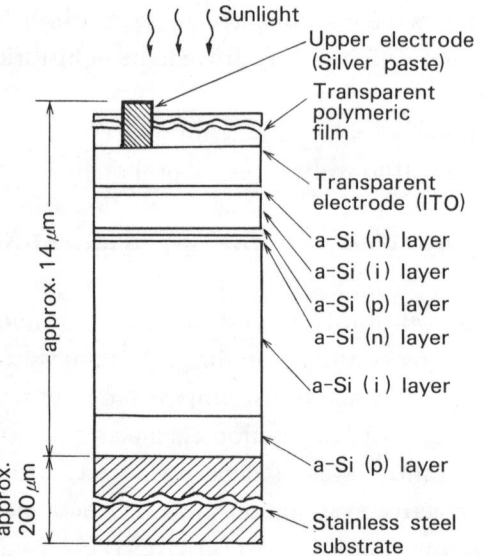

Fig. 5.3.2 Two-cell tandem structure on stainless steel substrate.

Fig. 5.3.3 Amorphous silicon alloy solar cells on stainless steel substrate after electrode and transparent polymeric film are applied.

Figure 5.3.3 shows a roll of solar cells with deposited electrodes and transparent polymeric film being transported to the next work station for further processing. At this

Fig. 5.3.4 Rolled amorphous silicon alloy solar cells and their applicability to pocket calculators.

work station, the solar cells are severed from the roll for use in consumer products such as pocket calculators. Note that, as shown in Fig. 5.3.4, the performance of the solar cell is not impaired by flexing the substrate. This represents an obviously novel feature of amorphous silicon solar cells vis-a-vis crystalline silicon solar cells or amorphous silicon cells deposited on non-flexible substrates.

5.3.3 Substrate Preparation

In this roll-to-roll process flexible substrates such as stainless steel, aluminum and polymer films are available. After considering (1) the mechanical and chemical strength of these substrates at temperatures under 300°C, (2) the cost and (3) the availability of the substrates, stainless steel was selected.

However, regarding the use of a stainless steel substrate, two areas required a closer examination. One problem was the presence of surface defects such as pits. The other

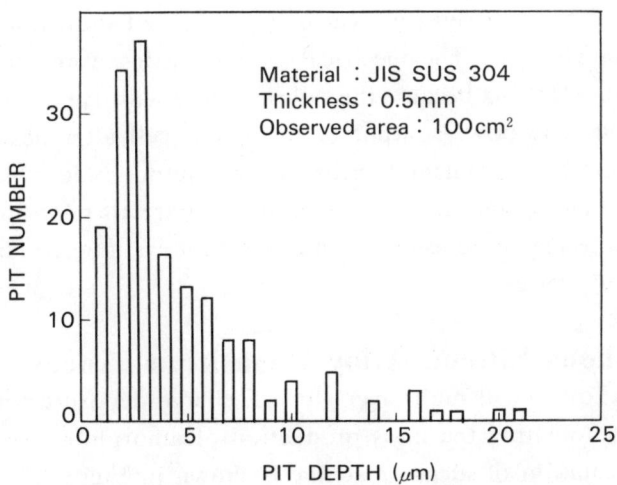

Fig. 5.3.5 Plot of the number of surface pits versus the depth thereof on stainless steel substrate. [after K. Tamiya[11)]]

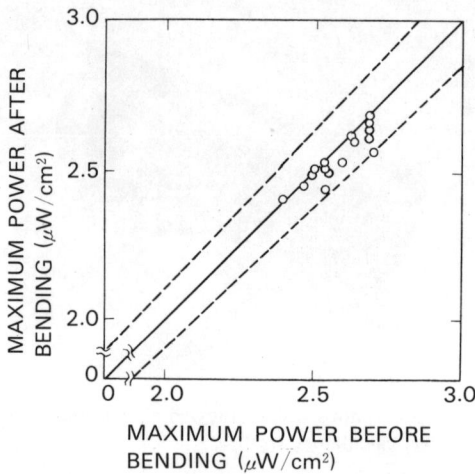

Fig. 5.3.6 Maximum power output of amorphous silicon alloy solar cells under fluorescent illumination before and after a thousand flexures to a radius of 50 mm.

problem was the stability of an amorphous silicon alloy solar cell deposited on the substrate after repeated flexure by the turning rollers in the automated deposition apparatus. The density of pits on the surface of a roll of stainless steel is show in Fig. 5.3.5. In this graph, the abscissa denotes the pit depth and the ordinate denotes the number of pits with a given depth in a 100 cm² sample. Since the depth of the pits can be as great as 20 μm, surface polishing to a depth greater 20 μm is necessary in order to remove all of the pits. Conventional buff polishing methods cannot be used due to its low polishing speed. Recently, an electrolitic buff polishing method has been proposed by Hitachi Zosen.[11]

The flexure problem was investigated through the use of a bending test machine. Figure 5.3.6 shows the maximum power solar cell output under fluorescent lamp illumination. The abscissa and ordinate denote the results before and after, respectively, of a thousand flexures of the solar cell (fabricated as per the cell shown in Fig. 5.3.2) about a roller having a 50 mm radius. The electrical characteristics were measured under low intensity illumination of 150 lux because the cells are more sensitive to shunt currents at low intensity illumination. The power output of each solar cell after flexure is of the same magnitude, within ± 5% experimental error, as the output before flexure. The results thereby demonstrate that an amorphous silicon alloy solar cell deposited atop a stainless steel substrate is unaffected by repeated flexure; therefore it is well adapted for use in a continuous roll-to-roll process.

5.3.4 Amorphous Silicon Alloy Deposition Process

A continuous roll-to-roll multilayer glow discharge deposition machine represents the key element in automating the mass production of amorphous silicon solar cells. A diagrammatic representation of such a machine is shown in Fig. 5.3.7. A roll of stainless steel substrate is continuously advanced from the pay-off chamber, through the gas gate and into the reaction chamber. The substrate is driven by a winding motor and a pay-off motor

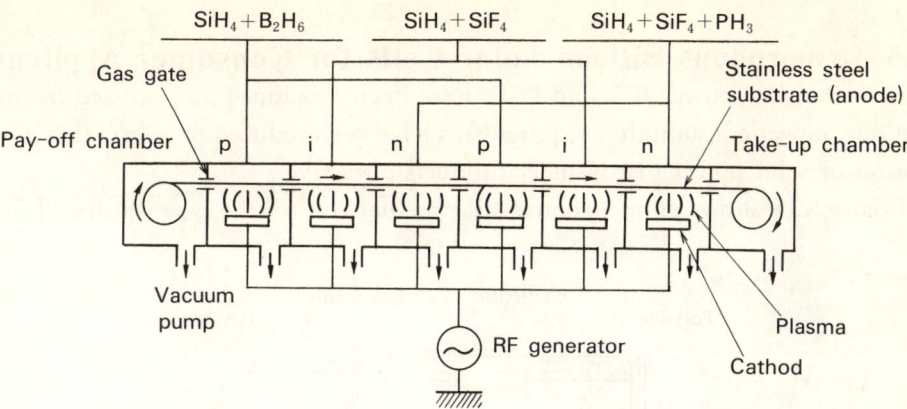

Fig. 5.3.7 Roll-to-roll continuous multilayer deposition machine.

located in the take-up and pay-off chambers, respectively. An RF plasma is generated by a 13.56 MHz RF generator for developing plasma to deposit each of the p, i and n semiconductor layers in discrete, isolated reaction chambers. The atmosphere of the reaction gases in adjacent reaction chambers are isolated by a gas gate. Measurements of the boron concentration profile obtained through the use of an ion microprobe analyzer in an n-i-p/s.s. structure are shown in Fig. 5.3.8. These measurements indicate that complete isolation between the process gas atmospheres in adjacent reaction chambers can be achieved using this gas gate. In this manner, many isolated plasmas are sustained during the continuous deposition of solar cells onto the elongated roll of substrate material resulting in the production of solar cells characterized by stability and homogeneity over the entire 300 m long substrate.

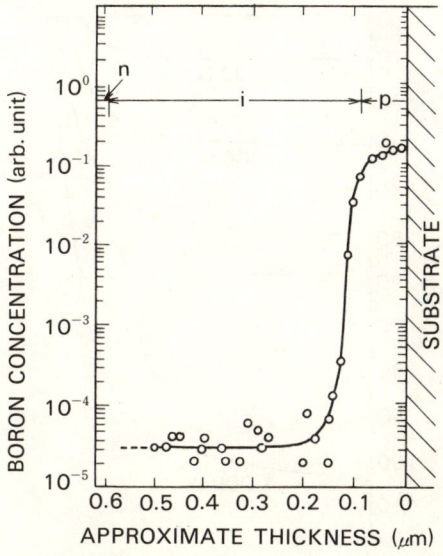

Fig. 5.3.8 Boron concentration profile in an n-i-p/s.s. structure obtained with an ion microprobe analyzer.

5.3.5 Amorphous Silicon Solar Cells for Consumer Applications

Now that low power IC's and LSI's have been developed and utilized in consumer applications, power consumption requirements have been reduced to a level that makes the application of solar power to consumer products attractive.

Figure 5.3.9 shows an amorphous silicon solar cell (of the type illustrated in Fig. 5.

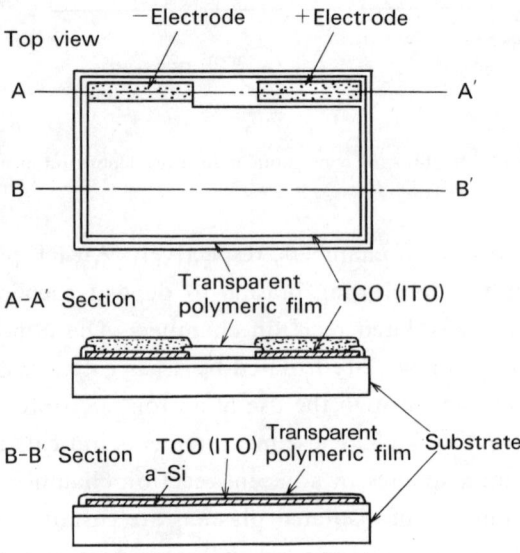

Fig. 5.3.9 An amorphous silicon alloy solar cell adapted for consumer power applications.

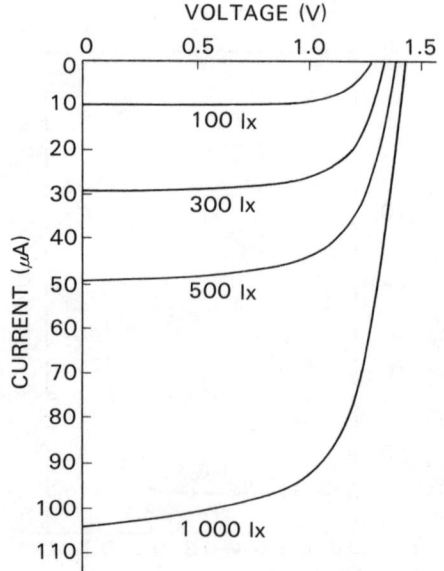

Fig. 5.3.10 Current-voltage characteristics of an amorphous silicon alloy solar cell under fluorescent illumination. [after H. Morimoto and Y. Yamauchi[12]]

3.2) adapted for consumer use. The electrodes of the cell are composed of silver paste, a negative electrode directly contacts the ITO, and a positive electrode electrically contacts the conductive substrate. The transparent polymeric film encapsulates the solar cell in order to prevent the cell from being harmed by the assembly elements.

Figure 5.3.10 is a curve illustrating the current-voltage characterisitics of a tandem amorphous silicon solar cell under a fluorescent lamp, which is the usual room ligth in Japan, as a function of intensity of illumination.[12] The short circuit current I_{sc} increases linearly in proportion to the intensity of illumination. The open circuit voltage V_{oc} is higher that 1V even under the lowest intensity of illumination and is almost twice as great as that of a single p-i-n cell. Accordingly, a pair of tandem solar cells are sufficient to drive a pocket calculator.

5.3.6 Amorphous Silicon Solar Cells for Power Applications

Large area solar cells have been successfully fabricated using the roll-to-roll process shown in Fig. 5.3.1. Figures 5.3.11 and 5.3.12 show large area solar cells produced in these roll-to-roll processes.

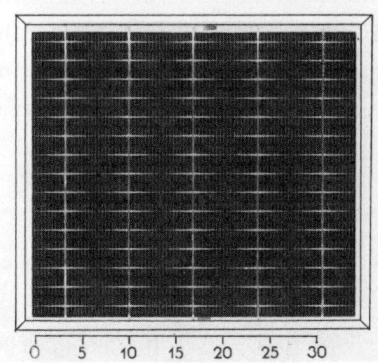

Fig. 5.3.11 Large area solar module; Rigid module.

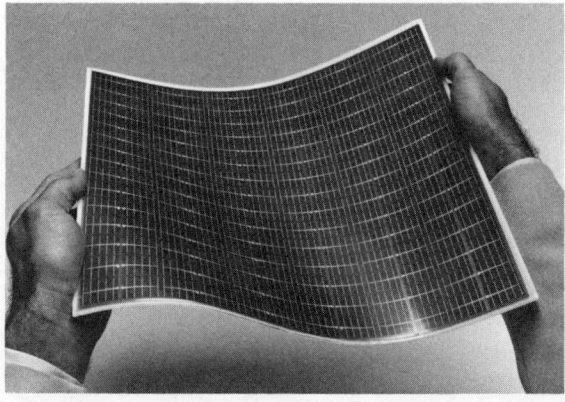

Fig. 5.3.12 Large area solar module; Flexible module.

A continuous roll-to-roll process is especially suited for the mass production of large area solar cells for power applications due to the following advantages:

(1) An automated mass production line can be adapted for mass production with a minimal cost outlay;
(2) Once the technology is established, the scale-up to high volume production is straightforward;
(3) Rolls of substrate material can be used, thereby reducing the cost of the substrate and handling costs, and minimizing the amount of edge loss. A range of materials for use as substrates is also available, and ;
(4) Solar cells may be stored and transported in a compact roll form and may be to size from the roll. This provides flexibility in the size and shape of the product.

5.3.7 Model of a Future Roll-to-Roll Mass Production Process

In order to further minimize the cost of producing amorphous silicon alloy solar cells, it is necessary to fully-automate the mass production process, thereby reducing associated labor costs. Figure 5.3.13 shows two projected models of future roll-to-roll processes; the first is a system for producing a conventional single junction type cell and the second is for producing an integrated type cell.[13] In the single junction type cell process, a TCO/a-Si/metal substrate photovoltaic structure is fabricated and cut into unit cells, which are then electrically connected to form a series or parallel array and laminated so as to form a module. In the integrated type cell process, interconnected arrays of solar cells are fabricated directly upon an insulated substrate and are laminated to form an encapsulated array.

5.3.8 Conclusions

Successful operation of the roll-to-roll process at the Sharp-ECD manufacturing

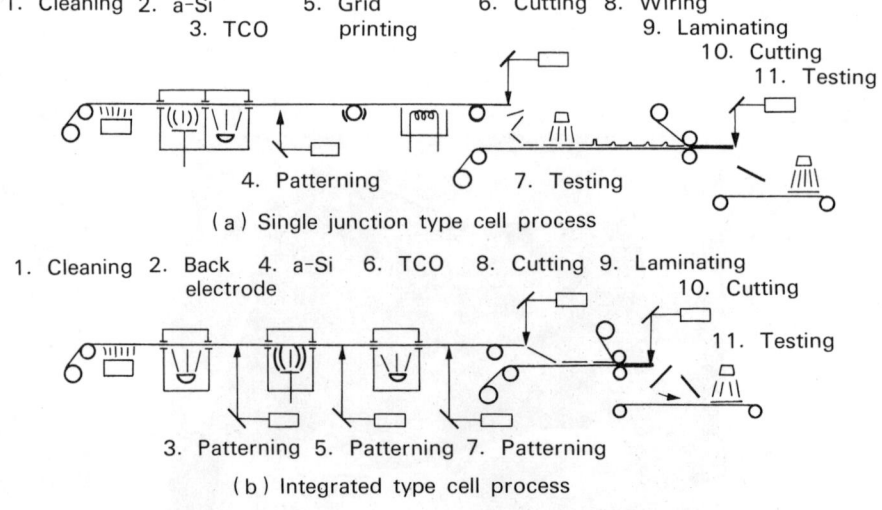

Fig. 5.3.13 Future models of amorphous silicon alloy solar cell mass production lines.

plant demonstrates the effectiveness of the roll-to-roll process. Its success confirms that all the major technical obstacles towards the commercialization of mass-production technology utilizing a roll-to-roll process have been eliminated.

Acknowledgements

The authors are grateful to Dr. Tadashi SASAKI and Mr. Minoru MIYUKI of Sharp Corporation, and Mr. Stanford R. OVSHINSKY of Energy Conversion Devices, Inc. for their thoughtful advice and encouragement during the course of this work. Thanks are also due to Dr. Yoshiharu NAKAJIMA for his help on the IMA measurements.

References

1) H. Morimoto, A. Yokota, K. Nomoto, M. Kaneiwa, Y. Yamauchi, and R. Yokota: Abst. 43rd Autumn Meeting of the Jpn. Soc. of Appl. Phys. (1982) 30p-W-9 [in Japanese].
2) M. Izu and S. R. Ovshinsky: Proc. Soc. Photo-Optical Instrumentation Engineering, Washington, Aprill (1983); and U.S. Patent Serial Nos. 4, 410, 558 and 4, 400, 409 covering a roll-to-roll process.
3) K. Nakatani, M. Yano, K. Suzuki, and H. Okaniwa: Proc. 10th Int. Conf. on Amorphous and Liquid Semiconductors, Tokyo (1983) 827.
4) M. Konagai, M. Kumata, and K. Takahashi: Abst. 43rd Autumn Meeting of the Jpn. Soc. of Appl. (1982) 30a-W-3 [in Japanese].
5) Y. Minamino, Y. Nitta, H. Kubo, T. Iwasaki, T. Minato, K. Tomita, and K. Ishibitsu: Abst. 30th Spring Meeting of the Jpn. Soc. of Appl. Phys. and the Related Soc. (1983) 6p-O-18 [in Japanese].
6) H. Sakai, M. Kamiyama, M. Ueno, Y. Uchida, and H. Haruki: Proc. 5th EC Photovol. Solar Energy Conf., Kavouri (Athens) (1983) 808.
7) Y. Ohashi, J. Kenne, M. Konagai, and K. Takahashi: Appl. Phys. Lett., 42 (1983) 1028.
8) J. J. Hanak: U. S. Patent No. 4, 292, 092 (1981).
9) S. Nakano, M. Ohnishi, H. Kawada, H. Nishiwaki, S. Tsuda, and Y. Kuwano: Proc. 5thEC Photovol. Solar Energy Conf., Kavouri (Athens)(1983).
10) Y. Kuwano, M. Ohnishi, S. Tsuda, Y. Nakashima, and N. Nakamura: Jpn. J. Appl. Phys. 21 (1983)413.
11) K. Tamiya: Amorphous Silicon Handbook edited by K. Takahashi and M. Konagai, Science Forum, Ltd., (1983) 283 [in Japanese].
12) H. Morimoto and Y. Yamauchi: Sharp Technical Journal 27 (1983) 7 [in Japanese].
13) Y. Kuwano and M. Ohnishi: JARECT Vol. 6, Amorphous Semiconductor Technologies and Devices, edited by Y. Hamakawa OHMSHA, LTD. and North-Holland Publishing Co. (1983) 204.

5.4 a-Si Solar Cell Industrialization and Systems

Yukinori KUWANO* and Shoichi NAKANO*

Abstract

The world's first industrialization of a-Si solar cells was accomplished by Sanyo and Fuji Electric in 1980. Two separate new technologies for the production of the a-Si solar cell were developed. The first is two new cell structures which utilize the characteristic features of the plasma reaction. They are, the integrated type and the tandem type structure. The second is a new fabrication method, the consecutive, separated reaction chamber method.

To apply the a-Si solar cell as an electric power source, model systems using a-Si solar cells are constructed and R & D is in progress. a-Si power generating systems, such as a 2 kW system, a 400 W system and solar cell roofing tile are also described.

5.4.1 Introduction

The development of solar cells which are capable of obtaining electrical energy directly from sunlight has become extremely active today, and one of the main goals is the development of a low-cost solar cell. The a-Si solar cell differs entirely from the conventional single crystal solar cell. Compared with the conventional single crystal solar cell, (1) the energy required for fabrication is small, (2) the fabrication process is simple, and (3) the amount of material necessary is much less. These and several other attractive features have brought attention to the a-Si solar cell as a promising device for the low-cost solar cell. In 1980, the Sunshine Project also began to focus its efforts on amorphous silicon, and the pace of R & D has been accelerated. Then, the "Amorphous Solar Cell Technology Development Program" of the New Energy Development Organization (NEDO) was begun in 1983.

An integrated type a-Si solar cell, which utilizes the natural features of the a-Si solar cell, was developed by Sanyo in 1979.[1] Various consumer products powered by this solar cell were then developed in 1980, marking the start of practical solar cell applications.[2] The a-Si solar cell continued to make advancements in the field of consumer products until its use eventually surpassed that of the conventional single crystal solar cell, and its annual production for use in consumer products reached the scale of approximately 2 MW/year by 1983, as shown in Fig. 5.4.1.

To apply the a-Si solar cell as an electric power source, which is the main goal, model systems using a-Si solar cells were constructed and R & D is under way.

---

* Reserch Center, SANYO Electric Co., Ltd., 1-18-13 Hashiridani, Hirakata City, Osaka 573.

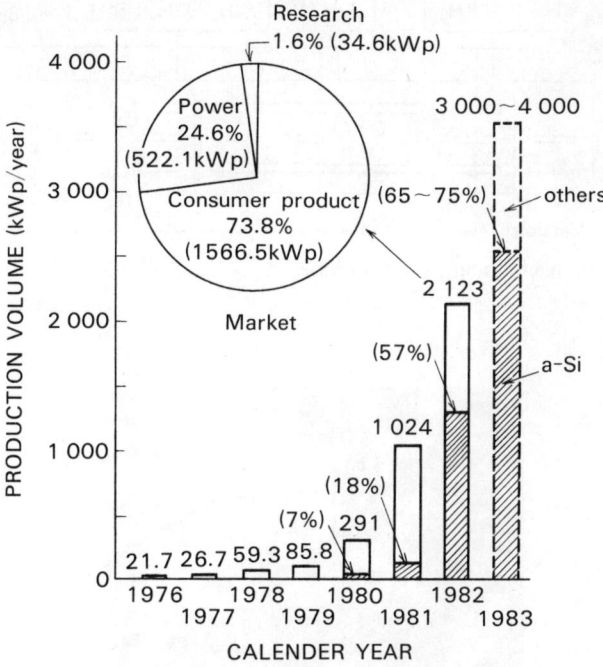

Fig. 5.4.1 Progress in production volume of solar cells in Japan (This survey was carried out by the Optoelectronic Industry and Technology Development Association. Overall production volume of a-Si solar cells, and that in 1983, are estimated by the author.)

This report describes the processes for producing a-Si solar cells on an industrial scale and power generating systems.

5.4.2 Industrialization of a-Si Solar Cells

One of the key technologies for the mass production of the a-Si solar cell is a fabrication method, the consecutive, separated reaction chamber method. Another comprises two new cell structures, the integrated type and the tandem type structure.

Fabrication method for mass production

Industrial fabrication methods for a-Si solar cells are the single reaction-chamber method and the consecutive, separated reaction chamber method.[3] The consecutive, separated reaction chamber method, which is highly efficient for mass production, was developed by Sanyo. This fabrication method is shown in Fig. 5.4.2 (a).

In this method, the p, i, and n type layers are deposited in different reaction chambers, as shown in Fig. 5.4.2 (a). The features of this method are as follows.

(1) It is possible to avoid the influence of the residual gaseous dopants which remain on the surface of the electrodes and reaction chamber walls.
(2) It is possible to sufficiently control the amount of dopant in each layer.
(3) a-Si solar cells can be produced consecutively.
(4) It is possible to prevent the inside of the reaction chamber from being exposed to the

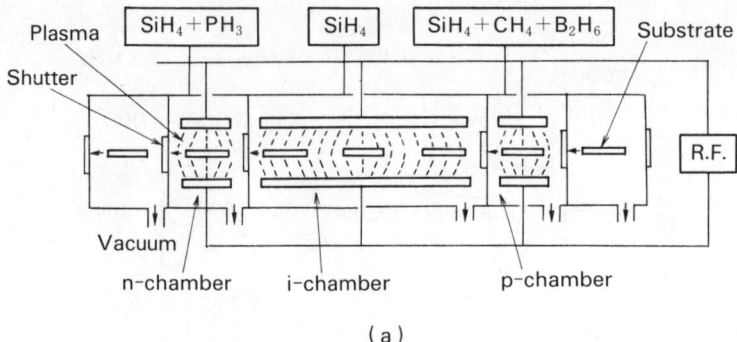

Fig. 5.4.2 (a) : Consecutive, separated reaction chamber apparatus. [Y. kuwano et al.[3)]]
(b) : Roll-to-roll plasma deposition machine.

open air.

(5) It is possible to reuse the raw material gases.

At present, the consecutive, separated reaction chamber method is the main fabrication method for mass production. And a similar roll-to-roll method has been developed by Sharp-ECD.[4)] This fabrication apparatus is shown in Fig. 5.4.2 (b).

For the industrial production of a-Si solar cells, a high deposition rate for the a-Si film is also important. Toward that end, fabrication methods using Si_2H_6 have been investigated.[5)]

New structures of a-Si solar cells

The development of new cell structures for a-Si solar cells played an important role for their mass production. The voltage obtained under sunlight by one unit (one wafer) of the conventional crystal Si solar cell is less than 1V. From the viewpoint of practical use, it is necessary to obtain higher voltages by series connection of lead wires.

For the practical industrial production of the a-Si solar cell, the integrated type structure, in which a-Si solar cell units are arranged in a cascade fashion on an insulated substrate, and the tandem structure, in which a-Si solar cell units are arranged in a vertical fashion, have been developed.

Integrated type a-Si solar cell   Employing the a-Si film characteristics of (1) formation in a gas reaction, (2) being a thin film, and (3) possessing high resistivity, this configuration comprises a cascade connection of numerous cells on one substrate, so that a cell capable of generating a high voltage is achieved. This solar cell was developed by Sanyo and is called the integrated type a-Si solar cell.[1] The construction of the integrated type a-Si solar cell is shown in Fig. 5.4.3. The integrated type a-Si solar cell is manufactured by the following method.

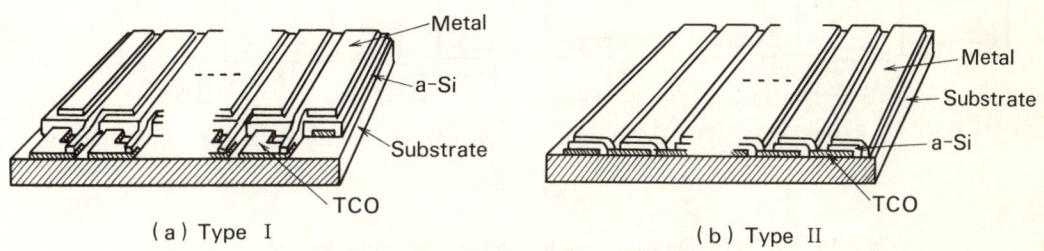

Fig. 5.4.3   Two kinds of integrated type a-Si solar cell modules (a) Type I, (b) Type II. [after Y. Kuwano et al.[1] © 1980   IEEE]

First, separate transparent electrodes are formed on an insulated substrate. Then, a-Si films are formed onto them through a mask pattern. Finally, metal electrodes are deposited onto the a-Si films through an appropriate metal mask. The metal electrode of the first cell is in contact with the transparent electrode of the second cell. In this way, each of the a-Si cell modules can be connected in series. It has become possible to achieve a high output voltage from a single substrate by this integration. This integration is also possible by using conventional photolithography techniques. The merits of this type are summarized as follows.

(1)   High voltage values can be achieved by connecting cells in series.
(2)   High voltage and low current operation makes the power loss due to the resistance effect small.
(3)   They are well suited to the assembly of large area devices.
(4)   The reliability of both the interconnections and the passivation is very high.

A new fabrication process for the integrated type a-Si solar cell was developed.[6] In this process, a transparent electrode film, an a-Si film and a metal electrode film, are selectively scribed by a laser beam, as shown in Fig. 5.4.4 (a). Figure 5.4.4 (b) shows a continuous manufacturing apparatus for the integrated type a-Si solar cell by adding the laser patterning method. From the viewpoint of mass production of the a-Si solar cell, it is helpful that the laser patterning method makes the conventional wet process unnecessary.

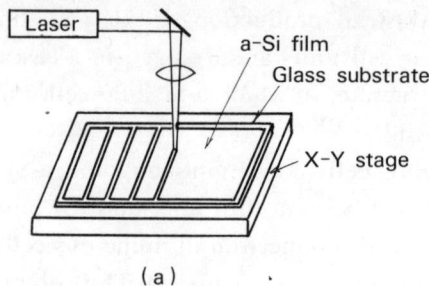

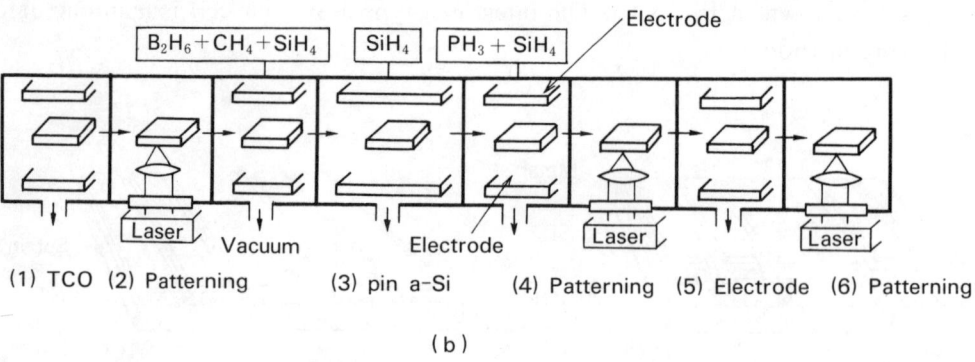

Fig. 5.4.4 (a) : a-Si film patterning by the laser patterning method.
(b) : A continuous manufacturing apparatus for the integrated type a-Si solar cell by adding the laser patterning method.

Tandem type structure a-Si solar cell   The tandem type structure a-Si solar cell is also mass produced. This new structure, which utilizes a large recombination current at the p-n junction, was developed by Hamakawa et al.[7] Cells of this structure are made by depositing continuously-doped layers and i-layers through a plasma reaction. The fabrication technique is similar to that employed for ordinary a-Si solar cells. For the formation of the internal p-n junctions they applied fixed doped layers of 500 Å thick p-layers

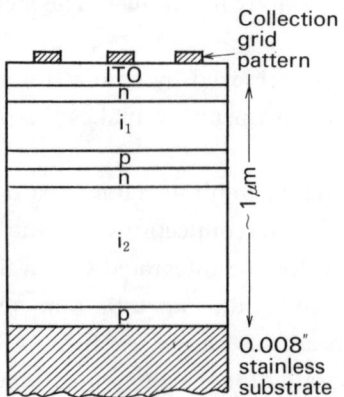

Fig. 5.4.5 The tandem type a-Si solar cell mass produced by Sharp & ECD. [after S. R. Ovshinsky[8]]

deposited from a silane mixture containing 0.1% diborane and a 100 Å thick n-layer from a mixed gas of $PH_3/SiH_4 = 0.2\%$. The tandem structure industrialized by Sharp & ECD is shown in Fig. 5.4.5.[8] The thickness of the first layer is different from that of the second layer in order to obtain the same current. The voltage obtained in this structure is about 1.6 V.

Application of a-Si solar cells

In the field of consumer application, the mass production of a-Si solar cells using both the integrated type structure and the tandem type structure is making rapid progress.

As shown in Table 5.4.1, pocket calculators powered by integrated type a-Si solar cells, which were marketed by Sanyo, represent the world's first practical use of a-Si solar cells.

Table 5.4.1 Development of industrial production for a-Si solar cell and systems.

Year	Development Activity
1975	Discovery of the ability to control the electrical properties of a-Si (Spear et al.)
1976	Development of a-Si solar cell (Carlson et al.)
1978	Development of tandem type a-Si solar cell (Hamakawa et al.)
1979	Development of integrated type a-Si solar cell (Kuwano et al.)
	Development of a-Si solar cell manufacturing process of consecutive, separated reaction chamber method (Sanyo)
1980	World's first electric calculator powered by an integrated type a-Si solar cell put on the market (Sanyo, Fuji Electric)
1981	Primary demonstration plant completed for application of the a-Si solar cell to electric power generating systems (2 kW system---Sanyo 400 W system---Fuji Eleetric)
	Watch powered by a-Si solar cell put on the market (Sanyo)
1982	a-Si photosensors put on the market (Sanyo)
	Color television set with automatic brightness controller using a-Si photosensor put on the market (Sanyo)
	2.5 kW electric power generating system completed (Sanyo, Fuji Electric)
1983	Ni-Cd battery chargers using large area integrated a-Si solar cell put on the market (Sanyo)
	Clock powered by a-Si solar cells put on the market (Sanyo)
	Electric calulator powered by a tandem type a-Si solar cell put on the market (Sharp & ECD)

Figure 5.4.6 (a) shows several consumer products, pocket calculators, watches, clocks, sensors and battery chargers, which were marketed by Sanyo. Other instruments, such as headphone type radios, tape recorders, and TV powered by a-Si solar cells, such as those shown in Fig. 5.4.6 (a), are being experimentally produced. Pocket calculators marketed by Fuji Electric and Sharp & ECD are shown in Fig. 5.4.6 (b) and Fig. 5.4.6 (c), respectively. In the field of consumer products, a-Si solar cells have surpassed the conventional single crystal solar cells.

(a)

(b)

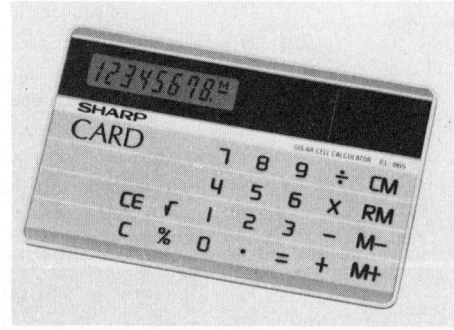

(c)

Fig. 5.4.6 (a) : Application of a-Si solar cells (Sanyo).
(b) : Pocket calculator (Fuji Electric).
(c) : Pocket calculator (Sharp & ECD).

5.4.3 Electric Power Generating System

A new a-Si solar panel for electric power was developed and applied to electric power generating systems. Figure 5.4.7 shows a picture and the structure of an a-Si solar panel developed by Sanyo. Figure 5.4.8 shows a picture and the structure of an a-Si solar module developed by Fuji Electric. This chapter describes 2 kW and 400 W photovoltaic power generating systems using a-Si solar cells.

2 kW system (Sanyo Electric Co., Ltd.)[9]

The world's first electric power generating system using a-Si solar cells was developed by Sanyo. A picture of an integrated type a-Si solar panel is shown in Fig. 5.4.7 (a). The

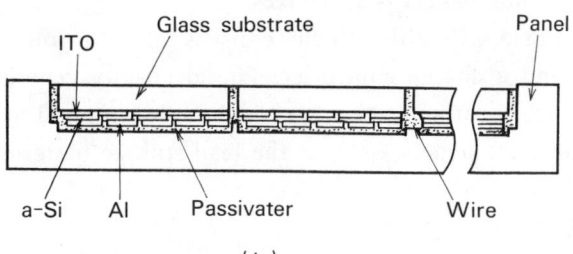

Fig. 5.4.7 (a) : a-Si solar cell panel consisting of 20 integrated cell modules (each 10cm × 10cm)(Sanyo).
(b) : a-Si solar cell panel structure (Sanyo). [after S. Nakano et al.[9)] © 1982 IEEE]

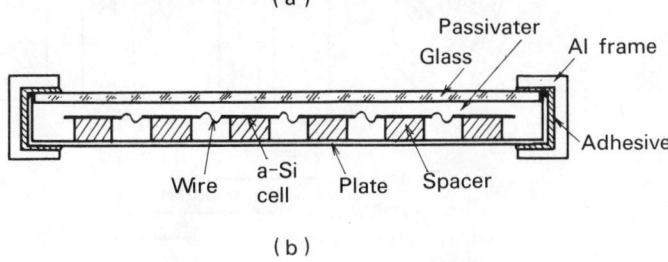

Fig. 5.4.8 (a) : a-Si power module (Fuji Electric).
(b) : a-Si power module structure (Fuji Electric).

solar panel is 45cm × 60cm and consists of 20 integrated type a-Si solar cell modules, each 10cm × 10cm. For the fabrication of the a-Si solar panel, the integrated type a-Si solar cell modules are constructed as shown in Fig. 5.4.7 (b). Epoxy resin is used for the passivator. The output power of the integrated type a-Si solar panel under AM-1 illumination of 100 mW/cm^2 was about 11 W.

The power generating system was set in an experimental model house, as shown in Fig. 5.4.9.

In this electric power generating system, there are 513 panels. Nine panels are connected in series as one unit, and 57 units are connected in parallel lines. Of the total 171 were installed on the main roof of the experimental model house, 189 on the awnings and 153 on an adjoining machine room roof. In accordance with an optimizing theory, the inclination of the a-Si solar panels is 15 degrees.

As shown in Fig. 5.4.10, this system consists of a-Si solar panels, lead storage batteries, an inverter and a charge controller. The electricity generated by the a-Si solar panels is stored in the lead storage batteries, or is directly supplied to the inverter. The DC output power from the solar panels, or from the lead storage batteries, is converted to AC

Fig. 5.4.9 View of experimental model house with solar photovotaic system. [after S. Nakano et al.[9] © 1982 IEEE]

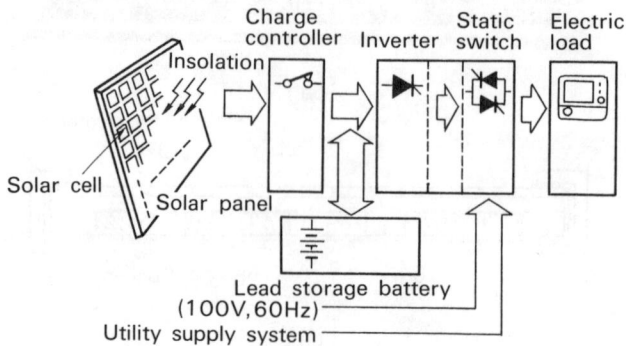

Fig. 5.4.10 Schematic diagram of a solar photovoltaic system. [after S. Nakano et al.[9] © 1982 IEEE]

output power of 100 V, 60 Hz, by the inverter and supplied to electric loads. Whenever the battery discharge is excessive and the charge becomes less than 50% of the full capacity of the battery, a utility power supply is automatically provided. The charge-discharge control is done by a charge controller. In this system, temperature, electric current, voltage, electric power, and solar radiation energy have been measured.

Figure 5.4.11 shows the relationship between the solar radiation energy and the photovoltaic characteristics from August 1 to August 31, 1981. The change in the output performance of the a-Si solar panels under sunshine was measured. As shown in Fig. 5.4.12, the conversion efficiency decreased about 10% within one month and then became stable. The open circuit voltage and short circuit current did not change significantly, but the change in the fill factor was similar to that of the conversion efficiency.

Although the conversion efficiency of the a-Si solar panels decreased slightly at first, their output power then became stable. This stability has continued untile the present time. This indicates that electric power generating systems using a-Si solar panels are already effective enough for practical use.

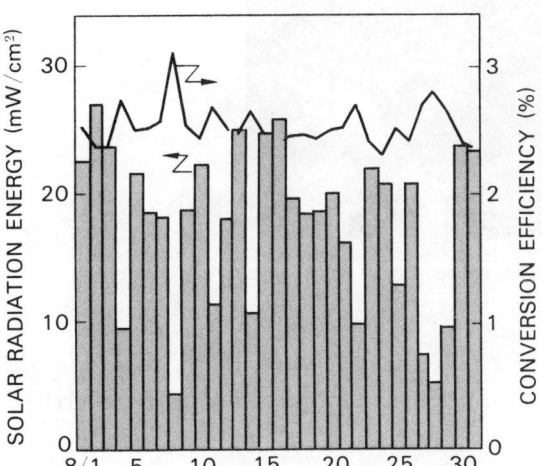

Fig. 5.4.11 The relationship between solar radiation energy and photovoltaic characteristics in August, 1981.

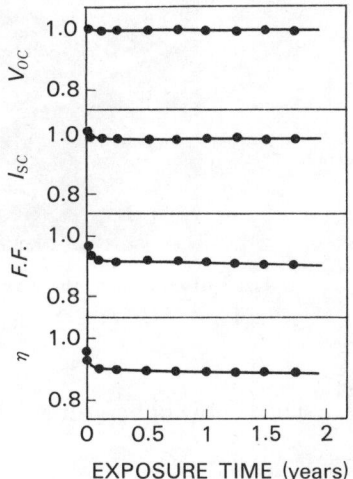

Fig. 5.4.12 The change in output performance of a-Si solar panels under sunshine. Esch value is normalized by the initial values.

400 W system (Fuji Electric)[10]

Figure 5.4.8 (a) shows the module made by Fuji Electric. This solar module consists of 36 a-Si solar cells, with stainless steel substrate, each 7cm × 7cm. This module is constructed as shown in Fig. 5.4.8 (b). A test system of an approximate 400 W a-Si solar array using 108 modules was installed on the rooftop of a building at the Engineering Research and Development Center of The Tokyo Electric Power Co., Inc. (TEPCO), as shown in Fig. 5.4.13, as a joint project between TEPCO and Fuji Electric in the spring of 1981. This system is composed of an a-Si array, an inverter, loads, a date acquisition system,

Fig. 5.4.13 400 W solar array installed as a joint project between The Tokyo Electric Power Co., Inc. and Fuji Electric.

Fig. 5.4.14 View of the 2.5 kW solar array on the roof of Tokyo Institute of Technology.

etc. This system is now undergoing field tests to evaluate its reliability and power generating capacity.

2.5 kW system

A 2.5 kW power generating system using the above-mentioned panels and modules developed by Sanyo and Fuji Electric was set on the roof of the Tokyo Institute of Technology in 1982. This system is shown in Fig. 5.4.14. Various data have been obtained from this system.

Solar cell roofing tile

In addition to the a-Si solar panel, the development of an a-Si solar cell roofing tile is in progress. Figure. 5.4.15 shows a picture of a solar cell roofing tile which was developed by Sanyo. This new home solar energy device combines the a-Si solar cell with a standard size unbreakable, transparent glass roofing tile. Formed into the backside of each tile, the cells are easily hooked up by simple waterproof lead wires. Perfectly substitutable for conventional roofing tiles, these solar tiles eliminate the high costs of panel stands, installa-

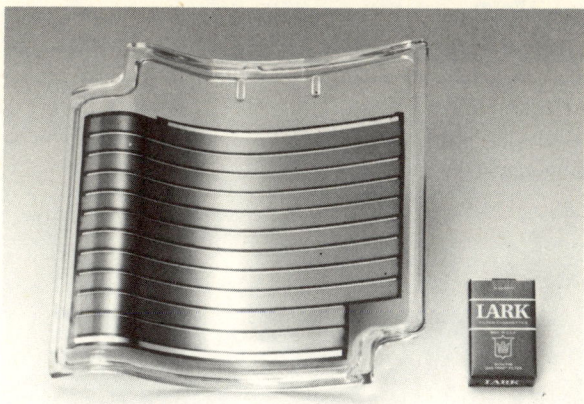

Fig. 5.4.15 Photograph of solar cell roofing tile.

tion and extra installation space. One solar cell roofing tile can generate 2W of electric power. In order to supply 2.7kWh of power, which is the amount utilized during daytime hours by the average Japanese household, a total of 510 solar cell roofing tiles is necessary.

5.4.4 Conclusion

The a-Si solar cell has the potential for realizing the desired low-cost solar cell power system. It is important, however, to secure basic production volume through the present applications in consumer products and to continue research on mass production methods and cost-reducing processes. At the same time, efforts must continue toward improving the conversion efficiency of the a-Si solar cell. Recent cell performance improvement indicates that the practical industrial production of a-Si solar cells for use as electric power sources will be achieved in the near future.

Acknowledgements

The authors wish to express their sincere appreciation to Prof. Y. Hamakawa of Osaka University for his kind guidance.

This work was supported in part by the Agency of Industrial Science and Technology under a contract of the Sunshine Project.

References

1) Y. Kuwano, T. Imai, M. Ohnishi, and S. Nakano: Proc. 14th IEEE Photovoltaic Specialists Conf. (1980) 1402.
2) Y. Kuwano and M. Ohnishi: Proc. 9th Inc. Amorphous & Liquid Semicon., (1981) C4-1155.
3) Y. Kuwano, et al.: Jpn. J. Appl. Phy., *21* (1982) 413.
4) M. Izu and S. R. Ovshinsky: Proceedings of SPIE—The International Society for Optical Engineering, *407* (1983) 42.
5) Scott. B. A., et al.: Apply. Phys. Lett., *37* (1980) 725.
6) S. Nakano, et al.: Proc. 5th EC Photovoltaic Solar Energy Conf. (1983).
7) Y. Hamakawa, H. Okamoto, and Y. Nitta:

Appl. Phys. Lett., *35* (1979) 187.
8) S. R. Ovshinsky: Proceedings of SPIE —The International Society for Optical Engineering, *407* (1983) 5.
9) S. Nakano, T. Fukatsu, H. Nishiwaki, H. Shibuya, K. Tsukamoto, and Y. Kuwano: Proc. 16th IEEE Photovoltaic Specialists Conf. (1982) 1124.
10) H. Haruki and Y. Uchida: Amorphous Semiconductor. Technologies & Devices, ed. Y. Hamakawa (OHM * North-Holland)(1982) 173.

CHAPTER 6

AMORPHOUS SEMICONDUCTOR ELECTRONIC DEVICES

6.1 Thin Film Diodes and Transistors

Masakiyo MATSUMURA*

Abstract

Recent theoretical and experimental results on hydrogenated amorphous-silicon diodes, field-effect transistors and charge-coupled devices are reviewed. Both pin and Schottky barrier diodes have a high rectification ratio of more than 10^9 and a high switching speed of more than 10 MHz. Similar high speed operation is expected in recently proposed field effect transistors having a vertical type structure. High speed operation of more than 200 kHz in charge-coupled devices is also described. And a self-alignment technology is reviewed.

6.1.1 Introduction

Hydrogenated amorphous-silicon thin-film field-effect transistors (a-Si FETs) are well known to hold promise as panel displays and image sensors. At the end of 1982, Sanyo demonstrated the a-Si FET addressed pocket TV set for the first time and clarified the usefulness of the device. However, various problems remain in a-Si FETs. The most serious one is low operation speed. In 1983, new device concepts and technologies for improving the operation speed have been presented at domestic and international technical meetings held in Japan. These included high speed switching[1] of pin and Schottky barrier diodes at more than 10 MHz, the proposal of a vertical type FET structure[2] where the channel length can be decreased to less than 1 micrometer, and 200 kHz operation[3] of charge-coupled devices (CCDs). Self-alignment technology[4] of the gate and source (or drain) will also be potentially useful. This review paper describes these recent developments.

6.1.2 pin and Schottky Barrier Diodes

Rectification ratios of amorphous-silicon pin[5] and Schottky barrier[1] diodes are improved to be more than 10^9, which will be sufficient for various electronic applications. Diode current, I, starts to increase exponentially with applied voltage, V, up to about $V = 0.5$ V and then increases slowly with V. At $V = 2$ V, I becomes more than $10 \, \text{A/cm}^2$. Reverse breakdown voltage of the diodes depends strongly on the active n^- a-Si thickness, W, and, in the case of $W = 0.5$ micron, it was more than 15 V. The dynamic characteristics of Schottky barrier diodes with molybdenum as a barrier metal are shown in Fig. 6.1.1. A

---

* Department of Physical Electronics, Tokyo Institute of Technology, Meguro-Ku, Tokyo 152, Japan.

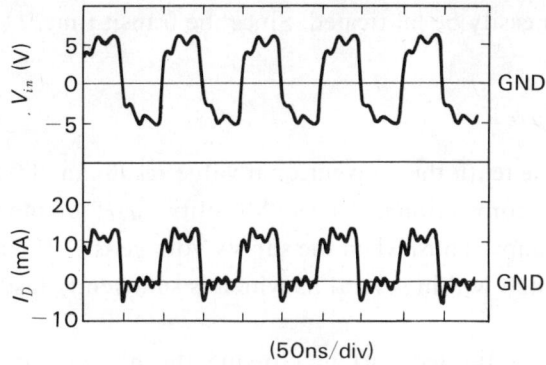

Fig. 6.1.1 Dynamic rectification characteristics of the a-Si Schottky barrier diode. The upper trace is the 10 MHz square wave input voltage. The lower trace is the diode current. [after Y. Nara et al.[1]]

10 MHz square wave voltage of ±5 V was applied to the diode. It can be seen that the rectification characteristics did not detoriorate even at 10 MHz, and that the charge storage effect was slight. Similar characteristics were obtained by Cr and Ta Schottky barrier diodes.

6.1.3 Vertical Field Effect Transistors

Though the fundamental reason for the low operation speed of a-Si FETs is low electron mobility, μ_{FE}, experimental results on a-Si diodes indicate that a sufficiently high operation speed of more than several MHz is attainable by shortening the channel length of the FETs to about 1 micron. However, in a conventional (horizontal) FET structure, where electrons flow along parallel to the substrate surface, delineation of the source and drain by photolithographic techniques make a decrease of the channel length to less than 10 micrometers to have no practical meaning. For breaking through this constraint, the Tokyo Institute of Technology proposed a vertical type FET[2] whose basic structure is shown in Fig. 6.1.2. The channel forms at the side-wall of the n^+-n^--n^+ structure. Since the thickness of the active undoped a-Si layer, n^-, is determined at the deposition, it can be decreased to below 0.1 micrometer. Thus, an FET with an effective channel length, L, of less

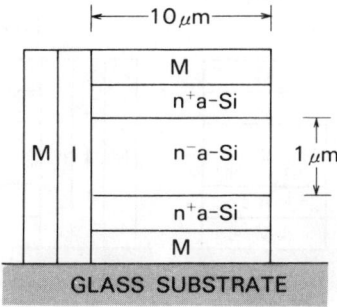

Fig. 6.1.2 Cross section of the basic vertical a-Si FET. The upper and lower metal electrodes are the source and drain electrodes. The left-hand-side metal electrode is the gate electrode. [after Y. Uchida et al.[2] © 1984 IEEE]

than 1 micrometer can easily be fabricated. Since the transit time, t_t, of electrons across the channel is given by

$$t_t = L^2/\mu_{FE} V$$

the decrease of L to one-tenth the conventional value results in 100 times higher operation speed compared with conventional FETs. Mobility, μ_{FE}, of more than $1\,\text{cm}^2/\text{V}\cdot\text{s}$ was reported by various groups. Thus, when the supply voltage is 10 V, an FET with a 1 micron channel length can switch within several ns which is sufficiently fast for panel displays and image sensors.

Since space-charge-limited current crossing the n⁻ layer increases rapidly with a decrease in the n⁻layer thickness, the on-off current ratio of the basic vertical FET will be detoriorated drastically. This drawback can be eliminated by the improved structure shown in Fig. 6.1.3. By changing the n⁻ layer to a p layer, as shown in (a), the space charge limited current is suppressed by the effect of the blocking characterisitics of the pn junction. Then, by dividing the p layer into a thinner p⁺ layer and a thick n layer, as shown in (b), the effective channel length can be decreased even further and the on-current will also be improved. However, it is well known that the introduction of boron into a-Si detoriorates carrier transport and, as a result, inversion type n-channel FET operation is difficult. Thus, the FET structure shown by (c) will be the most useful one. The channel is formed at the surface of the undoped n⁻ a-Si layer deposited at the etched side-wall of the n-I-n structure. Plamsa deposited silicon nitride or silicon dioxide can be used as the insulating layer, I.

Since electrons cross the active n⁻ layer near the source and the drain, the residual resistances arising from the n⁻ layer will play an important role. It must be noted that these

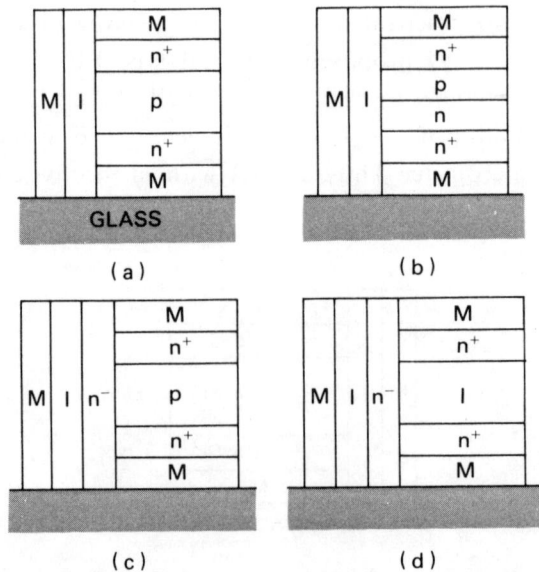

Fig. 6.1.3 Cross section of the improved vertical FETs. (a) n-p-n structure. (b) DSA (n-p-n⁻-n) structure. (c) Improved n-p-n structure. (d) Improved n-I-n structure. [after Y. Uchida et al.[2)] © 1984 IEEE]

space charge resistances are non-linear and that their values decrease either when the current density is increased or when the n⁻ layer thickness is decreased. Under the off-condition, the space-charge-limited current also flows through the n⁻ layer. Thus the on and off currents can be summerized as shown in Fig. 6.1.4.[3] The 1 micron FET will have an on-off current ratio of more than 10^4. We must take, however, the two dimensional effects into account, because all the size parameters of the device become approximately the same as the channel length. As the electric field which starts from the drain to the source is shortened by the gate, the two-dimensional effect acts to suppress the off current to a very small value compared with that in the FET structure shown by Fig. 6.1.2.

Preliminary experimental results for the vertical type $L = 1$ micron FETs indicated that the current is inversely proportional to L, at least down to $L = 1$ micron, and that the on-off current ratio decreases slightly with a decrease in L.

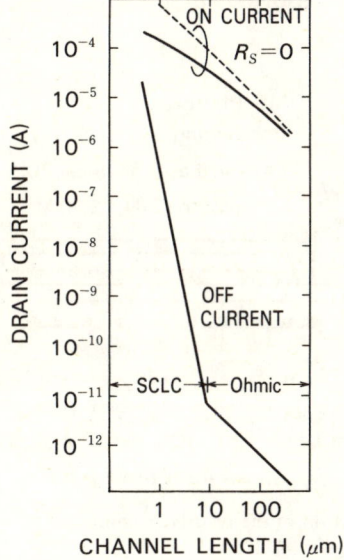

Fig. 6.1.4 Theoretical results of the on and off current for a-Si FETs with various channel lengths. [after M. Matsumura et al.[3]]

6.1.4 Self-Aligned Field Effect Transistors

Self alignment technology for the gate and source (or drain) is important because it not only decreases the residual capacitance but also decreases the fabrication steps a great

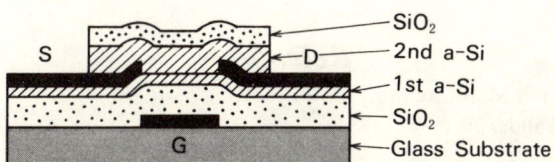

Fig. 6.1.5 Cross section of the self-aligned a-Si FET. [after T. Kodama et al.[4]]

deal. Figure 6.1.5 shows a cross sectional view of self-aligned FETs developed by Fujitsu.[4] The n^+ a-Si layer for the ohmic contact was deposited at a relatively low substrate temperature of 150°C because it must be deposited over the positive-type photo-resist layer above the gate electrode. This photo-resist layer lifts off the deposited n^+ a-Si and Ni-Cr layers on the source-drain spacing. Field-effect mobility measured was higher than $1\,cm^2/V\cdot s$.

6.1.5 Resistively Connected Gate Charge-Coupled Devices

Figure 6.1.6 shows the improved CCD structure with resistively connected transfer gates.[5] Since the narrow metalic electrode scarecely shields the electric field along the channel, a strong drift field for electrons is generated in the active n^-layer. As a result, the device can operate with sufficient speed. The device having 24 transfer gates with a 10 micron pitch and driven by a 3-phase clock pulse was operated at 200 kHz with a transfer efficiency of more than 99%/transfer. Parallel to serial transformation using a 40 micron pitch device at 10 kHz was also reported.[6]

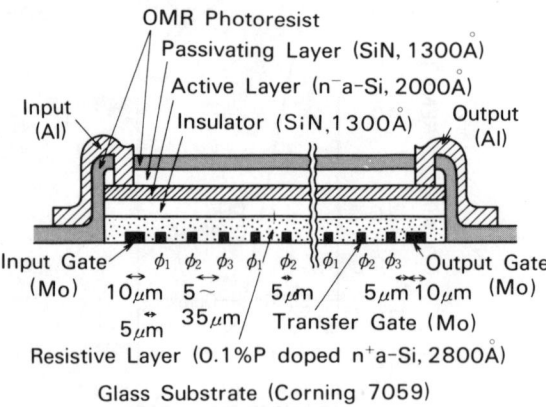

Fig. 6.1.6 Cross section of the resistively connected gate a-Si charge coupled devices. [after S. Kishida et al.[5]]

6.1.6 Conclusion

New device concepts and fabrication technologies for improving a-Si electronic devices have been proposed. Although the results are in a preliminary stage, they indicate that the a-Si devices can operate with sufficient speed. Thus the most serious problem for a-Si devices, low operation speed, will be overcome in the near future.

References

1) Y. Nara, Y. Kudoh, and M. Matsumura : J. of Non-Crystalline Solids, *59 & 60* (1983) 1175.
2) Y. Uchida, Y. Nara, and M. Matsumura : IEEE Elect. Device Lett., *EDL-5* (1984) 105.
3) M. Matsumura and Y. Uchida : IECE of Japan Digest of Technical Papers, *ED 83-67* (1983).
4) T. Kodama, S. Kawai, Y. Nasu, and N.

Takagi: Digest of Technical Papers on Fall Meet. of Jpn Soc. of Appl. Phys., *8 p-W-17* (1983) 320.
5) S. Kishida, Y. Naruke, Y. Uchida, and M. Matsumura: J. of Non-Crystalline Solid, *59 & 60* (1983) 1281.
6) Y. Uchida, S. Kishida, and M. Matsumura: Electronics Lett., *20* (1984) 422.

6·2 Microfabrication Technology Using Multicomponent Amorphous Alloy

Katsumi MURASE* and Yoshihiko MIZUSHIMA**

Abstract

Technologies for fabricating miniaturized electrode structures are developed on the basis of characteristic properties of a-Si-Ge-B, a tetrahedrally-bonded multicomponent amorphous alloy.

By using the large difference in the oxidation rate between a-Si-Ge-B and c-Si, a self-aligning process for miniaturized electrode structures is composed. This process can be performed at temperatures below 800°C owing to the extremely rapid oxidation of a-Si-Ge-B. A novel surface-electrode static induction transistor realized with this process exhibits sharp voltage-controlled negative resistance. This three-terminal negative-resistance device is promising for VLSI's with respect to high speed and small fabrication-scattering of device characteristics.

The good step-covering property of a-Si-Ge-B is applied to submicron electrode fabrication technology which needs no submicron photomasks.

6.2.1 Introduction

Since the first success in controlling the conduction property of amorphous silicon (a-Si : H),[1] tetrahedrally-bonded amorphous materials have intensively been studied from both physical-interest and practical-application viewpoints. So far, typical application fields of these materials have been those of optoelectronic devices and thin film transistors, which are reviewed in other sections. Amorphous chalcogenides, the counterpart of tetrahedrally-bonded materials, have also been applied to these devices. Besides, they have successfully been utilized as inorganic photoresists[2] in the microfabrication technology field. Application of tetrahedrally-bonded amorphous materials to such a field, however, has not yet been attempted to any great extent.

The subject of this section is miniaturized semiconductor-device technology using amorphous silicon-germanium-boron (a-Si-Ge-B) alloy, a tetrahedrally-bonded multi-component amorphous material.

Amorphous Si-Ge-B alloy has been introduced in the 1983 issue of this series,[3] where physical aspects of this amorphous alloy are presented, and positive utilization of localized

---

* Atsugi Electrical Communication Laboratory, Nippon Telegraph and Telephone Public Corporation, Ono 1839, Atsugi-shi, Kanagawa 243-01.
** Hamamatsu Photonics Co., Ltd., Ichino-cho, Hamamatsu-shi, Shizuoka 435.

states demonstrated. The following is characteristic features of this alloy particularly suitable for electrodes and connecting metallizations:
(1) High conductivity;
(2) High durability against acids and alkaline solutions;
(3) Mechanical stability;
(4) Strong adhesiveness to both silicon (Si) and silicon dioxide (SiO$_2$).

In addition, remarkable oxidation properties have been found[4]: a-Si-Ge-B is thermally oxidized much faster than crystalline Si (c-Si) and even faster than doped polycrystalline Si (poly-Si). Accordingly, a-Si-Ge-B can provide both conducting and insulating layers.

The microfabrication technology described first in 6.2.3 is based on the oxidation properties of a-Si-Ge-B, reviewed in 6.2.2. In this application, both the conducting layer and insulating oxide of a-Si-Ge-B are used as device elements. Device characteristics of a vertical MESFET or static induction transistor (SIT)[5] fabricated with this technology are also presented in 6.2.3.

Amorphous Si-Ge-B is prepared by a low-pressure chemical vapor deposition (LPCVD) technique. Consequently, it provides exellent step coverage. The feasibility of applying this aspect to a simple and precise fabrication technique with good reproducibility for submicron gate-electrodes is briefly pointed out.

6.2.2 Oxidation of a-Si-Ge-B[4]

Preparation

Amorphous Si-Ge-B films were prepared by decomposing SiH$_4$-GeH$_4$-B$_2$H$_6$ mixtures diluted with helium in a LPCVD furnace at 500°C. Boron-doped $2\sim6\,\Omega\cdot$cm (100)-oriented c-Si wafers were used as substrates. The pressure in the furnace was kept at 0.2 Torr.

Oxidation was carried out in an ambience of oxygen, bubbled through water at 90°C. The partial pressure of water vapor, estimated from its consumption, was 430 Torr.

Oxidation characteristics

Figure 6.2.1 shows oxide thickness versus oxidation time for wet oxidation of a-Si-Ge-B at 810°C. The flow-rate ratios between reactant gases for a-Si-Ge-B deposition are GeH$_4$/(SiH$_4$ + GeH$_4$) = 5×10^{-2} and B$_2$H$_6$/(SiH$_4$ + GeH$_4$) = 1×10^{-2}. Data for B-doped $5\,\Omega\cdot$cm (111)-oriented c-Si are also plotted in this figure for comparison. The oxide films grown from a-Si-Ge-B are about ten times as thick as the SiO$_2$ films grown from c-Si under the same oxidation conditions. The oxide growth-rate becomes larger as either the B content in the starting amorphous films becomes higher or the Ge content lower.

Oxide thickness x is related to oxidation time t in accordance with a linear-parabolic formula[6] as

$$\frac{x^2}{B} + \frac{x}{B/A} = t \qquad (6.2.1)$$

Figures 6.2.2 and 6.2.3 show parabolic rate constant B and linear rate constant B/A, respectively, derived by a least-square curve fitting method for various amorphous films.

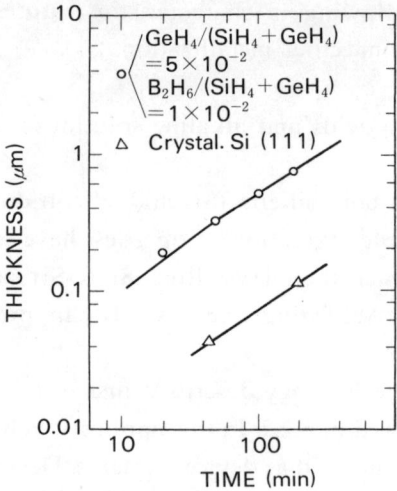

Fig. 6.2.1 Wet oxidation characteristics for B-doped 5 Ω·cm (111)-oriented c-Si and a-Si-Ge-B deposited with $GeH_4/(SiH_4 + GeH_4) = 5 \times 10^{-2}$ and $B_2H_6/(SiH_4 + GeH_4) = 1 \times 10^{-2}$. Oxidation temperature is 810°C. [after K. Murase et al.[4]]

The solid curves in Fig. 6.2.1 are given by Eq. (6.2.1), using the rate constants determined. As seen in Figs. 6.2.2 and 6.2.3, both these constants are remarkably enhanced in comparison with the case for c-Si, while the activation energies are reduced. It should be noted that although the activation energy of the linear rate constant for the amorphous film containing Ge is smaller than that for films without Ge, the value of the rate constant is nevertheless smaller for the former.

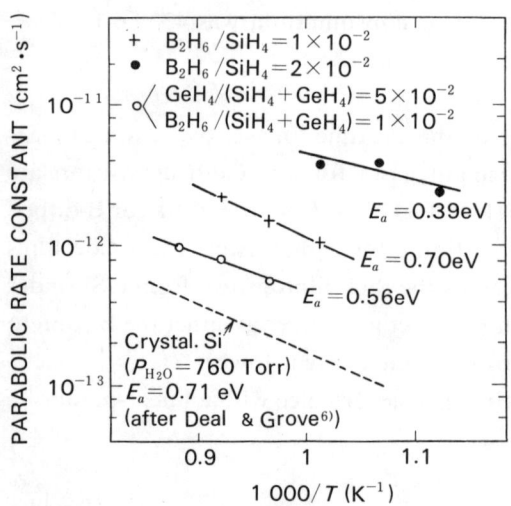

Fig. 6.2.2 Parabolic rate constant as a function of reciprocal absolute temperature. [after K. Murase et al.[4]]

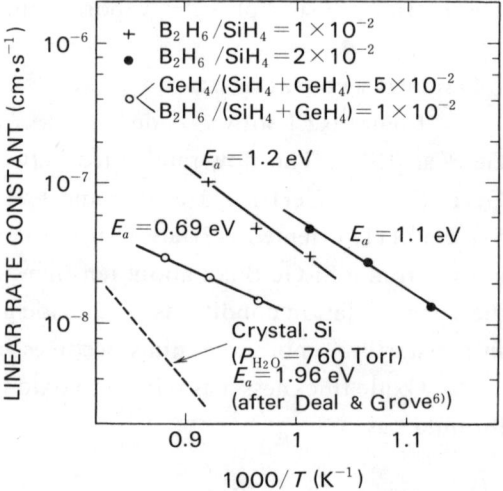

Fig. 6.2.3 Linear rate constant as a function of reciprocal absolute temperature. [after K. Murase et al.[4]]

Oxidation mechanism

According to Deal and Grove,[6] the linear rate constant for oxidation of c-Si is proportional to the chemical reaction rate for oxidation at the interface between SiO_2 and c-Si, and the parabolic rate constant is proportional to the diffusivity of the oxidizing species in SiO_2. This is presumably also the case for oxidation of a-Si-Ge-B.

The activation energy of the linear rate constant for oxidation of c-Si has been reported to be about 2 eV.[6] This value has little relation to whether oxidation is performed in a wet oxygen ambient or in a dry one. Moreover, it is hardly influenced by the impurity concentration in c-Si.[7]

The remarkable reduction in the activation energy of the linear rate constant for oxidation of a-Si-B and a-Si-Ge-B implies that the rate-determining process of their oxidation reaction is quite different from that for c-Si. First consider oxidation of a-Si-B without Ge. Table 6.2.1 presents the compositions of the amorphous films studied. The B

Table 6.2.1 The deposition condition and composition of the starting amorphous films. [after K. Murase et al.[4]]

Sample number	Deposition condition		Composition (mol %)		
	$\dfrac{GeH_4}{SiH_4+GeH_4}$	$\dfrac{B_2H_6}{SiH_4+GeH_4}$	Si	Ge	B
1	0	1×10^{-2}	73	0	27
2	5×10^{-2}	1×10^{-2}	61	16	23
3	5×10^{-2}	5×10^{-2}	40	11	49

content is more than 20 mol% for all films. This means that at least one of four bonding-orbitals of every Si atom forms a bond with a B atom. Such a situation never comes about in c-Si, even when c-Si is doped with B atoms to the solid solubility limit. Hence, it is probable that the reactions between Si-B and oxidizing species, water and oxygen, participate in the rate-determining process. Boron reacts with water and with oxygen much more readily than Si does.[8] Therefore, by analogy, it is expected that the reaction between Si-B and water or oxygen is more rapid, requiring lower activation energy, than that between Si-Si and water or oxygen. The reaction between Si-B and water (or oxygen) will result in either Si-O-B formation or creation of a B-O bond and Si with a free bonding-orbital. In the latter case, the free bonding-orbital will easily be attacked by the oxidizing species to form an Si-O bond.

Next, consider the effects of Ge on the oxidation reaction. Amorphous Si-Ge-B used for this study contains Ge to about 16 mol%, as seen from Table 6.2.1. Thus, Ge-Si and Ge-B bonds should play an important role in the oxidation reaction. It is known that Ge is more readily oxidized than Si, and Ge oxides are volatile.[9,10] Therefore, it is very likely that Ge-Si and Ge-B are more easily oxidized than Si-Si and Si-B, respectively, and subsequent Ge volatilization will take place. Thus, oxygen atoms will not introduce much strain in the oxide atomic-network when they occupy vacant sites created by Ge volatilization to combine with Si and/or B atoms. This can cause further reduction in the activation energy of the linear rate constant for films containing Ge. This also causes reduction in the effective oxide growth-rate. The validity of this model has been confirmed from infrared absorption measurements and Auger-electron spectroscopic measurements.[4]

As for enhancement of the parabolic rate constant, since diffusivities of various gases such as neon[11~13] and argon[11,12] are larger in B_2O_3 than in SiO_2, it is assumed by analogy that the diffusivities of water and oxygen increase as the B content in the oxide increases.

Properties of Si-Ge-B oxides

As a result of Ge volatilization, Si-Ge-B oxide contains only a small amount of Ge and can thus be regarded as borosilicate glass. Borosilicate glass prepared by CVD at 300 ~400° C is reported[14,15] to be strongly hygroscopic. However, Si-Ge-B oxides are very stable with respect to humidity.

Table 6.2.2 Electrical properties of various Si-Ge-B oxides. [after K. Murase et al.[4]]

Deposition condition		Oxidation temperature	Resistivity	Electric field at dielectric breakdown	Dielectric constant	tan δ
GeH_4 / $SiH_4 + GeH_4$	B_2H_6 / $SiH_4 + GeH_4$	(°C)	($\Omega \cdot$cm)	(V/cm)	(at 1 MHz)	(at 100 kHz)
0	1×10^{-2}	810	5×10^{16}	9.7×10^6	3.6	6.6×10^{-3}
5×10^{-2}	1×10^{-2}	620	4×10^{16}	1.0×10^7	3.8	7.7×10^{-3}
5×10^{-2}	1×10^{-2}	715	8×10^{16}	8.7×10^6	3.9	8.5×10^{-3}
5×10^{-2}	1×10^{-2}	810	6×10^{16}	8.6×10^6	3.7	4.2×10^{-3}
5×10^{-2}	2×10^{-2}	810	4×10^{16}	1.1×10^7	3.6	7.7×10^{-3}

Electrical properties are listed in Table 6.2.2 for various Si-Ge-B oxide films. All these properties are comparable to those of high quality SiO_2.[16] In particular, the electric field at dielectric breakdown is high when compared with that of oxidized poly-Si, $3 \sim 4 \times 10^6$ V/cm.[17] It should be emphasized that such high values result from uniform and smooth oxide textures due to the amorphous structures of the starting films.

6.2.3 Application of a-Si-Ge-B Films to Microfabrication Technologies

Self-aligning fabrication of fine electrode structures[4,18]

Process In many semiconductor devices, performance is improved as the structure is miniaturized. Self-aligning techniques are the most promising for miniaturizing device structures without reducing the production yield.

The following describes a self-aligning fabrication process for an electrode structure where two electrodes are formed very close to each other. This process utilizes the large difference in the oxidation rate between a-Si-Ge-B and c-Si. The processing steps are illustrated in Fig. 6.2.4. Amorphous Si-Ge-B is used as one of the two electrodes on the substrate surface, and its oxide as an insulating layer between it and the other electrode.

After a-Si-Ge-B is defined, the wafer is subjected to thermal oxidation. At this step, an oxide layer also grows on c-Si in the contact hole area. However, this SiO_2 layer is much thinner than Si-Ge-B oxide and is etched about 4 times as fast as Si-Ge-B oxide, as shown in Fig. 6.2.5. Therefore, no significant thickness change occurs after SiO_2 etching. As is easily seen, the contact hole for the second electrode is automatically formed at this stage, and is self-aligned with respect to the first or Si-Ge-B electrode. It should be stressed that the oxidation step for insulating the two electrodes can be performed at temperatures lower than 800°C owing to the high oxidation rate of a-Si-Ge-B. Consequently, no significant impurity diffusion occurs during oxidation. In other words, a "surface-electrode" structure is realized: no diffused layer is formed under the Si-Ge-B electrode. Distinctive device characteristics due to this structure are presented in the following.

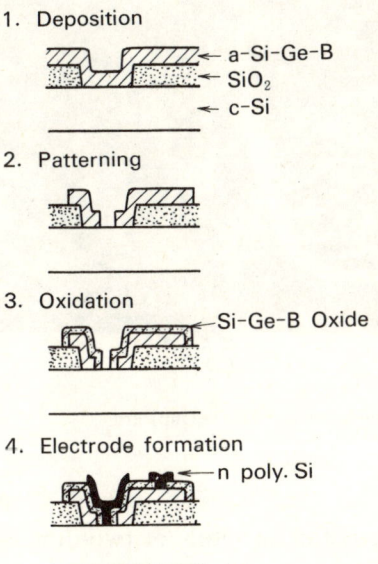

Fig. 6.2.4 Processing steps for miniaturized electrode structure. [after K. Murase et al.[4]]

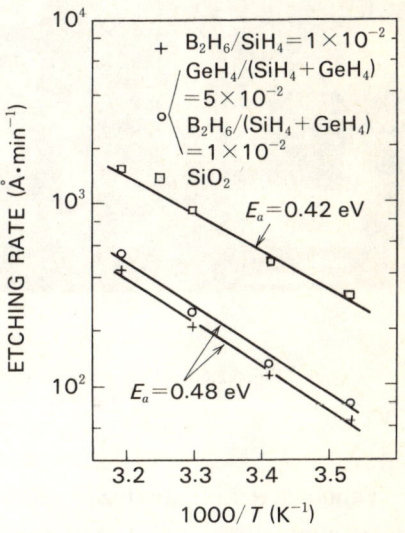

Fig. 6.2.5 Etching rates for various oxides in a solution of 1 part HF (49%) and 10 parts NH_4F (40%) as a function of reciprocal absolute temperature. [after K. Murase et al.[4]]

Surface-electrode SIT[19]   The device structure shown in Fig. 6.2.4 is that of a vertical MESFET or static induction transistor (SIT). The a-Si-Ge-B electrode, forming a bipolar-mode Schottky barrier[20] against n-type c-Si,[21] acts as the gate electrode. However, SIT's fabricated by the self-aligning process proposed above have several conspicuous structural features never observed in conventional SIT's. Namely, in the former, a surface-electrode structure is realized, and the gate-source spacing is greatly reduced to a submicron dimension. So far, the gate region has comprised an impurity-diffused layer. Its depth has normally been several micronmeters so that a high transconductance g_m could be obtained. Because of the deep diffusion of gate regions, the gate spacing cannot be reduced to less than 5 μm.

Owing to its structural features, surface-electrode SIT's exhibit negative-resistance operation. Figure 6.2.6 shows typical negative-resistance characteristics of an SIT in which the gate-source spacing, i. e., the thickness of Si-Ge-B oxide, is 0.2μm, and the gate spacing 1.0μm. An n-on-n$^+$ (111)-oriented c-Si wafer is used as the substrate. The resistivity and thickness of the n epitaxial layer are $20 \Omega \cdot$cm and 2.0μm, respectively. The second electrode is made of n$^+$ poly-Si. In Fig. 6.2.6, the "downward mode" denotes the operation mode in which the n$^+$ poly-Si is used as the source electrode and the wafer-backside electrode as the drain electrode. In the "upward mode", the source and the drain electrode are used inversely. A sharper negative-resistance is obtained in the upward mode than in the downward.

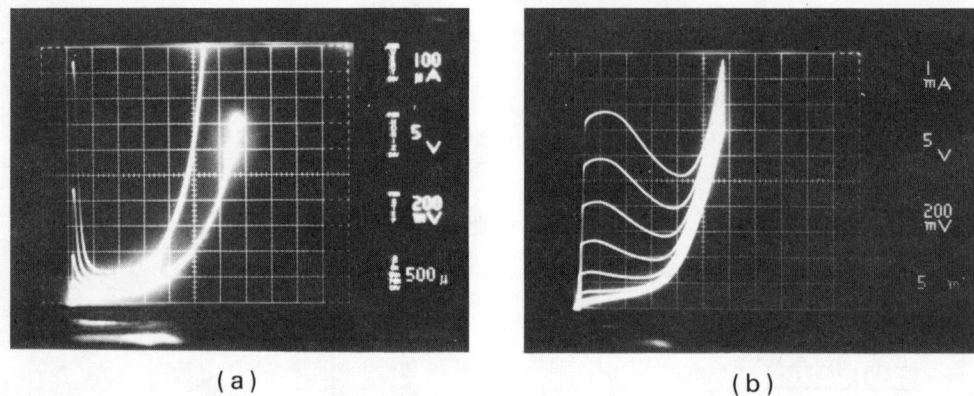

Fig. 6.2.6 *I-V* characteristics of a novel surface-electrode SIT: (a) upward mode, (b) downward mode. [after T. Tamata et al.[19]]

Tamama et al.[19] analyzed such device characteristics in terms of two-dimensional computer simulation, and found the conditions necessary to bring about the negative resistance. The origin of the negative resistance is interpreted as follows. First consider the upward mode. In this operation mode, the drain and the gate lie very close to each other on the same surface of the n epitaxial layer. When the gate is biased in the forward direction with respect to the n layer or source, hole injection from the gate occurs, increasing the conductivity in the epitaxial layer through the conductivity modulation. Thus, a rapid increase in the drain current appears in a low drain-voltage region. As the drain voltage is raised, the electric field caused by the drain bias penetrates into the area below the gate electrode, and the potential under this electrode exceeds the gate bias voltage. Hole injection ceases at this point. Consequently, the device plunges into a unipolar-mode state inducing negative resistance. The current in the unipolar-mode condition is space-charge-limited. The negative resistance in the downward mode is due to the same mechanism. However, in this operation mode, the negative resistance appears at higher drain voltages, since penetration of the drain field into the gate area is weakened on account of the larger distance between the gate and the drain electrode. Moreover, in the downward mode, penetration of the drain field into the gate area is difficult when the source-gate spacing is very small. Thus, in this operation mode, the source-gate spacing should be larger than some minimum value, which depends on the epitaxial-layer thickness, for the negative resistance to appear.

Negative-resistance characteristics can never be obtainded in conventional SIT's. Surface-electrode SIT's have several advantages over conventional ones. In the latter, operating in the bipolar-mode, holes which have accumulated at the near-surface part of the source-gate junction are protected against the drain field by deeply diffused gate regions, so that these holes must disappear through recombination, requiring a long time. In the present surface-electrode SITs, holes can be controlled through the drain electric field. Moreover, when the electrode-window area is constant, the surface-electrode structure has the lowest depletion-layer-capacitance. Therefore, surface-electrode SIT's are expected to operate faster than conventinal bipolar-mode SIT's. Another great advantage of the surface-electrode SIT is that it is promising for VLSI's. This is due to its large fabrication margin. Fabrication scatterings in both the gate junction depth and the source-gate spacing cause a large scattering in the drain current of the SIT. In this respect, conventional bipolar-mode SIT's are rather unfavorable to VLSI's as pointed out by Stork et al.[22] On the other hand, problems of scattering in the drain current rarely arise in surface-electrode SIT's, since the gate junction depth is almost zero and the source-gate spacing is determined by the Si-Ge-B oxide thickness which can be controlled precisely.

In this article, only the surface-electrode SIT is presented as an application of the self-aligning fabrication technique. However, this technique can effectively be applied to various other devices.

Submicron gate-electrode fabrication technology

In MOS and MES integrated circuit technology, there is a growing demand for higher operation speed. One key to higher device performance lies in precise and easy definition of submicron gate electrodes.

Here, a very simple fabrication technology designed for this purpose is presented. In this technology, the good step coverage of a-Si-Ge-B, prepared by LPCVD, is utilized in conjunction with directional etching techniques. A striking advantage of this technology is that no submicron-dimension photomasks are necessary. The processing steps are illustrated in Fig. 6.2.7. First, a rather steep step is formed. Then, a-Si-Ge-B is deposited on the step. The side of the step is uniformly covered with the amorphous film. When the amorphous film is etched by its thickness from the direction indicated in the figure using directional etching techniques such as reactive ion etching, only a portion of the film on the step wall remains. This portion can be used as gate electrodes of MOS or MES transistors. As is obvious, the thickness of the gate electrode is determined by the film thickness which can be controlled precisely and easily.

An example of a submicron electrode structure fabricated with this technology is shown in Fig. 6.2.8, where the width of the electrode is $0.2\,\mu$m and the height $1\,\mu$m. This fine electrode is successfully formed over a long distance.

It should be noted that the work function of this alloy can be tailored over a wide energy range by controlling the composition of the alloy. Work functions for various alloy

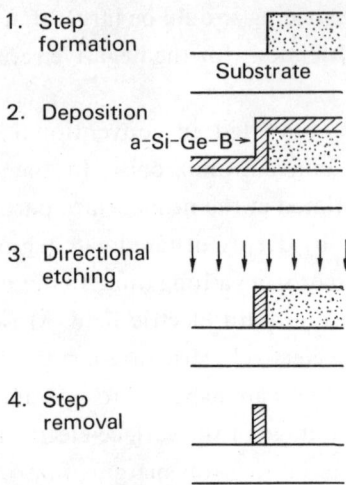

Fig. 6.2.7 Processing steps for submicron electrodes. [after K. Murase et al.[18)]]

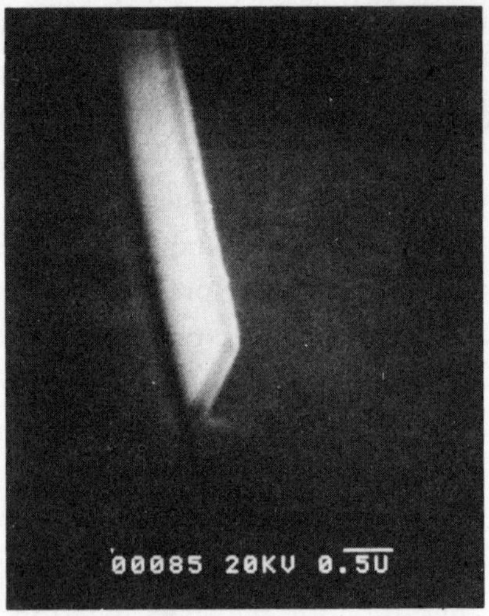

Fig. 6.2.8 Scanning electron micrograph of submicron a-Si-Ge-B electrode. The width is 0.2 μm, and the height 1.0 μm. [after K. Murase et al.[18)]]

Table 6.2.3 Work functions of various a-Si-Ge-B films. [after K. Murase et al.[18)]]

Gas flow ratio		Work function
$\frac{GeH_4}{SiH_4+GeH_4}$	$\frac{B_2H_6}{SiH_4+GeH_4}$	eV
0	1×10^{-2}	5.28
5×10^{-2}	1×10^{-2}	4.77
5×10^{-2}	2×10^{-2}	5.00
3×10^{-1}	1×10^{-2}	4.50

compositions are listed in Table 6.2.3. The work function increases with increasing B content and/or decreasing Ge content. This composition dependence is in accordance with the electronegativity of the constituent atoms. The variable work function property of a-Si-Ge-B is advantageous to barrier-height or threshold-voltage design.

6.2.4 Conclusions

Recent attempts at applying a tetrahedrally-bonded multicomponent amorphous alloy, a-Si-Ge-B, to the microfabrication technology field are reviewed in this section.

Amorphous Si-Ge-B is oxidized in a practical processing period at temperatures lower than 800°C, and converted into a high quality insulator. An oxide growth-rate more than ten times larger than that for c-Si can be obtained.

On the basis of these enhanced oxidation properties, a self-aligning process for miniaturized-electrode-structure fabrication is proposed. One of the advantages of this process is that it is performed at temperatures below 800°C. A novel surface-electrode SIT realized by this process shows striking voltage-controlled negative resistance. This three-terminal negative-resistance device is promising for VLSI's.

The good step-covering property of LPCVD a-Si-Ge-B is utilized for another microfabrication technology, with which submicron electrodes are easily and precisely formed without using submicron-dimension photomasks.

Acknowledgements

The authors are grateful to Dr. Teruo Tamama, Dr. Toshio Ogino, Dr. Masahiro Sakaue and Dr. Yoshihito Amemiya for valuable discussion.

References

1) W. E. Spear and P. G. LeComber: Solid State Commun., *17* (1975) 1193.
2) A. Yoshikawa, O. Ochi, H. Nagai, and Y. Mizushima: Appl. Phys. Lett., *29* (1976) 677.
3) K. Murase, Y. Amemiya, and Y. Mizushima: JARECT *Vol. 6*, Amorphous Semiconductor Technologies & Devices (Y. Hamakawa ed.) (OHMSHA, North-Holland, 1983) 274.
4) K. Murase, T. Ogino, and Y. Mizushima: Jpn. J. Appl. Phys., *22* (1983) 1771.
5) J. Nishizawa, T. Terasaki, and J. Shibata: IEEE Trans. Electron Devices, *ED-22* (1975) 185.
6) B. E Deal and A. S. Grove: J. Appl. Phys., *36* (1965) 3770.
7) E. A. Irene and D. W. Dong: J. Electrochem. Soc., *125* (1978) 1146.
8) J. W. Mellor: A Comprehensive Treatise on Inorganic and Theoretical Chemistry, *Vol. V* (Longmans, Green and Co., London, 1956) 14.
9) L. N. Dennis, K. M. Tressller, and F. E. Hance: J. Am. Chem. Soc., *45* (1923) 2033.
10) R. B. Bernstein and D. Cubicciotti: J. Am. Chem. Soc., *73* (1951) 4112.
11) K. N. Woods and R. H. Doremus: Phys. Chem. Glasses, *12* (1971) 69.
12) W. W. Brandt, T. Ikeda, and Z. A. Schelly: Phys. Chem. Glasses, *12* (1971) 139.
13) W. G. Perkins and D. R. Begeal: J. Chem. Phys., *54* (1971) 1683.
14) E. A. Taft: J. Electrochem. Soc., *118* (1971) 1985.
15) A. S. Tenney and J. Wong: J. Chem. Phys., *56* (1972) 5516.
16) R. M. Burger and R. P. Donovan ed.: Fundamentals of Silicon Integrated Device Technology, *Vol. 1* (Prentice-Hall, Englewood Cliffs, New Jersey, 1967) 102.
17) R. B. Marcus, T. T. Sheng, and P. Lin: J. Electrochem. Soc., *129* (1982) 1282 and references cited therein.
18) K. Murase, T. Tamama, M. Sakaue, T. Ogino, Y. Amemiya, and Y. Mizushima: Proc. 10th Int. Conf. of Amorphous and Liquid Semiconductors, held in Tokyo (1983) (J. Non-Cryst. Solids, *59 & 60*) 1211.
19) T. Tamama, K. Murase, and Y. Mizushima: submitted to Solid State Electron.
20) Y. Amemiya and Y. Mizushima: IEEE Trans. Electron Devices, *ED-31* (1984) 35.
21) K. Murase, Y. Amemiya, and Y. Mizushima: Jpn. J. Appl. Phys., *21* (1982) 1559.
22) J. M. C. Stork and J. D. Plummer: IEEE Trans. Electron Devices, *ED-28* (1981) 1354.

6.3 Anodic Oxidation and Its Device Applications

Hideki HASEGAWA,* Hidekazu YAMAMOTO,*
Satoshi ARIMOTO* and Hideo OHNO*

Abstract

Anodic oxidation allows formation of uniform thin and thick SiO_2 and Al_2O_3/SiO_2 insulating layers on amorphous silicon films at room temperature. The process is stable, reproducible and electrically controllable. Oxidation process, insulator properties and application to material assessment and device fabrication are reviewed here.

Anodic oxidation has been applied to MIS and p-i-n type a-Si solar cells for efficiency enhancement and material defect passivation. a-Si MOS FFTs and PSDs (position sensitive detectors) have also been fabricated for the first time with good device performance, indicating feasibility of new-type integrated planar functional devices based on anodic oxidation.

6.3.1 Introduction

Hydrogenated amorphous silicon (a-Si : H) films are of current technological interest for applications to low cost, large-area solar cells, integrated image sensors and large area displays. Judging from the tremendous success of the oxidation and MOS technology in single crystal Si (c-Si), establishment of a similar technology for a-Si : H films appears to be an interesting possibility to be explored. Since a-Si : H films are weak to any of high-temperature processings, thermal oxidation is obviously not suitable for the purpose. In fact, a-Si : H MIS FETs, recently reported by authors, employ chemical vapor deposited (CVD) films of SiO_2[1] and Si_3N_4[2] rather than using native oxide because of this difficulty. However, it is well known in the c-Si technology that the properties of deposited oxide-Si interface are inferior to those of native oxide-Si interface, particularly when the deposition temperature is low. In addition, the electrode structure of those FETs is usually staga type and not suitable for planar integration.

The purpose of this review is to show that uniform, thin and thick, native oxide layers and Al_2O_3/SiO_2 double-insulator layers can be grown on a-Si films at room temperature by anodic oxidation using electrolytes. The process is highly reproducible and electrically controllable. The oxidation process, insulator properties and applications to device fabrication are reviewed here.

---

* Department of Electrical Engineering, Faculty of Engineering, Hokkaido University, Kita-13, Nishi-8, Kita-Ku, Sapporo 060.

Anodic oxidation has been applied to MIS and p-i-n type a-Si solar cells for efficiency enhancement and material defect passivation. It is shown that anodization in the dark selectively passivates material defects by local defect oxidation, thereby enhancing the fabrication yields of successfully operating cells.

Planar a-Si MOS FETs using anodic Al_2O_3/SiO_2 double-insulator gate structures have been fabricated with reasonably good device performance.

Large area, low-cost and semi-transparent a-Si PSDs (position sensitive detectors) have also been fabricated, using a MIS type photo-sensing planar. Good position linearity has been obtained.

6.3.2 Anodic Oxidation Process and Properties of Oxides

Growth of native oxides of a-Si

a-Si : H films to be anodized are prepared either on stainless-steel or on c-Si substrates by rf glow-discharge decomposition of $SiH_4 + H_2$ gas mixture ($SiH_4 : H_2 = 1 : 9$) at a pressure of 2~5 Torr with a flow rate of 50 SCCM. The substrate temperature is 270°C and the rf power (13.56 MHz) is 20 W. Typical growth rate is 2.2 Å/s. n^+ and p^+ layers are obtained by adding PH_3 and B_2H_6 into the gas mixture, respectively.

The set-up for anodization is schematically shown in Fig. 6.3.1. The electrolyte is an ethylene glycol solution of KNO_3 with a concentration of 0.04 mol/l KNO_3. The anodization can be done either in constant voltage or in constant current modes or in their mixture, although only the result in the constant current mode is reported here. One of the key finding of the present study is that thick oxidation of a-Si : H i-layers can be made by providing sufficient illumination during anodization, whereas n^+ and p^+ layers are readily oxidized by anodization in the dark. The anode is thus illuminated by a tungsten lamp at 70000 lx for i-layer anodization. It is also found that anodization of i-layer in the dark

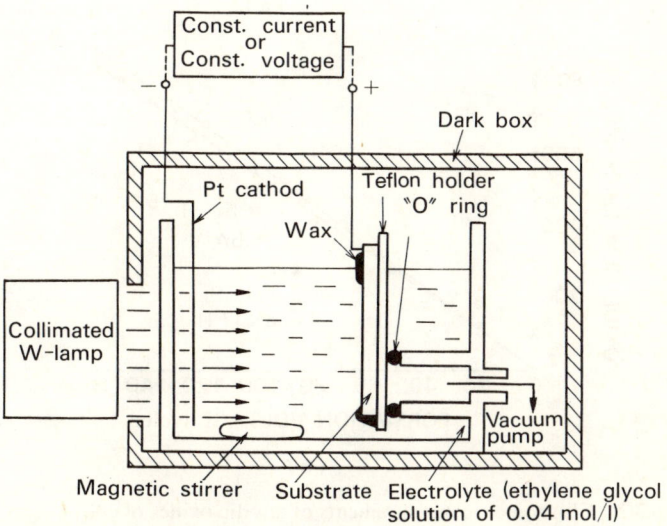

Fig. 6.3.1 Experimental set-up for anodic oxidation. [after S. Arimoto et al.[15]]

results in selective passivation of "weak" spots of the i-layer where material defects allow the current to flow in the dark and cause selective oxidation as described later.

Cell voltage vs. time curves for constant-current anodization of c-Si and a-Si : H are similar, showing linear increase of voltage with time, except for the initial portion where a more gradual increase of cell voltage is observed up to 20~30 V. The rate of the cell voltage increase vs. time in the linear region, is proportional to the current density for c-Si and a-

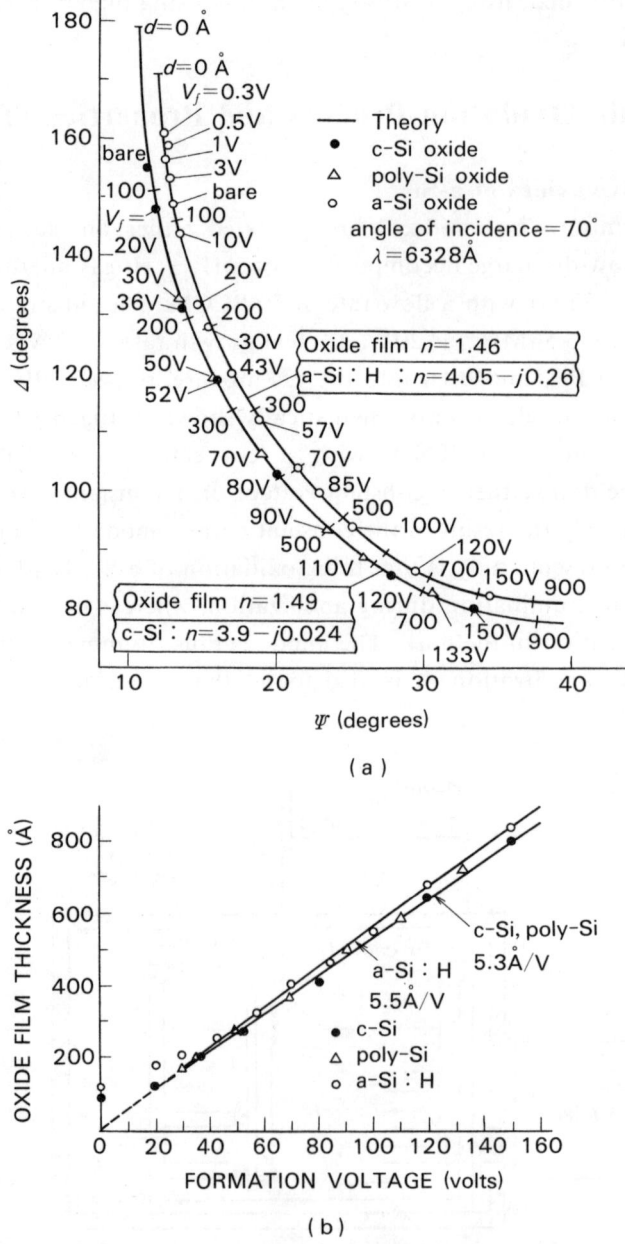

Fig. 6.3.2 Ellipsometric measurements of anodic oxides of c-Si, poly-Si and a-Si : (a) ellipsometric data at $= 6328$ Å (He-Ne laser) and (b) oxide thickness vs. formation voltage. [after H. Yamamoto et al.[3]]

Si : H. However, for a given current density the rate for a-Si : H surface is found to be smaller than that for c-Si, suggesting current concentration at material defects in a-Si : H.

Figure 6.3.2 (a) shows ellipsometric data and Fig. 6.3.2 (b) plots the oxide film thickness vs. formation voltage V_f for anodization of c-Si, poly-Si and a-Si : H.[3] The a-Si : H oxide film thickness changes linearly with V_f for $V_f \geq 50$ V, giving a thickness-voltage growth rate of 5.5 Å/V.[4] The value is close to the value for c-Si and poly-Si of 5.3 Å/V. A maximum thickness of a-Si : H oxide film of 2500 Å was obtained.

Formation of Al₂O₃/SiO₂ layer on a-Si

Al₂O₃/native SiO₂ double insulating layer was grown by anodization of an Al/a-Si : H structure. Al was deposited in vacuum onto the grown a-Si : H film at substrate temperature range of 80~150°C in order to enhance adhesion of the metal to the a-Si : H film. After Al deposition, the major part of the source and drain electrodes was covered by photoresist to avoid anodization. The electrolyte for anodization is AGW electrolyte,[5] which is a mixture of 3% aqueous solution of tartaric acid and propylene glycol (solution : glycol = 1 : 9). The anodization was done in constant current mode with current density of 3 mA/cm², which was supplied from an electrical contact made on deposited Al film. Since illumination was necessary for the anodization of a-Si : H i-layer, the anode was illuminated by W-lamp (70000 lx) during anodization. Resulting Al₂O₃ and SiO₂ thicknesses were 1200 Å and 60 Å, respectively.[6]

Properties of oxides

The properties of anodic native oxide and Al₂O₃ are summarized in Table 6.3.1.

Table 6.3.1 Properities of anodic oxides on a-Si : H films.

	SiO₂	Al₂O₃
Voltage growth rate (Å/V)	5.5	13~14
refractive index ($\lambda = 6328$ Å)	1.46	1.62
relative permitivity	3.2~3.5	8.1~8.7
resistivity (ohms − cm)	10^{12}	10^{15}
breakdown field strength (V/cm)	9~10 × 10^6	2~3 × 10^6

Material assessment by MOS structures

MOS structures allow standard C-V and DLTS measurement for interface state and bulk deep trap measurements. G-V characteristics in MOS FET structures also provide useful information concerning them. Figure 6.3.3 (a) shows examples of MOS C-V curves of a-Si MOS diodes, and the measured gapstate distributions, using both C-V and MOS

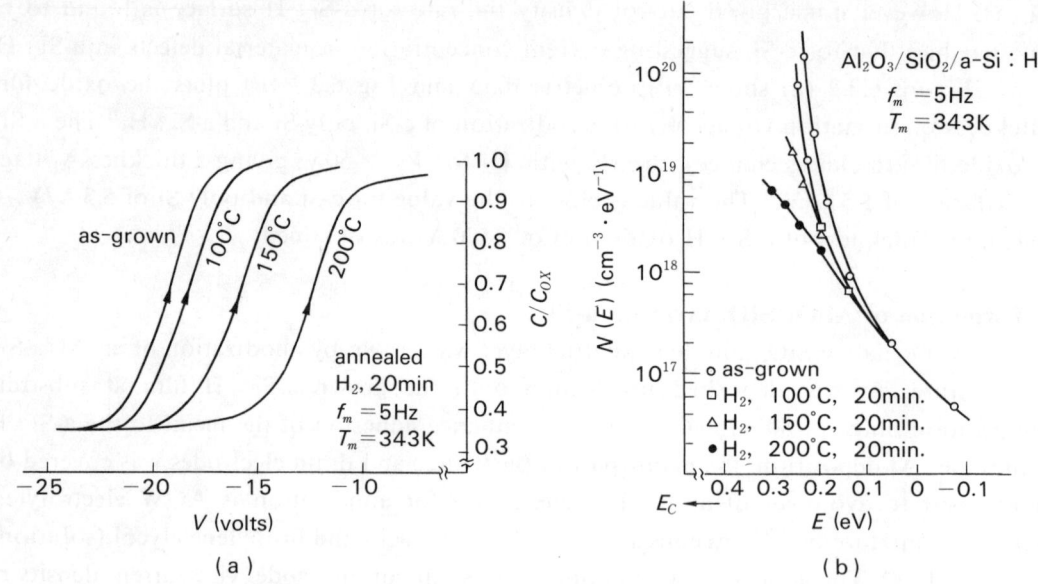

Fig. 6.3.3 Material characterization using anodic MOS structures: (a) MOS C-V curves and (b) measured gap state distribution. [after H. Yamamoto et al.[3]]

FET G-V data are shown in Fig. 6.3.3 (b).[3] Further work of this type may turn out to be useful for the understanding and control of the Staebler-Wronski effect.[7]

6.3.3 Solar Cells and Material Defect Passivation

a-Si MIS solar cells

MIS type solar cells are attractive because of their device simplicity and low processing cost.[8~10] Thin insulator films were grown on poly-Si and a-Si films, using the above described anodic oxidation process. For the purpose of comparison, Schottky cells on chemically etched surfaces using $HF + HNO_3 +$ acetic acid, were also prepared. Thin gold or aluminum barrier films (approximately 50 ohms/sq.), and thick metal comb patterns for current collection, were formed by vacuum deposition.

Amorphous Si films were produced by glow-discharge decomposition of SiH_4. Stainless steel substrates were used for a-Si films, and thin n^+-layers were grown by phosphine gas doping before deposition of i-layers, were deposited.

It was found that considerable increase of open-circuit voltage V_{oc} can be obtained by anodization in polycrystalline and amorphous cells without reduction of photocurrent.[11] For the anodization, there exists an optimum formation voltage, as shown in Fig. 6.3.4 (a). When the formation voltage is too high, anomalous reduction of photocurrent takes place. Otherwise, no systematic dependence of photocurrent on the process was observed.

Material defect passivation by anodization

Anodized poly-Si and a-Si surfaces were found to give much better yields of successfully operating MIS cells than as-grown or chemically polished surfaces. In the case of poly-Si cells, this can be attributed to selective oxidation of grain boundaries. For a-Si Schottky cells, the situation was much more severe, often resulting 0% yield. However, it was found that this yield can be dramatically enhanced to 100% by applying anodization in dark prior to cell fabrication.[11] The relevant mechanism was found to be passivation of pinholes or other defects in the i-layer owing to selective oxidation by dark leakage current. Since the i-layer is highly resistive in dark, leakage current is due to the n^+-layer being exposed through pinholes or other defects. By dark anodization, this current showed a rapid decay indicating selective oxidation. With the use of this technique, a large area a-Si MIS solar cell (3cm $\times$ 3cm) was successfully constructed, and its performance is shown in Fig. 6.3.4 (b).

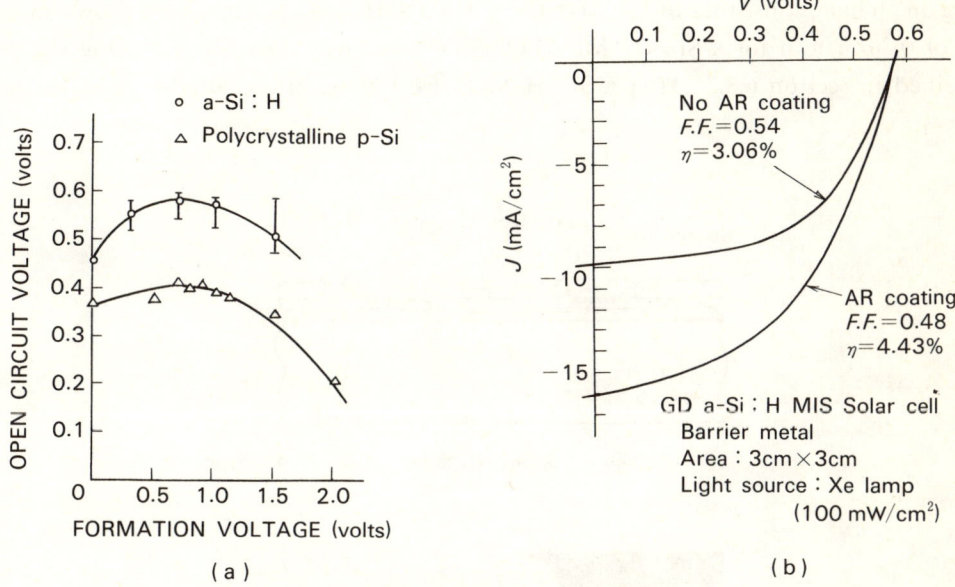

Fig. 6.3.4 Performance of a-Si : H anodic MIS solar cells : (a) V_{oc} vs. formation voltage and (b) characteristic of a large area (3cm $\times$ 3cm) MIS cell by defect passivation.

This type of process was found to also be useful for defect passivation in p-i-n a-Si solar cells. 36 p-i-n cell arrays were made on Corning 7059 glass substrate, each having a 0.5cm $\times$ 0.5cm area. Without the passivation process, the typical yield of operating cells was about 70%. By removing the metal electrode patterns and applying dark anodization, it was found that all the cells started to work under the optimized anodization condition.

6.3.4 Anodic Oxide Gate a-Si : H MOS FETs

Introduction

Recently, hydrogenated amorphous silicon field effect transistors (a-Si : H FETs) are receiving great interest because they can be used as integrated switching devices for image sensors and driving devices for large area displays.

Coplanar electrode structure a-Si : H MOS FETs using native oxide at the insulator/a-Si : H interface have been fabricated using anodic oxidation. Al_2O_3 (1200 Å)/native SiO_2 (60 Å) double layer structure grown by anodic oxidation process is used as the gate insulator in these FETs.

Fabrication process of a-Si : H MOS FETs

Figure 6.3.5 (a) shows a cross-sectional view of the fabricated FET. a-Si : H films were deposited on the glass (Corning 7059) by rf glow-discharge decomposition of a mixture of $SiH_4 + H_2$. Phosphine gas was added in the gas mixture to obtain an n^+-layer, which was used to make ohmic contacts between the source/drain electrodes and the a-Si : H film. The n^+-layer on the channel and device separation region was chemically etched off using an etchant consisting of $HF : HNO_3 : CH_3COOH = 3 : 5 : 15$, which shows an etching rate of $0.3 \mu m/min$ for a-Si : H. An Al_2O_3/SiO_2 layer was then formed using the process described in section 6.3.2. After a-Si : H MOS FET fabrication, annealing in H_2 gas at a

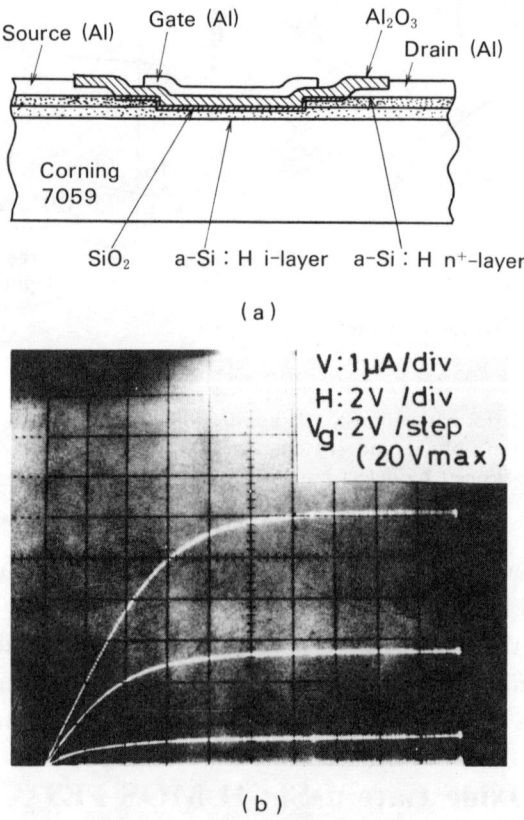

Fig. 6.3.5 Structure and performance of a-Si : H MOS FETs : (a) cross-sectional view, [after H. Yamamoto et al.[6]] and (b) output characteristics. [after H. Yamamoto et al.[3]]

temperature range of 160~230°C was done to improve the FET characteristics. The gate length and the gate width of the fabricated FET were 110 μm and 3.9 mm, respectively. The gate width was designed to be wide enough to obtain a reasonable drain current level, which is typical for a-Si : H FETs due to the low conductivity of the material.

FET performance

Figure 6.3.5 (b) shows the drain current, I_d, versus drain voltage, V_{ds}, characteristics of the fabricated a-Si : H MOS FET after annealing. The annealing was done in H$_2$ gas at 230°C for 15 min. The effective mobility μ_{eff} calculated from $\mu_{\text{eff}} = (L/W)(1/C_{ox})(I_d/V_g)(1/V_{ds})$ was typically 0.1~0.2 cm^2/V·s. The annealing process greatly enhanced the effective mobility by a factor of 2~10, which was typically 0.002~0.05 cm^2/V·s prior to annealing. It is believed that this improvement is due to the reduction by annealing of interface state densities at the insulator/a-Si : H interface.[3] FETs with higher mobility should be obtained by further optimization of the fabrication processes.

6.3.5 a-Si : H MIS Position Sensitive Detector

Introduction

Position sensitive detectors (PSDs) are photodetectors using a lateral photoeffect to detect the position of an incident light beam on the detectors. PSDs using crystalline materials, mainly crystalline silicon, have already been realized and used since the first report by Wallmark.[12] Crystalline materials are, however, unsuitable for large-area PSDs. Advantages which arise from the use of amorphous silicon for the PSD material compared with PSDs using crystalline materials include (1) low-cost, (2) large-area capability, and (3) flexibility of the choice of the substrate shape. In addition, the semi-transparent nature of the thin a-Si films on glass substrates makes it possible to fabricate an angle detecting PSD using one PSD on another to detect the angle of the incident light. An angle detecting PSD has a variety of potential applications such as sensors for manufacturing robots.

The anodic oxidation process has been employed to fabricate MIS PSDs as well as to enhance the fabrication yield of the detectors. The MIS structure offers high break-down voltage compared to the p-i-n structure, which results in higher speed, increased noise immunity, and enhanced design flexibility.

Fabrication processes of a-Si : H MIS PSDs

Cross-sectional views of the one-dimensional and two-dimensional dual-axis duolateral type PSDs are schematically shown in Fig. 6.3.6 (a) and (b), respectively. Corning 7059 glass was used as the substrate. Two parallel extended lateral Al electrodes at opposite sides and a thin Au-Cr film, which is the lower part resistive layer of a two-dimensional PSD, were deposited by vacuum evaporation. ITO films were deposited by rf sputtering to prevent diffusion of Au or Cr to the a-Si : H films. The sheet resistivity of ITO films is high enough so that there is only negligible effect on the sheet resistance of the resistive layer. In the case of the one-dimensional PSD, ITO films were deposited directly on the substrate to form ohmic contacts with the a-Si : H n$^+$-layer grown on top of them. The thickness of the i-layer was 6000 Å and that of the n⁺-layer was 400 Å for both the one- and the two-dimensional PSDs. The PSD to be anodized was fixed on a teflon holder with high-quality wax. The electrolyte is an ethylene glycol solution of 0.04 mol/l KNO_3. Two step anodic oxidation in the electrolyte was done prior to Au deposition. The first oxidation, which utilizes the current crowding enhanced oxidation of the defects,[11,13] was done in dark to passivate the

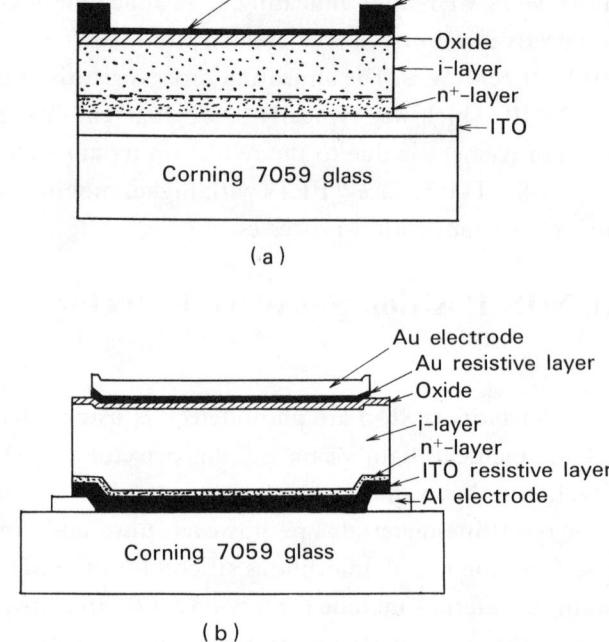

Fig. 6.3.6 Structure of a-Si : H PSDs : (a) one-dimensional PSD [after S. Arimoto et al.[14]] and (b) two-dimensional PSD. [after S. Arimoto et al.[15]]

material defects. This oxidation was done in the constant voltage mode to provide enough passivating effect. The second oxidation was done under illumination (W-lamp : 70000 lx) to grow a thin oxide layer of about 60 Å to form an MIS structure. This oxidation was also done in constant voltage mode and the formation voltage was chosen to be 1.0~1.5 V. Last, thin Au films and two parallel extended lateral Au or Al electorodes at opposite sides were deposited by vacuum evaporation. The thin Au films function not only as barrier metal but also as resistive layers.

PSD performance

The anodic oxidation process in the dark greatly enhanced the fabrication yield of the PSDs from 0~10% to 80~90%. It is found that this process is very useful for fabricating large-area a-Si : H MIS PSDs. The MIS structure by anodic oxidation offers high breakdown voltage, more than 10 V (best data : 15 V) in the present a-Si : H MIS PSDs.

Current I_{x-} and I_{x+} in Fig. 6.3.7 (a) were used to calculate the light position. For

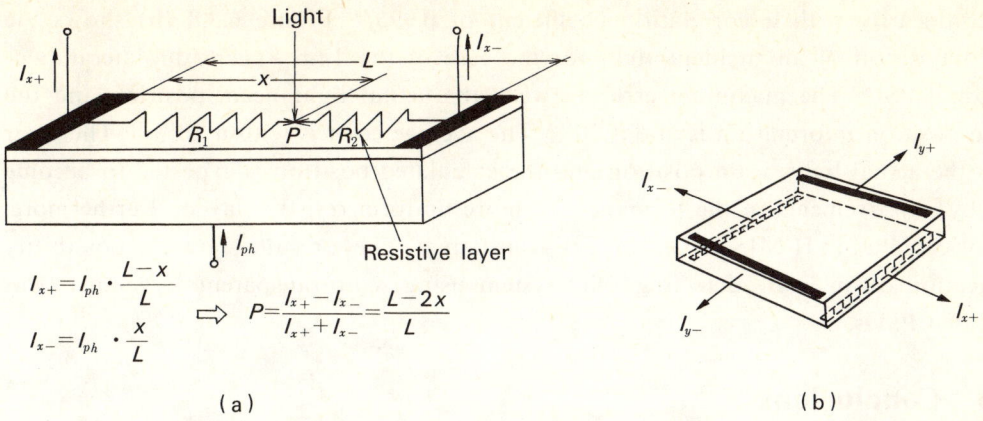

Fig. 6.3.7 Detection of an incident light position: (a) one-dimensional PSD, [after S. Arimoto et al.[14]] and (b) two-dimensional PSD. [after S. Arimoto et al.[15]]

two-dimensional PSDs, additional current I_{y-} and I_{y+} in Fig. 6.3.7 (b) were also used. The incident light beam position was calculated by using the following equations.

$$P_1 = (I_{y+} - I_{y-})/(I_{y+} + I_{y-})$$
$$P_2 = (I_{x+} - I_{x-})/(I_{x+} + I_{x-})$$

This way, the light position is stably detected because fluctuation of the light beam intensity and background illumination intensity have no effect on the detected position. A GaAlAs LED (1.7 mW/cm²) was used as the incident light source. Figure 6.3.8 (a) shows the lateral photocurrent response of the fabricated 3 mm × 26 mm one-dimensional PSD, which shows

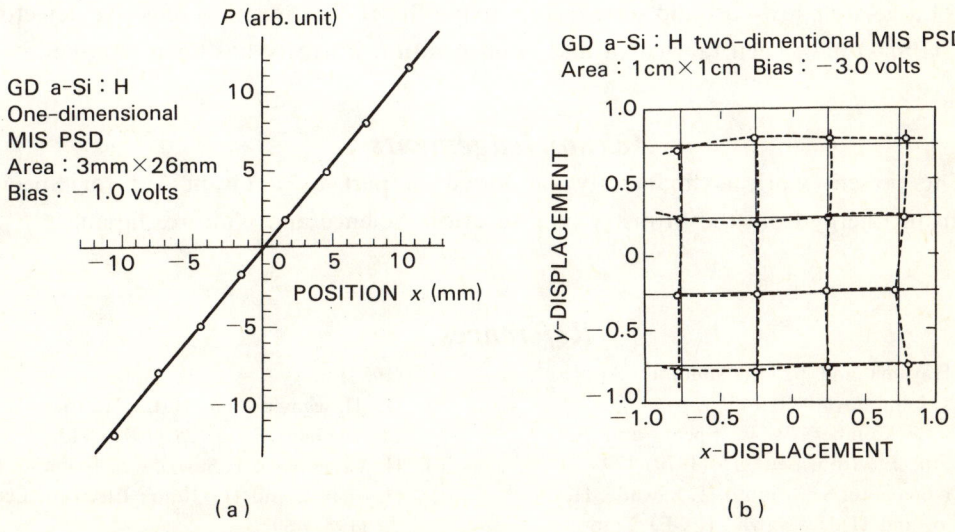

Fig. 6.3.8 Performance of a-Si : H PSDs : (a) one-dimensional PSD, [after S. Arimoto et al.[14]] and (b) two-dimensional PSD. [after S. Arimoto et al.[15]]

excellent linearity with a correlation coefficient of 0.996.[14] Figure 6.3.8 (b) shows the measured position of an incident light for the case of the 1cm × 1cm duolateral two-dimensional PSD. The maximum error between the actual light beam position and the obtained position information is about 20%. The average error is within 10%.[15] The error between the actual light beam position and the calculated position is expected to become smaller by improvements in the formation of more uniform resistive layers. Furthermore, these fabricated a-Si : H MIS PSDs are semi-transparent. These results show the possibility of fabrication of an angle detecting PSD system using semi-transparent large-area two-dimensional PSDs.

6.3.6 Conclusions

Insulator formation on a-Si films by anodic oxidation and its application to device fabrication are reviewed, and the following conclusions are obtained :

(1) By anodic oxidation, uniform thin and thick (up to 2500Å) SiO_2 and Al_2O_3/SiO_2 layers can be grown on a-Si films at room temperature in a reproducible and controllable way.
(2) Anodic MOS structures are useful for material assessment of a-Si : H films.
(3) Insertion of a thin anodic native oxide layer can significantly increase the open-circuit voltage of Schottky-type a-Si solar cells without reducing photocurrent.
(4) Anodization in the dark can selectively passivate material defects in a-Si : H films and therefore enhance the area and fabrication yield of normally operating MIS and p-i-n a-Si solar cells.
(5) Planar a-Si : H MOS FETs can be fabricated using anodic Al_2O_3/SiO_2 gate insulator. A maximum channel mobility of $0.1 \sim 0.2 cm^2/V \cdot s$ has been obtained. Such a technology seems promising for fully planar integration of functional devices on a-Si films.
(6) Large-area, low-cost and semi-transparent a-Si : H MIS position sensitive detectors (PSDs) have been fabricated and good position linearity has been obtained.

Acknowledgements

The present work is financially supported in part by a Grant-Aid for Special Research on Energy from the Ministry of Education, Science and Culture, Japan.

References

1) H. Hayama and M. Matsumura : Appl. Phys. Lett., *36* (1980) 745.
2) P. G. Le Comber, W. E. Spear, and A. Ghaith : Electron. Lett., *15* (1979) 179.
3) H. Yamamoto, S. Arimoto, T. Sawada, H. Ohno, and H. Hasegawa : IECEJ Trans., *ED 83-68* (1983) 35 [in Japanese].
4) H. Yamamoto, S. Arimoto, H. Hasegawa, H. Ohno, and J. Nanjo : Electron. Lett., *19* (1983) 6.
5) H. Hasegawa and H. L. Hartnagel : J. Electrochem. Soc., *123* (1976) 713.
6) H. Yamamoto, T. Sawada, S. Arimoto, H. Hasegawa, and H. Ohno : Electron. Lett., *19* (1983) 607.
7) D. L. Staebler and C. R. Wronski : Appl. Phys. Lett., *31* (1977) 292.
8) H. Hasegawa, S. Tamori, and T. Sawada :

Proc. 11th Conf. Solid State Devices, Tokyo, 1979, Jpn. J. Appl. Phys., *19* (1980) Suppl. 19-1, 1089.

9) J. Nanjo, H. Yamamoto, and H. Hasegawa: Bulletin of Fac. Eng., Hokkaido Univ., No. 1055 (1981) 65 [in Japanese].

10) H. Yamamoto, M. Moniwa, T. Sawada, and H. Hasegawa; Proc. 2nd Photovoltaic Science and Engineering Conf. in Japan, 1980, Jpn. J. Appl. Phys., *20* (1981) Suppl. 20-2, 87.

11) H. Yamamoto, M. Moniwa, and H. Hasegawa : Proc. 3rd Photovoltaic Science and Engineering Conf. in Japan, 1982; Jpn. J. Appl. Phys., *21* (1982) Suppl. 21-2, 53.

12) J. T. Wallmark : Proc. IRE, *45* (1957) 474.

13) M. Moniwa, S. Arimoto, and H. Hasegawa : Proc. 3rd Photovoltaic Science and Engineering Conf. in Japan, 1982; Jpn. J. Appl. Phys., *21* (1982) Suppl. 21-2, 57.

14) S. Arimoto, H. Yamamoto, H. Ohno, and H. Hasegawa : Electron. Lett., *19* (1983) 628.

15) S. Arimoto, H. Yamamoto, H. Ohno, and H. Hasegawa : Extended Abstracts of 15th Conf. on Solid State Devices and Materials, Tokyo, 1983, 197.

CHAPTER 7

OPTO-ELECTRONIC APPLICATIONS OF AMORPHOUS SEMICONDUCTORS

7.1 Optical Memories

Mutsuo TAKENAGA* and Masanari MIKODA*

Abstract

This review is related to recent Japanese developments in the field of optical mass memories applicable to optical disc memories. Research on memory materials has been centered on lessening the oxidation degradation of recording thin films and developing highly sensitive materials or disc structures for laser light.

Tellurium thin film containing Se has been analytically investigated and it has been found that the TeO_2 surface layer and Se-rich inner layer will prevent further oxidation degradation of the film. A double layer SbSe/BiTe film showing great changes in optical constants without any mechanical deformation resulting from laser heating has been developed. Some disc structures using amorphous thin film for multilevel recording or optical tracking have also been presented.

A reversible change in the reflectivity of $Te-TeO_2$ film containing a small amount of Ge and Sn was observed, and it offered the potential to achieve an erasing facility. Magneto-optical films, amorphous GdTbFe, TbFe and also CrO_2 film, all with erasable memories, are presented. Newly developed disc structure enhancing magneto-optic rotation is reviewed.

7.1.1 Introduction

In recent years, the first stage of the wide application of optical disc memories capable of storing very large quantities of information along with extremely rapid random access, such as video files, document files, data files and so on, has started. At the present time, the write-once type of recording medium for archive-like storage of information is commercially available. There are two types of write-once medium applicable to laser heat-mode recording. Using laser beam heating, one type makes pits on thin film through a process of melting or ablation, while the other changes its optical constants, refractive index and reflectivity without any mechanical deformation.

Recent research and developments concerning optical disc memories have been focused on improving material life shortened by oxidation degradation of the recording thin film and sensitivity to the laser diodes. Many trials have been made to lessen the humidity degradation of Te based recording film by adding some stabilizing additives such as carbon,[1,2] selenium[3] and CS_2.[4] The fundamental characteristics of these materials were reviewed last year.[5]

---

* Central Research Laboratories, Matsushita Electric Industrial Co., Ltd.

The mechanism of resistance to oxidation of Te-Se film and the effects of the additives have been clarified.[6] To obtain a new high performance medium, new materials and disc structures which reach sensitivities high enough to be applicable to video files have been studied. A double layered SbSe/BiTe film[7,8] shows great changes in optical constants. Heat increases the refractive index of the SbSe layer, while that of the light-absorbing layer, BiTe, decreases. By combining these films the optical change should be intensified. A multi-layered medium composed of an amorphous coloring reagent, light absorber and coupling reagent has possible multilevel recording applications.[9] An amorphous As-Se-S-Ge film[10] is found to be applicable to optical guide tracking because it has fewer drop-outs, instead of a conventional plastic groove.

Furthermore, erasable media have attracted great interest because of their possibility of replacing magnetic disc media. Tellurium oxide based film containing a small amount of Ge and Sn was found to have excellent erasability using irradiation by laser beams with different power densities.[11,12] An optical head having two laser beams, one for recording and playback and one for erasing, has been developed.

Descriptions of magneto-optical materials[13~15] are also included because of their importance as erasable optical memories.

7.1.2 Optical Memory Material

Te-Se film

It has already been reported that recording thin film composed of Te and Se has high performance. Here, details of the tests on Te-Se thin film resistance to oxidation and the effects of additional metallic elements such as Pb, Bi, Sb, Sn, In and As on microcrystal growth[6] will be discussed.

The Te-Se thin film, 30 nm thick, was deposited on a disc substrate rotating at a high speed. Component elements of the recording film were evaporated from several evaporation boats. The deposited film did not have a layered structure, but had a uniform one because the rotational speed of the substrate was high.

The change in light transmittance due to thin film oxidation was studied by changing the ambient temperature and humidity for various compositions of the Te-Se alloy. The results are shown in Fig. 7.1.1. The experiments were performed by making the temperature-humidity conditions severer step by step. In the pure Te thin film, light transmittance change resulting from oxidation gradually occurred even at 25°C and 70% relative humidity. On the other hand, in the case of thin films which contain 14% or more Se, no change in light transmittance was observed even at 60°C and 95% RH.

The depth profile of elements in Te-Se thin films was studied using Auger electron spectroscopy (AES) and X-ray photo-electron spectroscopy (XPS). The result of the AES is shown in Fig. 7.1.2. The Te oxides are formed near the surface, and a relatively Se-rich part exists on the inner side of the oxide layer.

The role of Se in oxidation prevention is considered to be as follows. Tellurium atoms near the surface are ionized in the process of oxidation and move toward the surface, leaving Se atoms behind. As a result, a layer which has a relatively high density of Se is

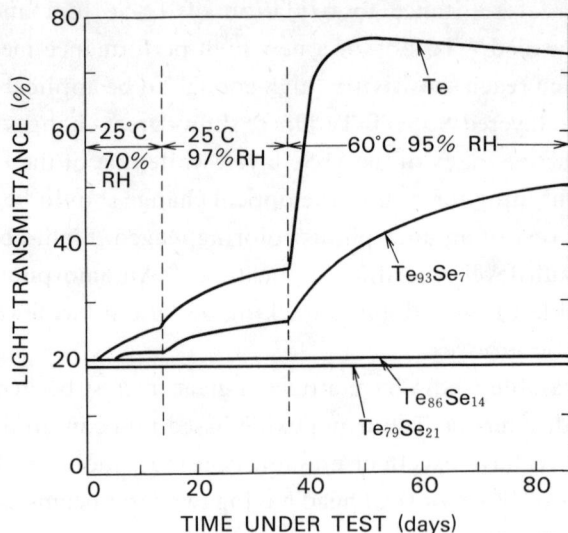

Fig. 7.1.1 Relation between the composition of the thin film and the light transmittance change. [after M. Terao et al.[6)]]

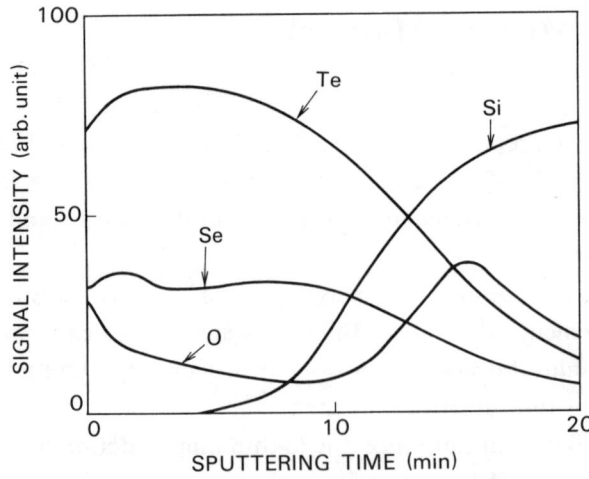

Fig. 7.1.2 Depth profile of elements in a Te-Se thin film. [after M. Terao et al.[6)]]

formed on the inner side of an oxidized Te layer. In Te oxides, diffusion of Se presumably is difficult, though mutual diffusion of Te and Se occurs easily. It can be considered that the Se-rich layer and the Te oxides prevent Te ions from moving toward the surface of the thin film.

More than a 10 atomic % addition of either Pb or Sb or Bi to the Te-Se thin film had the effect of making the crystal grains smaller, and as a result decreased disc noise. When more than 10 atomic % of Sn or In was added to the Te-Se thin film, the thin film did not crystallize when heated at 80°C for 10 hours. The Te-Se thin film containing As was amorphous just after deposition, but in many samples, large crystal grains appeared when heated at 60°C.

SbSe/BiTe film

A new optical recording material made of vapor-deposited Sb_2Se_3 film has been developed.[7,8] This film shows great change in optical characteristics without any mechanical deformations from laser irradiation or thermal treatment below 200°C. Since the physical disc form is never affected by heating, the recording material can be directly sealed by a protective layer, which ensures higher reliability.

The absorption of Sb_2Se_3 film is very low at a wavelength of 830 nm. Recording sensitivity was enhanced by a unique double-layer structure consisting of a recording layer (Sb_2Se_3) and a light-absorbing layer (Bi_2Te_3).

The as-deposited Sb_2Se_3 thin film has an amorphous phase. After annealing above 170°C, it reveals the growth of microcrystals with a grain size of about one micron in diameter. In order to determine the amorphous-crystalline transition temperature, thermal analysis was carried out. The result is shown in Fig. 7.1.3. The crystallization temperature is 170°C, where an exo-thermic reaction took place. The two broad peaks which appeared at higher temperatures were identified with the oxidation of Sb_2Se_3 since these peaks did not appear in a nitrogen atmosphere. The TG curve in the figure shows no weight loss at the crystallization temperature, i. e., sublimation did not occur.

This physical phase change causes a change in the optical characteristics of the Sb_2Se_3 film in that the refractive index, $n + ik$, changes from $3.8 - i0.1$ to $4.7 - i0.8$ following crystallization.

On the other hand, light-absorbing Bi_2Te_3 thin film has optical characteristics opposite those of Sb_2Se_3. The refractive index of Bi_2Te_3 changes from $2.8 - i4.8$ to $2.6 - i3.0$ following heat treatment above 170°C. By combining these films to form the recording layer, the optical change should be intensified.

To investigate thermal transition in the reflectivity of the double layer, a sample

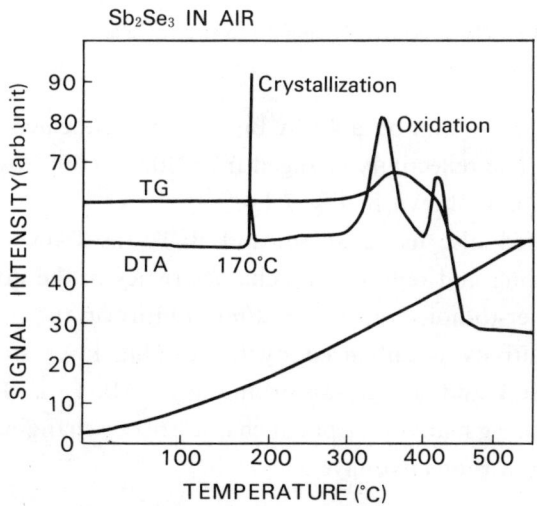

Fig. 7.1.3 Thermal analysis of the SbSe film. [after Y. Aoki et al.[8]]

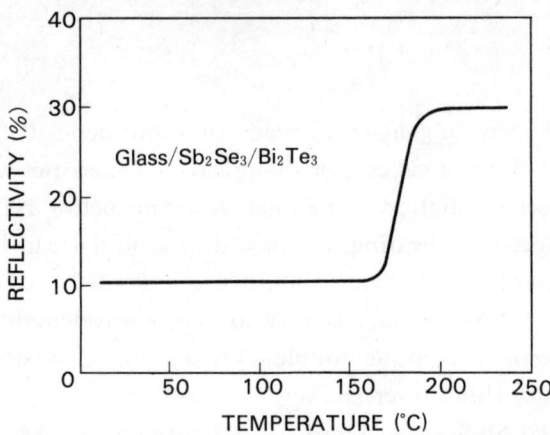

Fig. 7.1.4 Temperature dependence of reflectivity of a 400 Å Sb_2Se_3 film on a 400 Å Bi_2Te_3 film. [after Y. Aoki et al.[8]]

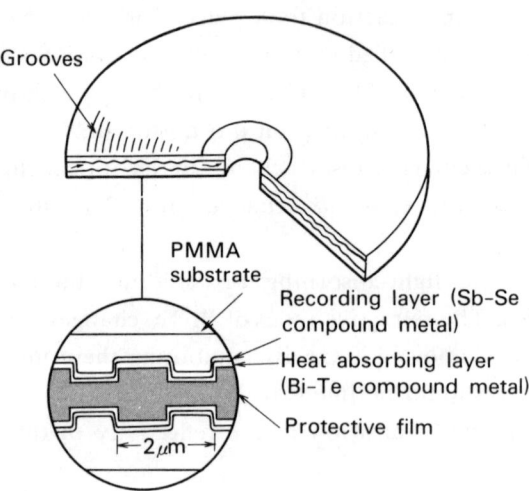

Fig. 7.1.5 The structure of a double-sided disc. [after Y. Aoki et al.[8]]

consisting of a 400 Å Sb_2Se_3 film on a 400 Å Bi_2Te_3 one was heated up to about 300°C at a constant heating rate. The reflectivity changed from 10% to 30% over a temperature range of 20°C (170°C~190°C), as shown in Fig. 7.1.4.

A double layered disc using Sb_2Se_3 and Bi_2Te_3 as shown in Fig. 7.1.5 has been developed. The recording and reproducing characteristics of the double layered disc were investigated. The carrier-to-noise ratio at a 60 mm radius on the disc was over 60 dB.

Recording sensitivity is enhanced by the double layer structure consisting of a recording layer (Sb_2Se_3) and a light-absorbing layer (Bi_2Te_3). Since disc recording is facilitated by the recording material's optical characteristics changing, the signal surface can be directly sealed with a protective layer.

Multilevel recording medium

A new recording medium, a multilayer medium with thermal coloration, has been developed.[9] This medium is applicable to multicolor micrographic hard-copy and multi-level optical memory systems.

An elementary unit of the medium is composed of three layers of a coloring reagent, a light absorber, and a coupling reagent. The coloring reagent and coupling reagent react with each other due to laser beam irradiation, and coloration occurs. Each layer is made by vacuum deposition. Each layer must be in an amorphous state so that recording and reading light can pass through the medium without scattering.

The reagents used here are tabulated in Table 7.1.1. Up to now, four colors (red, green, blue, and black) have been attained. The coloration mechanism of this medium is the

Table 7.1.1 Reagent for multilayer recording medium. [after S. Oikawa et al.[9]]

Coloring reagent	Leuco dye	Crystalviolet lactone	Blue
		QZ-1017 *	Red
		TH-107 *	Black
		3-ditoluilamino-7-diethylaminofurane	Green
Light absorber		Vanadilphthalocyanine	
		Te	
Coupling reagent	Phenolic acid	Phenolphthaleine	
		Thymolphthaleine	

* Trade name of Hodogaya Chemical Co. Ltd., Japan.

A_1 : leuco dye (red), 5 μm
D_1 : light absorber, 150Å
B : coupling reagent, 10 μm
D_2 : light absorber, 150Å
A_2 : leuco dye (green), 1 μm
substrate

$A_1 + B \rightarrow$ red, $A_2 + B \rightarrow$ green
D_1, D_2 : vanadilphthalocyanine
B : phenolphthaleine

Fig. 7.1.6 Structure of two color recording medium. [after S. Oikawa et al.[9]]

Fig. 7.1.7 Recorded pattern on multicolor recording medium. [after S. Oikawa et al.[9]]

same as that in thermal printing paper.

The structure of the recording medium for red and green two-color recording is shown in Fig. 7.1.6. Focusing a laser beam on the D_1 layer results in red, while for D_2 the result is green. An example of a recorded pattern on two-color medium is shown in Fig. 7.1.7. The recorded color is red or green depending on the focusing depth.

The above points are applied to recording media composed of the same kinds of light absorbers and coloration reagents. If several kinds of absorbers and coloration reagents are used, layers with different light absorbers or coloration reagents are independently recorded or read. In addition, if a visible light source, such as a tunable dye laser, is used as a recording light source, many narrower absorption peak dyes are available as light absorbers. This medium's wavelength degree of multiplex recording is shown in Fig. 7.1.8. Multilevel recording may be possible about 3000 times when the coloration contrast, ΔT, is 2.5%, the reading light multiplex degree, N_r, is 8, and the allowable reading light transmittance decrease, T, is 1%.

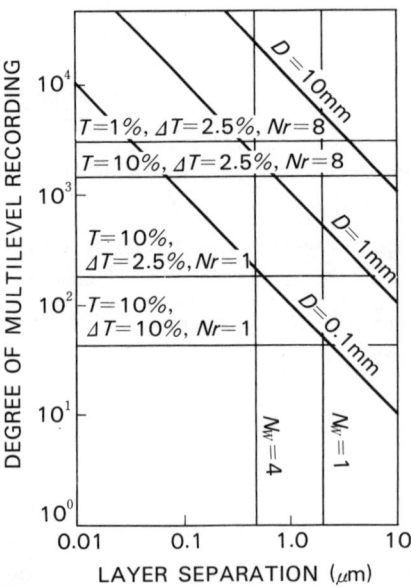

Fig. 7.1.8 Total degree of multilevel recording.
T : allowable reading light transmittance decrease
ΔT : coloration contrast
N_r : reading light multiplexing degree
N_w : writing light multiplexing degree
D : total layer thickness
[after S. Oikawa et al.[9)]]

Optical tracking guide using amorphous film

Optical disc tracking guides are considered to be indispensable for achieving high recording density and accurate random access. Optical discs with conventional tracking guides are manufactured through complex processes, namely, optical beam recording of a master disc, nickel electro-plating to make stampers, groove transfer to substrates by a compression or injection method, and deposition of a recording film. These complex processes often cause defects which lead to drop-outs or bit errors. A new tracking guide fabrication method using a photo-induced refractive index change accompanied by photo-darkening of the As-Se-S-Ge amorphous film has been developed.[10]

The structures of the conventional and new tracking guides are shown in Fig. 7.1.9. In the new method, the disc can be formed directly on a disc substrate without transferring processes. The surface of the disc is flat and optically equivalent to the conventional pre-grooved disc. Therefore, the new disc with photo-darkened As-Se-S-Ge tracking guides is expected to have fewer defects than in the conventional one.

The tracking operation of the disc shown in Fig. 7.1.9 (b) can be performed by detection of the phase difference between the reflected light from tracks and that from non-track areas. The reflected light from the film through the substrate suffers a complex phase change due to multiple interference in the As(10)Te(90) recording film and As-Se-S-Ge film. Therefore, the reflectivity and the phase difference depend on the refractive index and thickness of these films.

Using RF sputtering, As-Se-S-Ge film about 5000Å thick was deposited on a glass substrate. After heat-treatment of the film at 190°C for 20 min., circular tracking guides were recorded using an Ar laser ($\lambda = 5145$Å). Then, As-Te recording film 400Å thick was evaporated on it.

Tracking characteristics were measured by the differential phase detection method using a diode laser. The observed tracking signals were similar to that of pre-grooved discs, but the signal intensity was a little lower than that of pregrooved discs with a depth of 700 Å and a width of 0.8 μm. This may be caused by the fact that the thickness of the As-Se-S-Ge film was about half of the optimum obtained by calculation.

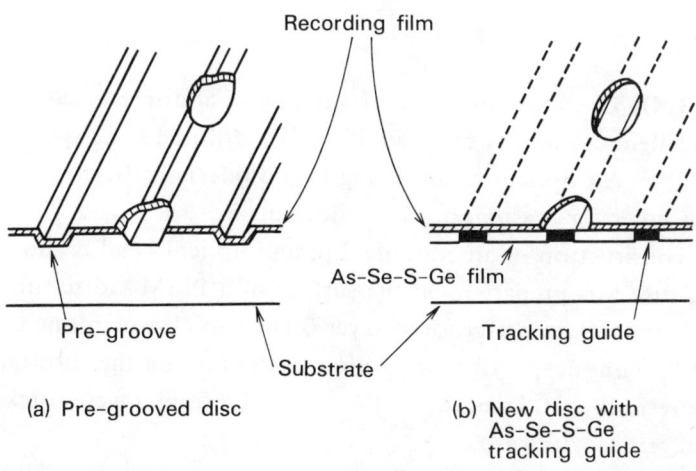

Fig. 7.1.9 Structures of conventional pre-grooved disc and new disc with As-Se-S-Ge tracking guide. [after M. Miyagi et al.[10]]

Erasable memory using TeO_2-Te

Oxide based thin films, such as Te-O, Ge-O, Sb-O[16,17] were already reviewed last year.[5]

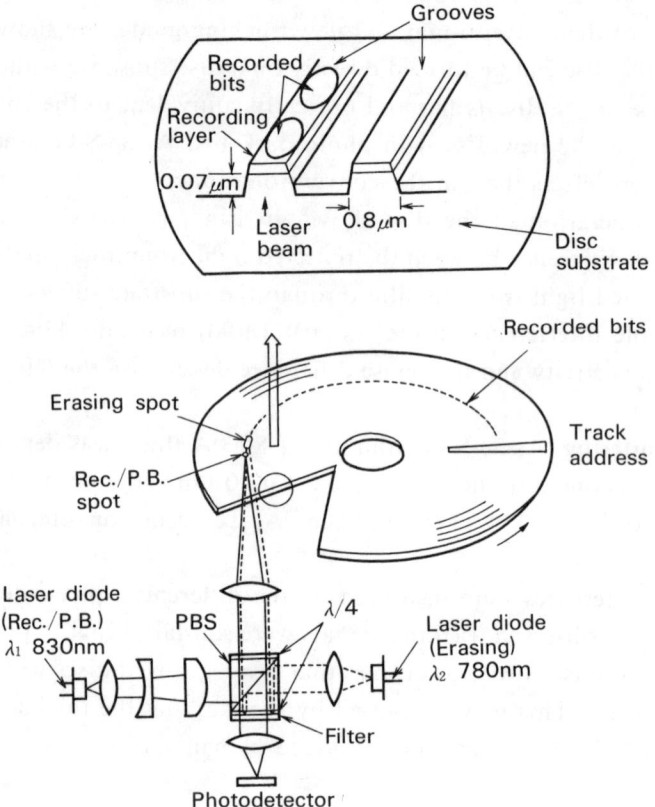

Fig. 7.1.10 Basic construction of an erasable optical disc and head. [after N. Yamada et al.[11]]

Recently, TeO_2-Te containing a small amount of additives such as Sn and Ge was found to have excellent erasability by irradiation of a diffused laser spot and a diffraction-limited laser spot.[11,12] An optical head having two diode laser beams, one for recording/playback and the other for erasing, has been developed.

The basic construction of an erasable disc and optical head is shown in Fig. 7.1.10. A reflection type disc was prepared. On the surface of a PMMA disc substrate 200 mm in diameter and 1.1 mm thick, a pre-grooved layer 0.1 mm thick was formed. The groove was 0.8 μm wide and 0.07 μm deep. After deposition of the film on the substrate, it was coated with a protective resin layer 0.1 mm thick. Two such disc frames were stuck together so that their protective layers face inwards.

A diffraction-limited laser beam 0.8 μm in diameter and $\lambda = 830$ nm was used for recording and playback. In addition, a diffused laser beam with an elliptical shape, 1×10 μm and $\lambda = 780$ nm, was used for erasing. The cw erasing beam irradiates the thin film, and it is immediately followed by the modulated recording beam.

A diffused laser beam having an elliptical shape as shown in Fig. 7.1.11 was used. This erasing beam causes a micro-area on the film to heat and cool rather slowly, and this

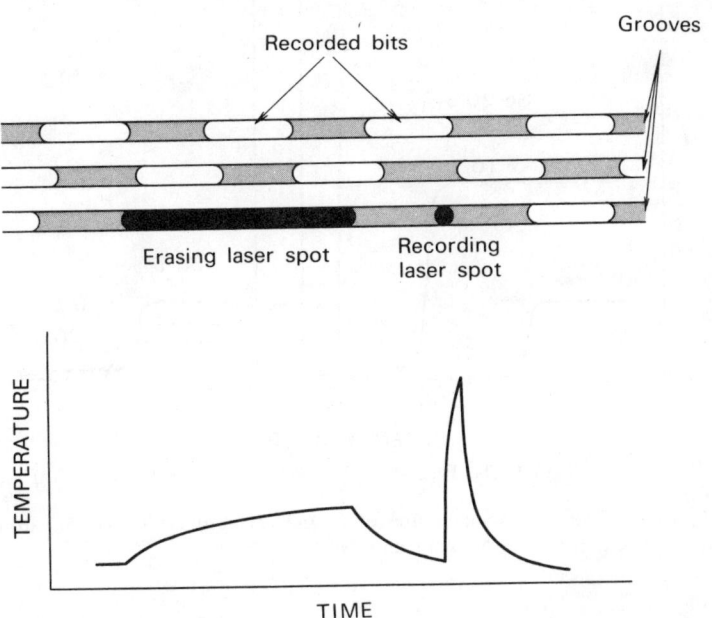

Fig. 7.1.11 Schematic model of heating-cooling by the two laser system. [after N. Yamada et al.[11]]

action results in an increase in the reflectivity of the area. The recording beam which follows must be focused to a diameter as small as possible in order to keep the recording at the highest possible density. Also, a short irradiation time causes sudden melting and rapid quenching, which results in a decrease in the reflectivity to the former level. The use of such a two laser optical head for erasing and recording permits real-time recording while erasing.

Samples of TeO_2-Te thin film containing a small amount of Sn and Ge were prepared by evaporation using several sources of TeO_2, Te and additives. Both the thermal property and erasability of the TeO_2-Te (Ge, Sn) thin film depend on the content of Sn and Ge in the film. The effect of the additives on these properties was examined.

Changes in optical transmission and reflectivity at a wavelength of 830 nm when the samples of TeO_2-Te (Ge) and TeO_2-Te (Sn) deposited on Pyrex substrate were heated up to 300°C at a rate of 100°C/min. are shown in Fig. 7.1.12. Each curve has a critical transition temperature, above which the transmission markedly decreases. At the same time, the reflectivity changes in intensity in the same way. This transition temperature depends on the Ge content and rises from 130°C for 5% Ge to 280°C for 20%. The addition of Sn to TeO_2-Te, in contrast, produces no significant change in the transition temperature.

As a result of this, adding Ge to TeO_2-Te produces a rise in the transition temperature. However, this results in a significant decrease in sensitivity to laser light, though it

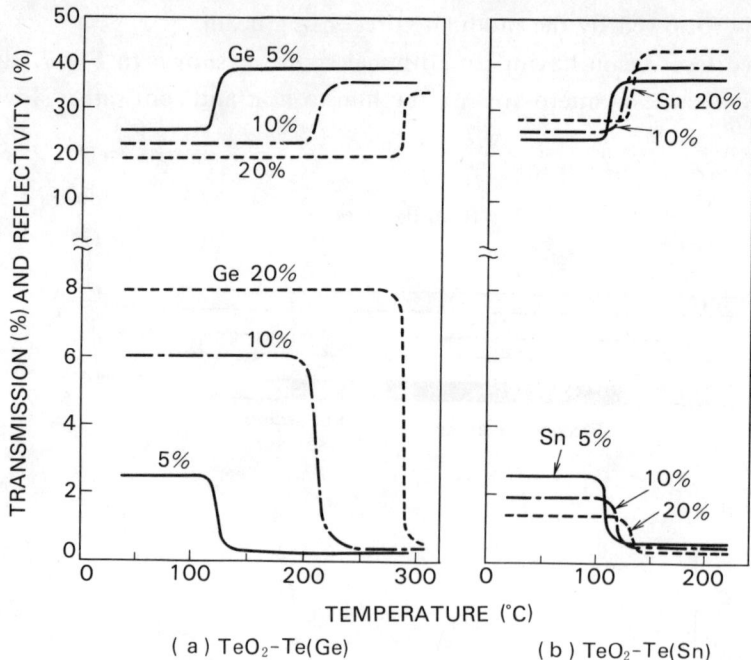

Fig. 7.1.12 Transition temperatures of TeO$_2$-Te (Ge) and TeO$_2$-Te (Sn) thin films. [after N. Yamada et al.[11]]

becomes much more resistant to heat degradation. The conclusion reached, therefore, was that 5% Ge is the most preferable from the viewpoint of both sensitivity and stability.

The relationship between the irradiation time of laser light on TeO$_2$-Te (Sn) thin films and changes in their reflectivities is shown in Fig. 7.1.13. The results when films with low reflectivity were irradiated by laser light with a power density of 0.8 mW/μm^2 and various pulse widths are shown in Fig. 7.1.13 (a). The reflectivities increased with the irradiation time. In addition, each curve shows saturation beyond a certain irradiation time. When the samples saturated in terms of their reflectivities were re-irradiated with a high power density laser pulse, as shown in Fig. 7.1.13 (b), the reflectivity decreased and returned to the starting level in the 5 and 10% Sn samples, with a slight recovery in the 20% Sn sample. The same threshold re-irradiation time, about 20 ns, was observed with various Sn content, in contrast to Fig. 7.1.13 (a). As mentioned above, TeO$_2$-Te thin film containing a small amount of Sn shows an increase in reflectivity with irradiation from a relatively weak and long laser pulse, and it shows a decrease with an intense and short pulse. As far as the results are concerned, the sample showing appreciable efficiency in both contrast in reflectivity change and sensitivity to short pulses contained 10% Sn.

A microscopic photograph of the recorded bits is shown in Fig. 7.1.14. The white lines indicate the grooves in the area as deposited. The three sets of black lines show the grooves irradiated by the erasing laser beam with the elliptical shape. Relatively wide zones are darkened due to the thermal diffusion of laser energy. The white points in the middle set of black lines show the recorded bits of a 2 MHz binary signal. The length of a bit is about 3 μm. In order to ascertain the real time erasing, a modulated erasing laser beam at

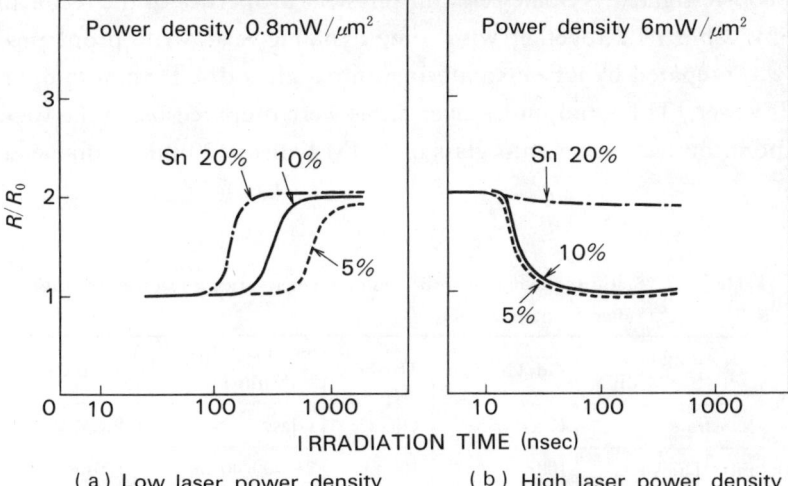

Fig. 7.1.13 Response characteristics of TeO$_2$-Te (Sn) thin films to laser pulse. (a) low power density irradiation, (b) high power density irradiation to the samples with saturated reflectivity by laser irradiation in (a). R_0 and R are reflectivities of starting level and respectively after being irradiated. [after N. Yamada et al.[11]]

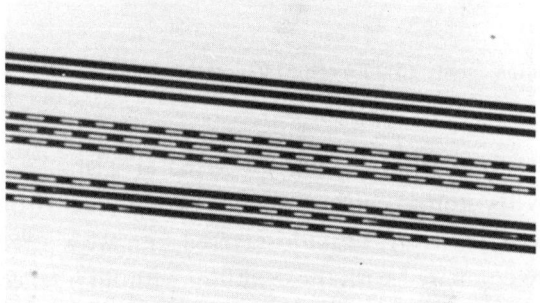

Fig. 7.1.14 Microscopic photograph of recorded bits on TeO$_2$-Te (Ge, Sn) thin film. [after N. Yamada et al.[11]]

200 kHz irradiated the recorded bits. As can be seen from the lower part of the photograph, there are completely erased bits and partially erased ones. It can be concluded that the latter are being erased by the edge of the erasing laser beam.

The carrier-to-noise ratio was measured with the conditions as follows: 5 MHz, a 30 kHz bandwidth, an 8 mW recording power and a 10 mW erasing power. The C/N ratio obtained was more than 55 dB. A reversible change in the reflectivity was observed over more than one million cycles of erasing and recording.

7.1.3 Magneto-optical Memory Material

Iron-based memory

Thin amorphous iron-based alloy film with the magnetization perpendicular to the film plane was investigated.[13] Some relevant physical properties of the recording media are summarized in Table 7.1.2 together with some dynamic read/write properties. Film made of GdTbFe was prepared by RF co-sputtering onto a glass disc 115 mm in diameter and 1.0 mm thick. However, TbFe and multi-layer films were prepared using the vacuum electron beam co-evaporation technique onto glass or PMMA discs 120 mm in diameter and 1.2 mm thick.

Table 7.1.2 Static and dynamic properties of magneto-optical recording media. [after N. Imamura.[13]]

Recording Medium	$Gd_7Tb_{13}Fe_{80}$	$Tb_{22}Fe_{78}$	$Tb_{22}Fe_{78}/SiO/Ag$ Multi-Layer	$Tb_{22}Fe_{78}$
Substrate	Glass	Glass	Glass	PMMA
Film Thickness	100 nm	100 nm	25/400/40 nm	100 nm
Laser Power	8 mW	6 mW	6 mW	5 mW
Bit Width	~1.2 μm	~1.55 μm	1.3~1.4 μm	———
T_c (°C)	160	140	140	140
θ_k (°)	0.41	0.43	0.45	———
H_c (kOe)	1.2	>5	>5	>5
C/N (dB)	40	40	44	40

Disc rotation speed : 1360 rpm, Bit rate : 2 Mbps

In Fig. 7.1.15, the thickness of TbFe was fixed at 22 nm and its composition, $Tb_{22}Fe_{78}$, was determined by its atomic percentage. The minimum writing laser power, P_m, and the effective Kerr rotation angle, θ_k, of multi-layered film with a reflective layer, measured with a He-Ne laser, were compared with those of film without a reflective layer as a function of SiO buffer film thickness. The reflectivity, minimum writing laser power and effective Kerr rotation angle in the case of multi-layered film with a reflective layer and that without a reflective layer were R_1, P_1, θ_{k1}, R_0, P_0, and θ_{k0}, respectively. From this experiment, the optimum thicknesses for the TbFe and SiO layers for low writing laser power and large effective Kerr rotation angle with a He-Ne laser are thought to be 22 nm and 250 nm, respectively. The Kerr rotation angle is three times as large as that of TbFe single film with an angle of 0.15.

The dynamic recording and readout properties were also measured both for the TbFe film and for the multi-layered film, and are summarized in Table 7.1.2.

Bit density of 15~20 Mb/cm^2 was attained with a C/N of more than 35 dB. In addition, a C/N of 44 dB was obtained for multi-layered film at a disc rotating speed of 1350 rpm and a bit rate of 2.0 Mb/s with a recording power of 4 mW. It was also confirmed that the recording power can be reduced using a PMMA substrate instead of a glass substrate without a change in the C/N readout value.

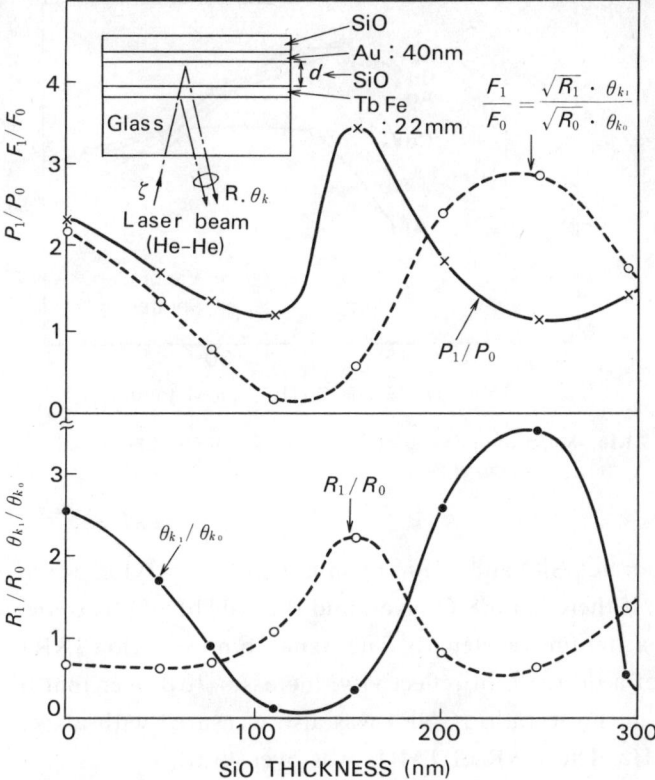

Fig. 7.1.15 The value of P_1/P_0, θ_{k1}/θ_{k0}, R_1/R_0 and F_1/F_0 as a function of SiO thickness. [After N. Imamura.[13]]

GdTbFe film with reflecting layer

It is well known that magneto-optic rotation is increased when a magnetic thin film is deposited on a reflector. It is because the Faraday effect is added to the Kerr effect. The former rotates a light's plane of polarization when it passes back and forth through the magnetic layer, and the latter rotates it at the surface of the layer.

It was found that the rotation angles of GdTbFe films prepared on reflecting layers such as Al, Cu, Ag and Au were increased.[15] The rotation ($\theta_{k'}$) as a function of GdTbFe thickness for each reflecting layer is shown in Fig 7.1.16. The $\theta_{k'}$ becomes 0.27° above a 60 nm thickness, which is not shown in the figure. The value 0.27° is the Kerr rotation angle of a thick GdTbFe film. The $\theta_{k'}$ increases up to 0.5° for each Cu, Ag and Au layer at a GdTbFe thickness of 20nm, and up to 0.4° for Al at 30nm. The samples were prepared as follows: The GdTbFe films were sputtered on glass substrates 0.5mm thick, and then the reflecting layers were deposited by sputtering (○) or by vapor deposition (●). The $\theta_{k'}$ was measured through the glass substrate using a light with a 630nm wavelength.

Dielectric overcoating on a magnetic film has been investigated as a means of Kerr rotation enhancement. It may also be employed in the reflecting layer structure. The $\theta_{k'}$ of the typical quadrilayer structure is increased to 1.75° at a 780nm wavelength. Its structure

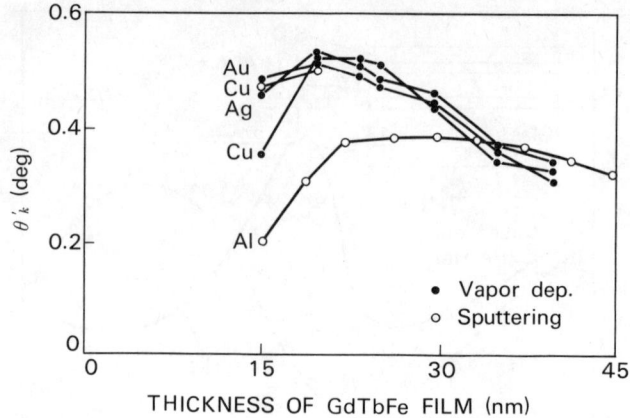

Fig. 7.1.16 Magneto-optic rotation as a function of GdTbFe thickness. [after K. Ohta et al.[15]]

has 120 nm of evaporated SiO and 15 nm of sputtered GdTbDyFe, 30 nm of SiO_2 and 40 nm of Cu accordingly. If there is no SiO layer, and the GdTbDyFe is optically thick, the Kerr rotation is 0.27° at a 780 nm wavelength. The signal to noise ratio (SNR) in this quadrilayer medium, in spite of a decrease in reflectivity, increases 7 dB over that of thick film.

The carrier to noise ratio (CNR) was also measured with a spectrum analyzer at a bandwidth of 30 kHz. The CNR of 1 MHz was over 40 dB.

CrO_2 film

A new type of magneto-optical recording system using a CrO_2 flexible disc as a storage medium and a magnetic garnet film as the magneto-optical readout medium has been developed.[14]

The Curie temperature of CrO_2 tape is approximately 130°C. Magnetic hysteresis measurement shows that as temperatures increase both saturation magnetization and coercivity ($H_c \simeq 500$ Oe at room temperature) decrease. The recording mechanism is based on thermoremanent magnetization.

Information can be recorded by raising the temperature of a small region of the continuously moving tape with a focused laser beam and applying a magnetic field produced by a magnetic head. This recording concept is shown schematically in Fig. 7.1.17.

In using CrO_2 tape for magneto-optical memories, the readout technique is somewhat different from conventional ones. The information recorded on the tape should be transferred to a magnetic thin film, from which the information can be read out magneto-optically. There are several ways of transferring information from one magnetic medium to another. Among these methods, noncontact transfer, which utilizes only the fringing field from the tape, is desirable because no special apparatus is needed for transfer. The basic configuration of this readout system is illustrated in Fig. 7.1.18. As the tape is brought close to the magnetic thin film, the bit pattern is transferred by the magnetic fringing field of the tape. A conventional magneto-optical Kerr or Faraday readout system is employed to read the transferred information.

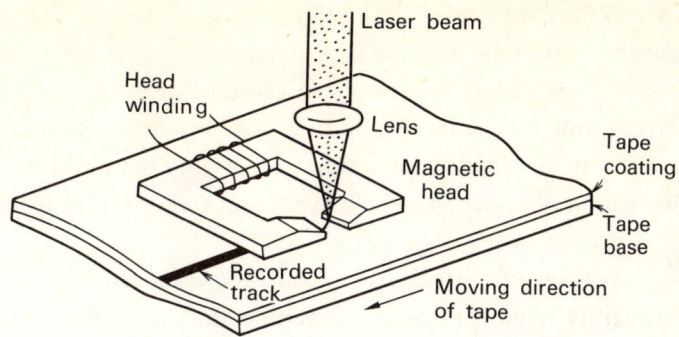

Fig. 7.1.17 Schematic representation of thermomagnetic recording on the CrO_2 tape. [after T. Nomura et al.[14]]

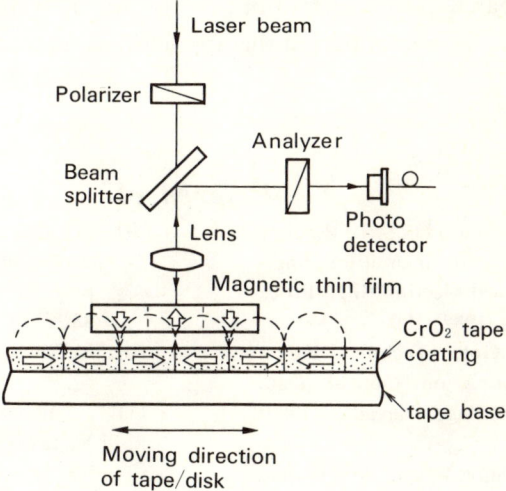

Fig. 7.1.18 Magnetic transfer process and related optics. The magnetic fringing field of the storage medium (tape or disc) switches the magnetization direction of the thin film according to the recorded bit patterns. [after T. Nomura et al.[14]]

The signal was recorded with 40 mW of laser power on the surface of the CrO_2 disc, which was rotating at a speed of 40 m/s. The magnetic garnet film used as the second magnetic medium for transferring the information had a thickness of $3.6 \mu m$, a saturation magnetization of 244 Gauss and domain structure period of $6.6 \mu m$. Measurements using a spectrum analyzer showed that the 4 MHz carrier-to-noise ratio of the readout signal was 38 dB, with a bandwidth of 100 kHz.

7.1.4 Conclusions

Recent research and developments into optical disc memories in Japan have been reviewed. Many recording materials and their structures have been studied and work on practical optical disc development has been carried out to achieve highly sensitive and stable memories. At the present time, write-once type disc memories are commercially available as video files, document files and data files. They are expected to have long term stability, more than 10 years, and the sensitivity to be recordable using laser diodes.

Work on memory materials has been centered around lessening oxidation degradation of the recording thin film. By adding some stabilizing additives such as carbon, selenium or TeO_2 to Te thin film, stable and highly sensitive media satisfying the requirements have been obtained. The role of the additives in resistance to oxidation has been clarified.

Further advances in optical disc memories are expected to result in the achievement of an erasing facility. Amorphous chalcogenide thin films triggered the optical memory studies. A new medium based on Te-TeO_2 thin film shows reversible changes in reflectivity using laser irradiation with different power densities. In addition, it has the possibility of achieving an erasable optical disc memory in place of magnetic media.

A lot of research into magneto-optical media as well as optical memories is in progress. Optical disc memories started the applications and are expected to result in many more.

References

1) M. Mashita and N. Yasuda : SPIE (Society of Photo-Optical Instrumentation Engineers), 12th Technical Meeting, SPIE Proc., *329* (Los Angeles, 1982) 190.
2) Y. Unno and K. Goto : Technical Digest of Topical Meeting on Optical Data Storage, Incline Village, Nevada, (1983) MB 2.
3) M. Terao, K. Shigematsu, M. Ojima, T. Taniguchi, S. Horigome, and S. Yonezawa : J. Appl. Phys., *50* (1979) 6881.
4) Y. Asano, H. Yamazaki, A. Morinaka, and K. Murase : Digest of SID, *81* (1979) 70.
5) M. Takenaga and T. Yamashita : JARECT, Amorphous Semiconductor Technologies and Devices (ed. Y. Hamakawa) (OHMSHA * North-Holland, 1982) 311.
6) M. Terao, S. Horigome, K. Shigematsu, Y. Miyauchi, and M. Nakazawa : Technical Digest of Topical Meeting on Optical Data Storage, Incline Village, Nevada, (1983) ThB 2.
7) K. Watanabe, T. Oyama, Y. Aoki, N. Sato, and S. Miyaoka : ibid., (1983) WA 4.
8) Y. Aoki, K. Watanabe, T. Oyama, N. Satou, and S. Miyaoka : Proc. of SPIE—The Intern. Society for Optical Engineering, *420* (Arlington, Virginia 1983) 313.
9) S. Oikawa, A. Morioka, and H. Yamazaki : Proc. of the 3rd Intern. Display Research Conference, (Kobe, 1983) 38.
10) M. Miyagi and S. Fukunishi : ibid., (1983), 42.
11) N. Yamada, S. Ohara, K. Nishiuchi, M. Nagashima, M. Takenaga, and S. Nakamura : ibid., (1983) 46.
12) M. Takenaga, N. Yamada, S. Ohara, K. Nishiuchi, M. Nagashima, T. Kashihara, S. Nakamura, and T. Yamashita : Proc. of SPIE—The Intern. Society for Optical Engineering, *420* (Arlington, Virginia 1983) 173.
13) N. Imamura : Technical Digest of Topical Meeting on Optical Data Storage, Incline Village, Nevada, (1983) ThA 1.
14) T. Nomura, H. Tokumaru, and S. Nakagawa : ibid., (1983) ThA 3.
15) K. Ohta, A. Takahashi, T. Deguchi, T. Huga, S. Kobayashi, and H. Yamaoka : ibid., (1983) ThA 4.
16) T. Ohta, M. Takenaga, N. Akahira, and T. Yamashita : J. Appl. Phys., *53* (*12*) (1982) 8497.
17) M. Takenaga, N. Yamada, K. Nishiuchi, N. Akahira, T. Ohta, S. Nakamura, and T. Yamashita : ibid., *54* (*9*) (1983) 5376.

7.2 A CCD Imager Overlaid with An a-Si : H Layer

Nozomu HARADA*

Abstract

An a-Si : H photoconductive film with high resistivity is the most suitable material for a "two-story (two-layer)" solid state image sensor which is composed of a conventional solid state image sensor and a photoconductive amorphous layer. In 1981, a two-story MOS imager using a reactively sputtered a-Si : H film first appeared. Recently, a two-story CCD imager overlaid with a glow discharge a-Si : H/a-SiC : H film was reported. This imager has excellent features, such as high sensitivity, low blooming, low smearing and low burn in and it aims to replace conventional vidicon camera tubes.

7.2.1 Introduction

Although remarkable progress has been achieved in monolithic solid state image sensors, including MOS,[1] CCD,[2,3,4] CPD[5] and CID,[6] each aiming to replace vidicon camera tubes, there are still persistent requirements for improving disadvantages in their photo-electric properties, such as photosensitivity, image smearing and a blooming phenomena associated with high light level illumination. Recently, new kinds of solid state image sensors, overlaid with photoconductive layers, have attracted interest as the possible next generation solid state image sensor. They are called "two-story (two-layer)" image sensors. In these image sensors, conventional image sensors are used as signal charge read-out scanners and a photoconversion is made in the photoconductor formed on the top of the scanner.

The first two-story image sensors, a MOS imager overlaid with a Se-As-Te amorphous chalcogenide glass[7] and a BBD imager overlaid with a ZnSe-Zn$_{1-x}$Cd$_x$Te heterojunction photoconductor,[8] were reported by Hitachi in December 1979 and by Matsushita in February 1980, respectively. In August 1981, Hitachi reported an MOS image sensor using a reactively sputtered a-Si : H photoconductive layer.[9]

In these photoconductive layers, the Se-Te-As film was found not suitable for color solid state image sensors.[9,10] The reasons reported are that the photoconversion requires supply voltages as high as 50 V. These voltages are rarely used for conventional driver ICs. Also, the film cannot survive the on-wafer color filter process after film deposition. The ZnSe-Zn$_x$Cd$_{1-x}$ Te film, though reported as successfully utilized with BBD scanners, CCD

---

* Toshiba VLSI Research Center, Toshiba Corp., 1, Komukai Toshiba-cho, Saiwai-ku, Kawasaki, Japan, 210.

scanners[11] and lately the MOS scanner,[12] is not ubiquitous nor familiar to conventional MOS-LSI fabrication technologies. In this sense, the a-Si H film has better prospects for use in accordance with the present Si technology. In July 1983, Hayashimoto et al. of Toshiba reported a 500(V) × 400(H) picture element CCD image sensor overlaid with an a-Si : H (intrinsic)/a-SiC (p-type) photoconversion layer produced by glow discharge.[13] This was the first sensor to be combined with a glow discharge a-Si : H photoconductive film and a CCD scanner.

Device structure fabrication and performance for the reported CCD imager using a glow discharge a-Si : H photoconductive layer are described in this paper.

7.2.2 Device Structure and Fabrication

The sensor unit cell structure reported by Hayashimoto et al. of Toshiba is shown in Fig. 7.2.1. The unit cell consists of an a-Si : H/a-SiC : H photoconversion layer, a storage diode and a half stage of vertical CCD (V-CCD). The CCD scanner for reading out signal charges generated in the photoconversion layer is constructed with an interline transfer CCD (IT-CCD) having double poly Si electrodes and two-level Al electrodes. In the previously mentioned sensor using a reactively sputtered a-Si : H film, an a-Si : H/ITO (Indium-Tin-Oxide) Schottky barrier junction has been employed to stop minority carrier (electron) injection from the ITO electrode. Hayashimoto et al. used an a-Si : H (intrinsic)/a-SiC : H (p-type) p-i junction for minority carrier injection stopping. The intrinsic a-Si : H film (undoped) and the p-type a-SiC : H film (B doped) are continuously deposited on the IT-CCD scanner. The ITO film for applying voltages to the amorphous p-i junction is formed on the p-type a-SiC : H layer by dc-magnetron sputtering. It acts as an overflow drain for excess charges under intense light illumination. Signal charges, electrons, generated in the a-Si : H film, are transfered to the storage diode by an electric field in the film, and are stored there. In the report, Hayashimoto et al. indicated that this device featured high sensitivity, low blooming, low lag and low burn in.

For the sensor using such a p-type amorphous layer, Harada of Toshiba analyzed that it was possible to obtain high blue sensitivity by using a thin a-SiC : H film of about 200 Å thickness.[14] Also he analyzed that μ_e (electron mobility)·τ_e (electron life time) product of more than $5 \times 10^{-7} \text{cm}^2/\text{V}$ was necessary in the a-Si : H photoconductive layer.

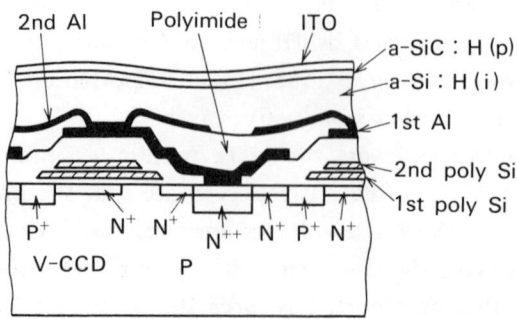

Fig. 7.2.1 Cross sectional view of the two-story CCD imager. [after Y. Hayashimoto et al.[13]]

Harada et al. of Toshiba reported a sensor fabricated by using 100 Å thickness a-SiC : H (p-type) film.[15] It showed high photosensitivity over the whole visible light spectral region. This will be described in the next chapter.

For the two-story imager, a smoothed scanner surface is required to decrease the dark current value and the number of image defects. Chikamura et al. of Matushita showed that through-hole taper with an inclination of 30 degrees or more led to a dark current increase in the sensor using a $ZnSe-Zn_xCd_{1-x}Te$ layer.[16]

Uya et al. of Toshiba suggested that glow discharge a-Si : H film on step regions in the CCD scanner was formed with less density than that formed on the flat regions.[17] In the previously mentioned CCD imager overlaid with a glow discharged a-Si : H layer, Harada et al. used a polyimide insulative layer for scanner surface smoothing.[15] Also, they used a p/p^+ epitaxial Si substrate. The p^+ substrate has a low life-time for minority carrier diffusion. With this substrate, those unnecessary charges, which are generated in the bulk or at the back of the chips and flow into a p-type active region, are successfully avoided. On the other hand, in the MOS imager overlaid with a sputtered a-Si : H film reported by Baji et al. at Hitachi, a p-well structure has been employed for unnecessary charge injection stopping.[9,18]

Harada et al. reported that resistivities for each film were $1.5 \times 10^{10} \Omega \cdot cm$ for a-Si : H film and $5 \times 10^6 \Omega \cdot cm$ for a-SiC : H film, respectively. Measured band gaps for each film were 1.8 eV for a-Si : H film and 2.13 eV for a-SiC : H film, respectively.

A diagram showing the reported device organization is shown in Fig. 7.2.2. The read-out elements from the device were 400(H) × 500(V). Effective picture elements in the imaging area were 378(H) × 486(V). Driving modes for CCDs used in this device were four phase operation for V-CCD and two phase operation for H-CCD, respectively. Also, its cell size was 22 μm(H) × 13 μm(V). Figure 7.2.3 shows a cross section SEM photomicrograph of the fabricated device.

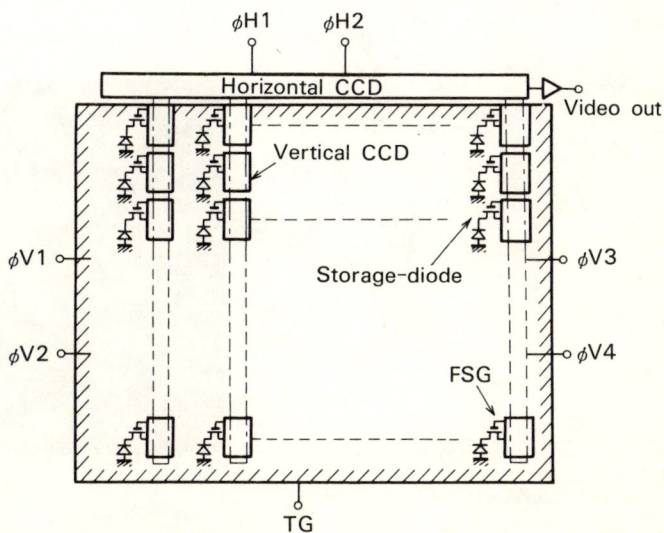

Fig. 7.2.2 Diagram showing the device organization. [after N. Harada et al.[15]]

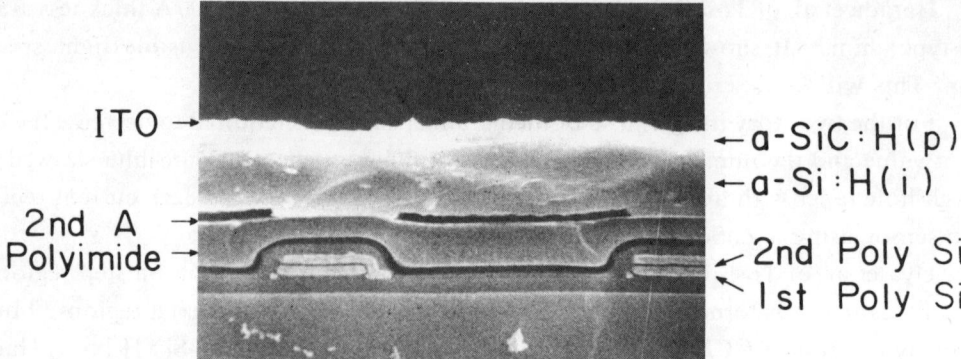

Fig. 7.2.3 Cross section SEM photomicrograph of the device. [after N. Harada et al.[19]]

7.2.3 Performances

One of the advantages of the two-story imager is that high sensitivity is obtained by using the whole area on the top surface of the CCD scanner as the photosensitive area. Harada et al. reported that the fabricated CCD imager overlaid with an a-Si : H/a-SiC : H photoconversion layer had high sensitivity over the whole visible light spectral region.[15,19] Figure 7.2.4 shows the reported spectral response curves for the sensor. As shown in this figure, an imager with 100 Å thickness a-SiC : H film has a flat spectral response thoughout

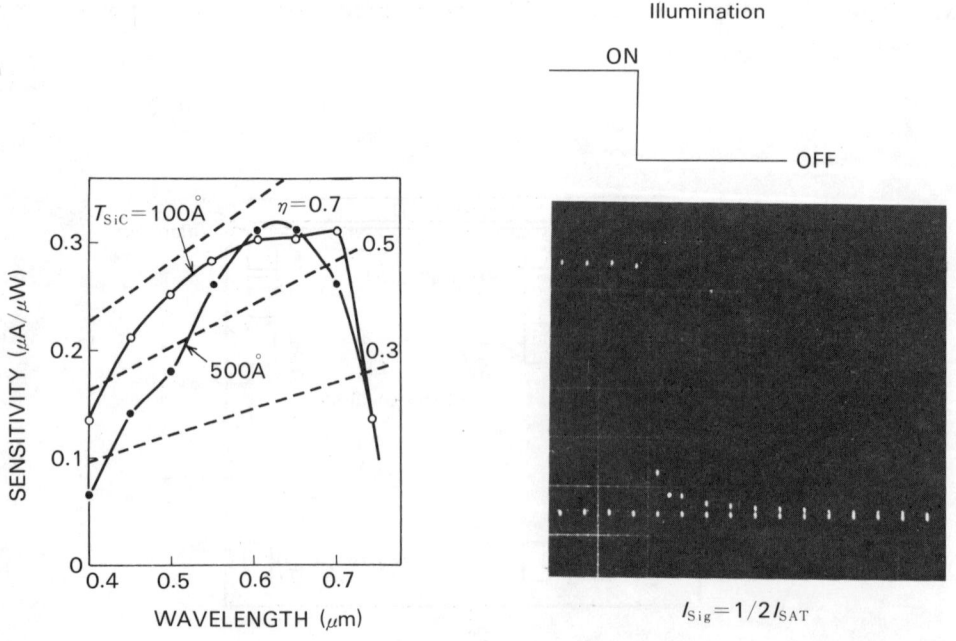

Fig. 7.2.4 Spectral response curves for the device. [after N. Harada et al.[19]]

Fig. 7.2.5 A decay lag characteristic. [after N. Harada et al.[15]]

the visible light region. Also, the sensitivity under 2856 K illumination was $0.14\,\mu A/lx$. On the other hand, the sensitivity of an imager with 500 Å thick film is low at the short wavelength region. The dark current was 2 nA at 27°C.

A decay lag characteristic reported by Harada et al. is shown in Fig. 7.2.5. The image lag value under a half of the saturation light illumination was 5% at the third field. They suggested in the report that, in this imager, the lag was mainly due to photoconductive phenomena in the a-Si : H film.

Kon et al. of Toshiba proposed a two-story CCD imager operation mode for high light lag characteristic improvement.[20]

They showed that the high light lag characteristic was markedly improved by applying a two-step-pulse to the ITO electrode during the vertical blanking period. Figure

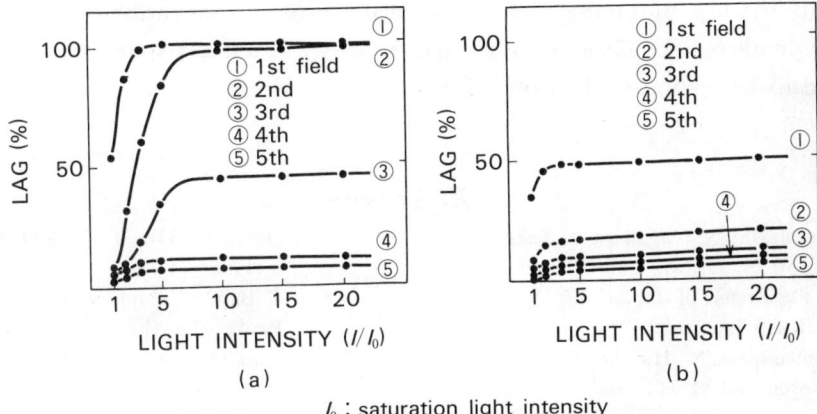

Fig. 7.2.6 High light lag characteristics (a) obtained by a conventional mode (b) obtained by the two-step-pulse mode. [after T. Kon et al.[20]]

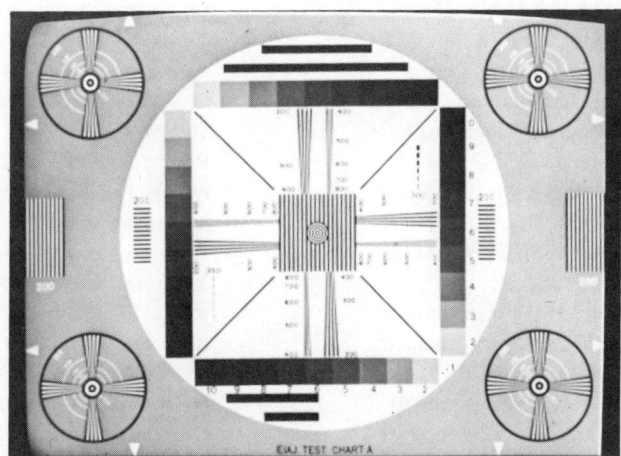

Fig. 7.2.7 An image picture of a RETMA resolution chart, taken by the two-story CCD imager. [after N. Harada et al.[19]]

7.2.6 shows the high light lag characteristic obtained by the two-step-pulse mode in comparison with that by a conventional operation mode.

An image picture of a RETMA resolution chart, taken by the two-story CCD imager, is shown in Fig. 7.2.7. Horizontal and vertical limiting resolutions were 280 TV lines and 400 TV lines, respectively. Harada et al. pointed out that, by the photosensitive area increase, the Moiré phenomena in the reproduced picture was much smaller than that obtained by a conventional CCD imager, giving an improved reproduced image quality.

7.2.4 Conclusion

The two-story CCD imager using a glow discharge a-Si : H photoconductive layer, which has recently been reported, was reviewed. This device is supposed to be one of the new three demensional ICs, aiming to replace vidicon camera tubes which are currently primary in various fields. It was demonstrated that the device has many advantages, such as high sensitivity, low blooming and low smearing, over a conventional solid state imager. These new imagers, including an MOS imager using a reactively sputtered a-Si : H film, are considered to be second generation solid stage imagers.

References

1) K. Takahashi, S. Nagahara, I. Takemoto, M. Aoki, N. Ozawa, and T. Suzuki : J. Inst. TV Engrs. of Japan, *37*, 10 (1983) 812-818 [in Japanese].
2) H. Matsumoto, Y. Hirata, H. Matsui, K. Takeshita, and M. Hamasaki : J. Inst. TV Engrs. of Japan, *37*, 10 (1983) 776-781 [in Japanese].
3) Y. Matsunaga and N. Suzuki : Digest of 1984 ISSCC, WAM 2.4 (1984).
4) E. Oda, I. Akiyama, Y. Ishihara, A. Kohno, K. Arai, and T. Kitagawa : Digest of 1983 ISSCC, FAM 18.7 (1983) 264-265.
5) S. Terakawa, Y. Matsuda, T. Kozono, T. Yamada, K. Senda, I. Murozono, Y. Hiroshima, K. Horii, T. Takamura, and T. Kunii : J. Inst. TV Engrs. of Japan., *37.10* (1983) 795-802 [in Japanese].
6) D. M. Brown, H. K. Burke, M. Ghezzo, P. McConnelee, G. Michon, and T. L. Vogelsong : Digest of 1980 ISSCC, WAM 2.3 (1980). 28-29.
7) T. Tsukada, T. Baji, H. Yamamoto, Y. Takasaki, T. Hirai, E. Maruyama, S. Ohba, N. Koike, H. Ando, and T. Akiyama : Digest of 1979 IEDM, 6-1 (1979).
8) Y. Terui, T. Wada, M. Yoshino, H. Kadota, T. Komeda, T. Chikamura, S. Fujiwara, H. Tanaka, Y. Ota, Y. Fujiwara, K. Ogawa, O. Kitahiro, and S. Horiuchi : Digest of 1980 ISSCC, WAM 2. 6 (1980) 34-35.
9) T. Baji, Y. Shimomoto, H. Matsumaru, N. Koike, T. Akiyama, A. Sasano, and T. Tsukada : Proc. 13th CSSD (Tokyo, 1981) 269-273.
10) T. Tsukada, T. Baji, Y. Shimomoto, A. Sasano, Y. Tanaka, H. Matsumaru, Y. Takasaki, N. Koike, and T. Akiyama : Digest of 1981 IEDM (1981) 379-482.
11) T. Chikamura, Y. Miyata, K. Yano, Y. Ohta, S. Fujiwara, Y. Terui, M. Yoshino, M. Nakayama, and M. Fukai : IEEE Trans. Electron Devices, *ED-30*, 10 (1983) 1386-1391.
12) O. Kyogoku, Y. Chatani, T. Kawashima, and M. Ozaki : 1983 National Convention Report of Inst. TV Engrs. of Japan, *37* (1983) 19-20 [in Japanese].
13) Y. Hayashimoto, N. Harada, S. Uya, Y. Komatsubara, K. Ide, O. Yoshida, T. Yoshino, K. Yano, M. Kakegawa, and T. Kon : 1983 National Convention Report of Inst. TV Engrs. of Japan, *3-5* (1983) 45-45 [in Japanese].
14) N. Harada : Technical Group on Electron Devices of Electronics & Communication Engrs. of Japan, *IE 83-43* (1983) 73-78 [in Japanese].
15) N. Harada, S. Uya, Y. Hayashimoto, Y.

Komatsubara, K. Ide, O. Yoshida, T. Kon, K. Yano, M. Kakegawa, and T. Yoshino: Proc. 15th CSSDM (Tokyo, 1983) 209-212.

16) T. Chikamura, S. Fujiwara, T. Shibata, Y. Terui, T. Wada, T. Yoneda, K. Ogawa, K. Ogura, H. Tanaka, and S. Fukai: Technical Group on Electron Devices of the Inst. TV Engrs. of Japan, *ED 490* (1980) 13-18 [in Japanese].

17) S. Uya, Y. Hayashimoto, Y. Komatsubara, N. Harada, and O. Yoshida: Autumn Meeting of Japan Society of Applied Physics, *28 a-k-10* (1983) 347 [in Japanese].

18) T. Tsukada, T. Baji, Y. Shimomoto, A. Sasano, Y. Tanaka, H. Matsumaru, Y. Takasaki, N. Koike, and T. Akiyama: Digest of 1981 IEDM, 19.5 (1981) 479-482.

19) N. Harada, Y. Hayashimoto, S. Uya, Y. Komatsubara, K. Ide, T. Kon, K. Yano, M. Kakegawa, T. Yoshino, and O. Yoshida: Technical Group on Electron Devices of Inst. TV Engrs. of Japan, *ED 760* (1983) 55-60 [in Japanese].

20) T. Kon, Y. Endo, N. Harada, and O. Yoshida: 1983 National Convention Report of the Inst. TV Engrs. of Japan, *3-6* (1983) 47-48 [in Japanese].

7.3 Amorphous Silicon Linear Image Sensors

Toshihisa TSUKADA*

Abstract

Recent technical efforts to develop amorphous silicon linear image sensors are reviewed. A contact linear imager can read a manuscript without a bulky lens system. The size of the imager, however, has to be as wide as the width of the manuscript. Amorphous silicon has been investigated as a material for this type of sensor because of its high photoconductivity and large area capability features. Contact imagers can be categorized into two types: direct address imagers, and matrix drive sensors. Direct address imagers can be made with a simpler process, but they have to have as many terminals as the total picture elements of the sensor. The matrix drive sensors have a smaller number of terminals, but they have to have switching elements such as diodes and transistors fabricated on the glass substrate. The drive methods, sensor structure, fabrication, and characteristics of these sensors are described.

7.3.1 Introduction

Hydrogenated amorphous silicon (a-Si) has such unique features as,
(1) Capability of large area deposition,
(2) High photoconductivity,
(3) Spectral response in the visible region, and
(4) Efficient doping, i.e., structure sensitive properties.

By virtue of these features, such applications as solar cells, xerographic printers, thin film transistors and image sensors, have been investigated in many research institutes. Among these devices, contact image sensors have attracted substantial research and development efforts. These devices can read documents and manuscripts without a bulky lens system. Therefore, the reading apparatus for such systems as facsimile and copiers can be very much reduced in volume, resulting in a scale reduction of these systems. In order to fabricate this kind of imager, the device has to be as wide as the manuscript width, to be sensitive to visible spectral range, and has to include a good rectifier. Since these requirements are met quite satisfactorily with amorphous silicon, this material is quite suitable for contact imagers.

For a linear sensor to be applied to the facsimile equipment, the sensor size has to be 216 mm (A 4 size) or more in length. Resolution should be equal to or higher than 8 bits/

---

* Central Research Laboratory, Hitachi, Ltd., 1-280, Kokubunji, Tokyo 185.

mm. Total pixel number of an A4 sensor is 1728 bits. These sensors can be classified into two categories. One is a straightforward way in which an analog switch is connected to every reading bit. The other is a matrix drive method in which the number of terminals from a sensor can be greatly reduced.

The next section describes direct address sensors, followed by a section describing matrix drive sensors. Drive method, sensor structure, fabrication and sensor characteristics are described.

7.3.2 Direct Address Sensor

In this type of sensor, amorphous silicon photodiodes are arranged in a linear array. All bits in this sensor are addressed by a combination of an individual analog switch and shift register. The structure of this amorphous silicon sensor and its plan view are shown in Fig. 7.3.1.[1] In most cases, glass plates such as 7059 or Pyrex are used as the sensor substrates but in some cases ceramic substrates are used. The amorphous silicon film is deposited by capacitive-coupled rf glow discharge by using 100% monosilane reaction gas. The major physical properties of the deposited film are as follows: Optical band gap is about 1.7 eV, ESR spin density is less than $10^{16} cm^{-3}$, IR absorption peaks at 2000 and 630 cm^{-1}. The film thickness is about 1.0 μm, and the thickness deviation for each run measured at the center of the substrate is less than $\pm 5\%$. The thickness variation in a 130 mm length is smaller than $\pm 5\%$, which indicates good deposition controllability. The final step in the sensor fabrication is the deposition of indium tin oxide (ITO) which forms a transparent upper electrode. This film is made by dc sputtering and the thickness of the film is 150 nm.

The dc I-V characteristics (photo- and dark currents) and their dependence on the lower electrode materials have been measured. In this experiment, negative bias is applied to an upper ITO electrode, while lower electrodes were grounded as shown in Fig. 7.3.1. A

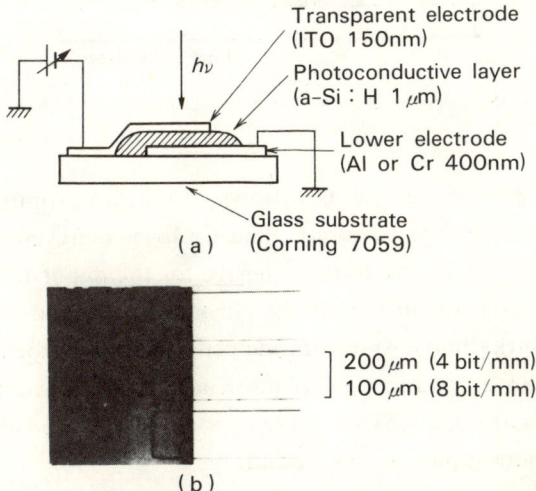

Fig. 7.3.1 Structure of direct address linear image sensor. [after T. Hamano et al.[1]]
(a) Schematic cross-section,
(b) microscopic plan view.

10 W fluorescent lamp with a peak intensity at 350 nm was used as a light source. Saturation levels of the photocurrent depend little on the lower metal electrodes, whereas the dark current largely depends on the metal electrodes. In the case of Al, a thin oxide film of a few nanometers is formed on the surface by a heat treatment in air. Because of the blocking characteristics of this film, photocurrent level has been greatly improved. The difference of the dark current levels between Au and Ni or Cr corresponds to the amount of injected hole current from each metal electrode, and is due to the difference in work functions of the metals used. It was found that chromium is the best choice.

The relation between light intensity and the photocurrent of the sensor was measured with fluorescent lamp mentioned above, the intensity of which was varied by inserting neutral density filters. The slope of the curve (γ) is almost unity in the range from 6 to 1250 lx. A 1056-bit test device which can read a document 132 mm wide with a resolution of 8 lines/mm was fabricated.

In an amorphous silicon linear sensor array, the reduction of switching noise in the driving circuit is very important to read out the signal charge from the sensor. A new 128-bit low noise multiplexer LSI has been developed to minimize switching noise. This LSI basically consists of 128-channel pre-amplifiers, C-MOS analog switches, and shift registers.[2] The gate sizes of both n- and p-type MOSFET are designed to have the same dimension, which is minimized until their on-resistance values significantly influence the read-out time. It has been proved that a sensor with this LSI can read an A4 size document within 0.8 ms/line with a resolution of 16 lines/mm. A cross-sectional view of the A4 sensor is shown in Fig. 7.3.2. A photosensitive area is formed at the center of the glass substrate.

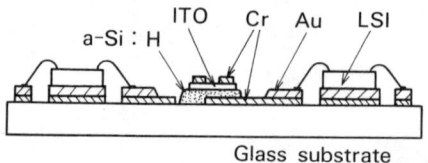

Fig. 7.3.2 Cross-sectional view of the A4 sensor. [after T. Saito et al.[2]]

The fabrication procedures are as follows: First, a chromium metal layer 200 nm thick is evaporated over the glass substrate. The Cr layer behaves as a barrier against the holes in the amorphous sillicon and as an adhesive for the deposition of gold. Amorphous silicon 1 μm thick is deposited on top of the chromium, and an ITO film 75 nm thick is deposited to make Schottky barrier photodiode cells. Then a chromium layer 200 nm thick is again formed to enhance conductivity of the lead electrode and mask the region where photosensitivity is not required. LSI's are then mounted on the substrate and electrically connected to the conducting pad by wire bonding.

7.3.3 Matrix Drive Sensor

As described in the previous section, a direct bit-address linear sensor has to have as many as 1728 terminals for an A4 size sensor. In order to reduce the number of terminals connected to the scanning circuit, the matrix drive method is preferable.

The equivalent circuit configuration of a matrix drive sensor is shown in Fig. .7.3.3. In this configuration, each photodiode is connected in series and in the reverse direction to a blocking diode.[3] Both the photodiode and the blocking diode are made of a-Si pin diodes. Each set of diodes is connected to common electrodes (Y) and to individual electrodes (X). The operational principle of this sensor can be described by way of equivalent circuits as shown in Fig. 7.3.4. In this figure, four states selected by pairs of X switches (SW_X) and Y switches (SW_Y) are shown. First, the photodiode is charged to V_T through the blockding diode (Fig. 7.3.4 (a)). Then the sensor is switched to V_B, reverse biasing the blocking diode (Fig. 7.3.4 (b)). Since the reverse bias current of a blocking

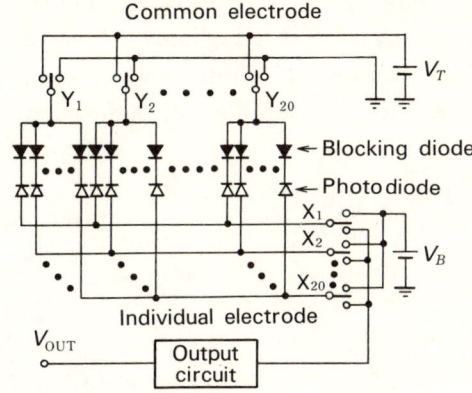

Fig. 7.3.3 Equivalent circuit configuration of the matrix drive linear imager. [after H. Yamamoto et al.[3]]

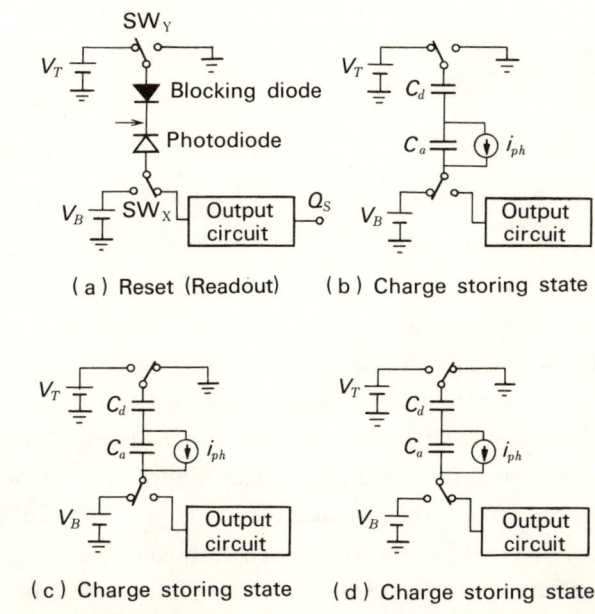

Fig. 7.3.4 Four states of the picture element consisting of a blocking diode and a photodiode. [after H. Yamamoto et al.[3]]

diode is extremely small, the photodiode is completely isolated from the others. Thus, crosstalk between photodiodes is prevented. Three states of charge storing are shown in Fig. 7.3.4 (b), (c), (d), depending on the reading condition.

The schematic diagram of this sensor is shown in Fig. 7.3.5. The fabrication procedure is as follows: First, $0.2\,\mu$m thick chromium stripe electrodes are formed on a glass substrate. A thin film of a-Si is deposited by an rf glow discharge method to a total thickness of $0.6\,\mu$m, with an n-layer first, to be followed by an i-layer and p-layer. The thickness of each layer is 0.03, 0.55, and $0.02\,\mu$m, respectively. This a-Si film is dry-etched to form a $100 \times 150\,\mu$m pattern. An isolation film of SiO_2 is then formed by an rf sputtering method. This serves as a cover to prevent the side walls of amorphous silicon pin diodes from shortening out as well as an insulator for double layer inter-connection. The size of contact holes in this SiO_2 film is $70 \times 100\,\mu m^2$. Indium-tin-oxide (ITO) transparent electrodes are then deposited by rf sputtering and Al is vacuum-evaporated to form connecting electrodes. In this sensor, the photo- and blocking diodes are made of the same amorphous silicon film. This results in a greatly simplified sensor fabrication process.

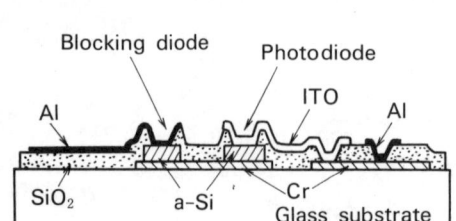

Fig. 7.3.5 Cross-section of the fabricated matrix drive sensor. [after H. Yamamoto et al.[3]]

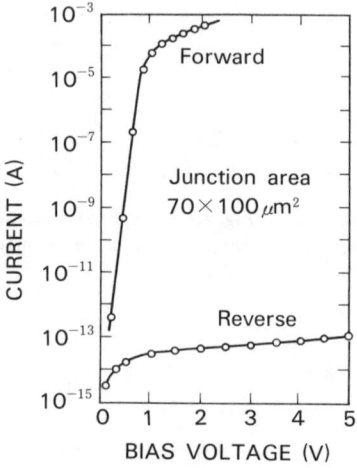

Fig. 7.3.6 I-V characteristics of a blocking diode of which the rectification ratio is greater than 10^{10}. [after H. Yamamoto et al.[3]]

The blocking diode plays the role of suppressing crosstalk between picture elements. Therefore, it has to display excellent rectifying properties. The typical I-V characteristics of a blocking diode fabricated for this purpose are shown in Fig. 7.3.6. A rectifying ratio as high as 10^{10} or more has been obtained at voltages higher than 2 V with a diode quality factor of 1.2. The typical irreversible breakdown voltage is about 25 V. The spectral sensitivity of the photodiode matches the visible spectrum very well, and peaks at about 550 nm where its quantum efficiency is 0.7. The ratio of the photocurrent under an illumination of 100 lx to the dark current is greater than 6×10^3. In addition, it has a high speed responsivity with a response time of less than $3\,\mu$s.

This sensor can read an A4 size document at a speed of less than 5 ms/line. The saturation exposure is 0.3 lx·s. Deviation of the sensitivity is in the range of ± 7%, and the S/N ratio is 20 dB. The noise component is primarily due to clock noise, and can therefore be reduced.

Another version of matrix drive linear image sensor utilizes a Schottky barrier diode between platinum and amorphous silicon.[4] The photodiode consists of indium oxide (In_2O_3) and amorphous silicon. The structure of photo- and blocking diodes and their current-voltage characteristics are shown in Fig. 7.3.7. In contrast to the pin diode version, the light is introduced through the glass substrate. First, an indium oxide film is deposited by electron beam evaporation at a substrate temperature of 300°C. The transmittance of the 50 nm thick indium oxide film in the visible region is greater than 90%. On top of this transparent electrode, a 0.5 μm thick layer of amorphous silicon is deposited using a capacitively coupled rf glow discharge from 10% monosilane diluted with argon gas. Deposition is made at a substrate temperature of 250°C, an rf power of 10 W, a gas flow rate of 65 sccm, and a gas pressure of 0.1 Torr. As shown in this figure, the indium oxide/amorphous silicon heterojunction diode exhibits a high photocurrent, and the platinum/amorphous silicon Schottky diode maintains a very low reverse current level.

The output voltage of the photodiode is shown in Fig 7.3.8 as a function of the integration time. The cell size of the diode is 100 μm square, and the bias voltage is 1 V. The signal-to-noise ratio at a saturation output voltage under 200 lx illuminance is 6.4 dB. Although this value is somewhat low, the reduction of the noise level will yield a high signal-to-noise ratio even at low illuminance levels. In this case, the dark current of the indium oxide/amorphous silicon diode is a major part of the noise level. When tin oxide

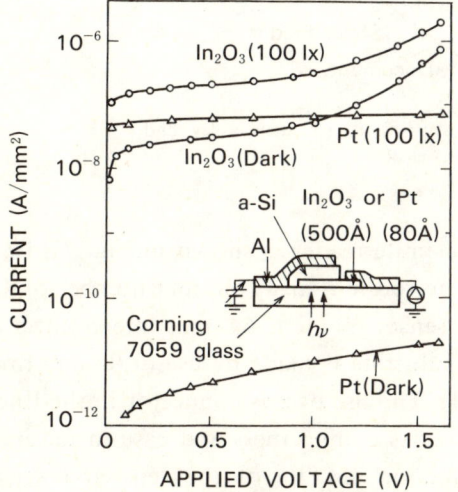

Fig. 7.3.7 I-V characteristics of indium oxide/amorphous silicon diode and platinum/amorphous silicon diode. [after K. Ozawa et al.[4]]

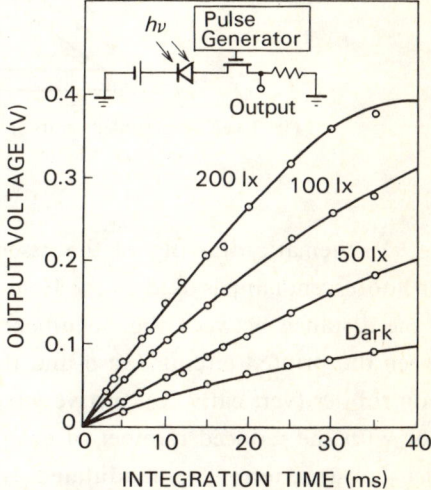

Fig. 7.3.8 Output voltage of indium oxide/amorphous silicon diode as a function of integration time which is used for Schottky type matrix drive sensor. [after K. Ozawa et al.[4]]

(50 nm thick) is used as the electrode of the photodiode, the signal-to-noise ratio is improved to 15 dB, while the output voltage decreaes to one half that of the case of indium oxide.

A cross-sectional view of the linear sensor is shown in Fig. 7.3.9. The pitch of the diode array is 145 μm, and the areas of the photodiode and blocking diode are 100 μm square. The thickness of the amorphous silicon and indium oxide layer are 0.5 μm and 50 nm, respectively. A NiCr layer assures the adhesion of the platinum layer to the glass substrate. A photosensitive layer of polimer (CBR) is used as an intermediate insulator to separate the upper individual electrodes (Al) from the lower ones (Au/Cr).

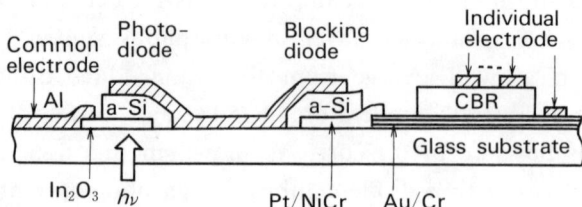

Fig. 7.3.9 Cross-sectional view of the amorphous silicon Schottky type matrix drive sensor. [after K. Ozawa et al.[4])]

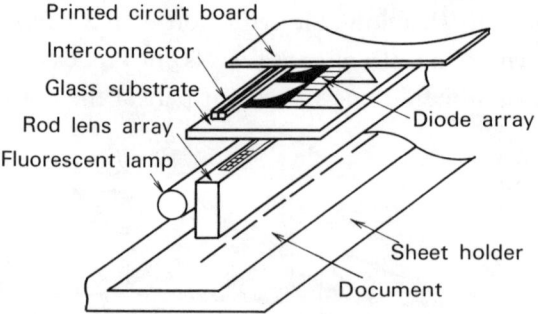

Fig. 7.3.10 Assembled unit of the linear sensor and rod lens array and fluorescent lamp. [after K. Ozawa et al.[4])]

A schematic drawing of the assembled facsimile reader is shown in Fig. 7.3.10. A green fluorescent lamp is used as the light source. The rod lens array has an f number of 4.8, and the distance between the document and the sensor plane is 64 mm. Interconnection between the printed circuit board and the sensor substrate is made by use of 0.8 mm thick silicon rubber (vertically conductive when pressed). The use of this connector in the linear sensor with the reduced number of electrodes provides compactness and ease in facsimile reader construction. The modulated transfer function (MTF) of the lens and sensor combined system is about 80% at 8 lines/mm when illuminated with 550 nm light, while it decreases to 65% with the white fluorescent lamp.

Instead of an amorphous silicon diode, an amorphous silicon transistor (TFT) can be used as a switching element of the linear photodiode array. This gives another version of the linear sensor reducing the number of terminals. This type of sensor is shown in Fig. 7.3.11, and consists of a photodiode array, TFT array, matrix circuit, and external circuit which includes transfer capacitance and analog multiplexer.[5] Each outlet terminal is connected to the discrete transfer capacitor which transfers signal charge from the photodiode to a detector circuit. However, in the case of a large size linear sensor, stray capacitance functions as a transfer capacitance.

A cross-sectional view of this linear sensor is shown in Fig. 7.3.12. The photodiode has a sandwich-like structure with silicon nitride and p-type amorphous silicon blocking layers on both sides of a photosensitive amorphous silicon layer. These blocking layers reduce dark current and provide high photosensitivity. The effective photosensitive area is $100\,\mu m \times 100\,\mu m$ (8 bit/mm). The amorphous silicon TFT configuration is an inverted stagger structure. The gate insulator is silicon nitride. A thin phosphorus-doped layer is deposited on top of the undoped layer as an ohmic contact to an evaporated Al electrode. Channel length and width are $20\,\mu m$ and 1 mm, respectively. Polyimide is used as an insulator for the double layer interconnection.

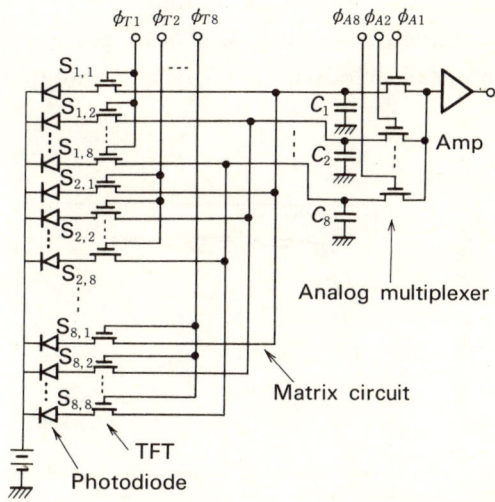

Fig. 7.3.11 Equivalent circuit for 64 bit amorphous silicon linear image sensor driven by TFT matrix array. [after F. Okumura et al.[5]]

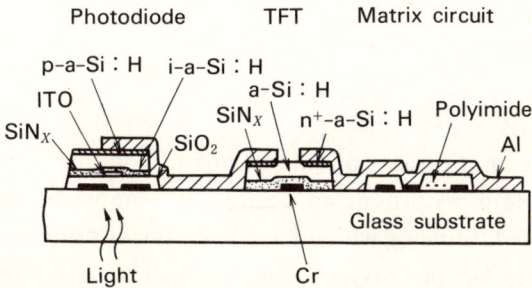

Fig. 7.3.12 Cross-sectional view of amorphous silicon linear imager operated by amorphous silicon TFT array. [after F. Okumura et al.[5]]

Fabrication procedures are as follows: A 100nm thick Cr film is first evaporated on a Corning 7059 glass substrate which acts as a light shield layer, gate electrode, and lower electrode for matrix configuration. Polyimide is coated and chemically etched by a photoetching process. A $0.5\,\mu m$ thick silicon dioxide film and 40nm thick transparent ITO film are deposited by rf sputtering. To fabricate TFT, $0.3\,\mu m$ thick silicon nitride, $0.3\,\mu m$ undoped amorphous silicon, and 50nm n-type amorphous silicon layers are deposited continuously. Both silicon nitride and amorphous silicon layers are deposited by capacitively coupled rf glow discharge technique. The final process in the sensor fabrication is an aluminum evaporation.

Typical characteristics of fabricated TFT are as follows: On resistance is $400 \sim 500$ kΩ at a gate voltage of 30 V. Mobility and threshold voltage are $0.6\,cm^2/V \cdot s$ and $6 \sim 7$ V, respectively. Drain current on/off ratio is 10^5 or more. The photocurrent of a photodiode under 100 lx (550nm) exposure is 9×10^{-10} A at a saturation voltage of 3 V. Dark current is less than 10^{-13} A at a voltage ranging from 0 to 15 V. Photorespones time of the photodiode is less than 0.1 ms. The storage capacitor in the photodiode is $10 \sim 20$ pF.

The charge transfer characteristics from the photosensor to the transfer capacitance (330 pF) are shown in Fig. 7.3.13. Charge transfer time is $20\,\mu s$ at a gate voltage of 20 V and $5\,\mu s$ at 30 V. After image is then less than 3%. These results agree with the test element estimates. The operation speed achieved is $1.3\,\mu s/bit$, and this satisfies the GIII mode facsimile spcifications.

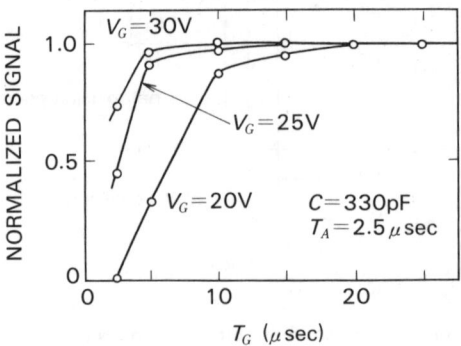

Fig. 7.3.13 Charge transfer characteristics of a TFT driven linear imager. [after F. Okumura et al.[5]]

7.3.4 Conclusion

Among the many applications of amorphous silicon, the contact image sensor could be counted as one of the most promising devices. A number of image sensors have been proposed and investigated, some of which have been reviewed in this article. Most of these sensors are test samples or prototypes, but they all satisfy the GIII mode facsimile specifications. Research and development efforts towards higher performance and higher yield will be accelerated.

References

1) T. Hamano, H. Ito, T. Nakamura, T. Ozawa, M. Fuse, and M. Takeuchi: Japan. J. Applied Physics, *21* (1982) Supplement 21-1, 245.
2) T. Saito, K. Suzuki, Y. Suda, S. Takayama, T. Nakai, K. Mori, and O. Takikawa: IEE Technical Report, Vol. ED 83-64, 1 [in Japanese].
3) H. Yamamoto, T. Baji, H. Matsumaru, Y. Tanaka, K. Seki, T. Tanaka, A. Sasano, and T. Tsukada: Ext. Abstracts 15th Conf. Solid State Devices and Materials, Tokyo (1983), 205.
4) K. Ozawa, N. Takagi, K. Hiranaka, S. Yanagisawa, and K. Asama: Japan. J. Applied Physics, *22* (1983) Supplement 22-1, 457.
5) F. Okumura, S. Kaneko, and H. Uchida: Ext. Abstracts 15th Conf. Solid State Devices and Materials, Tokyo (1983), 201.

7.4 Enhancement of Long Wavelength Sensitivity

Isamu SHIMIZU*

Abstract

A current survey is first outlined of the progress made in an a-Si : H photoreceptor of electrophotography, and a copier furnished with an a-Si : H drum and its performance are introduced.

Spectral tuning with a diode laser emitted near infra-red light(700~800nm) is expected for the a-Si : H photoreceptor. With respect to enhancement of sensitivity vis-a-vis long wavelength light, amorphous alloys, viz. a-SiGe, and a-SiSn and the recent advances achieved utilzing them are summarized. The device structures for this purpose are also described.

7.4.1 Introduction

Subsequent to the great success of a-Si : H solar cells,[1] some optoelectric devices comprised of a-Si : H have attracted attention as promising candidates for commercialization in the near future. The a-Si : H photoreceptor[2] of electrophotography is one of the products expected to be furnished in copying machines or laser line-printers. Other optoelectric devices, i. e., a linear photosensor for facsimile,[3] and image pick-up devices for TV cameras[4] are also promising.

In a laser-line printer, some diode lasers emitting light of near infra-red (700~800 nm in wavelength) are often adopted as the light source. Spectral tuning, therefore, with this light region becomes quite important in a-Si : H devices since the optical absorption of a-Si : H decreases rapidly in these wavelength regions.

In this report, we will first survey the current situation in the development of the a-Si : H photoreceptor of electrophotography and subsequently describe the preparation of amorphous alloys with a band-gap narrower than that of a-Si : H. In particular, the photoconductivity of a-SiGe alloy has been rapidly improved due to a dramatic reduction in the dangling bond.[5] We will introduce both the preparation technique and some characteristics of the a-SiGe alloy.

Care must be taken to prolong the dielectric relaxation time of the device when the materials with the narrower gap are adopted in order to enhance the sensitivity to long wavelength light. Some attempts to solve this problem will be discussed.

---

* Imaging Science and Engineering Laboratory, Tokyo Institute of Technology, Nagatsuta, Midori-ku, Yokohama 227.

7.4.2 a-Si : H Photoreceptor of Electrophotography

Since the first proposal presented in an article published in 1979,[6] considerable efforts have been made to develop an a-silicon photoreceptor, an outline of which was summarized in the latest volume of JARECT.[7] Consequently, announcements have been made by several groups in Japan pertaining to success in developing a-Si : H photoreceptors. With regard to performance, extremely attractive characteristics have been obtained in a-Si : H photoreceptor provided in copying machines or laser printers. Most difficulties from a technological aspect have been solved through enthusiastic efforts, but not on a mass-production basis as yet.

How to provide and retain a sufficient amount of acceptance voltage is one of the key factors in designing an a-Si photoreceptor because the charge on the surface provided by corona must be retained during the processing time (T_p) in the electrophotographic image-processes. Attempts were made at preparing highly resistive a-Si : H films by doping foreign elements (boron and oxygen)[7] and by sputtering in a H_2-Ar gaseous mixture.[8] Care must be taken during the procedures to minimize the deterioration of photoelectric properties.

The blocking electrodes to prohibit carriers from being injected from them are essential in this device to establish a "charge depletion condition" in the photoconductive film under a high electric field (5×10^5 V/cm). Either homophase or heterophase blocking is applicable at the contact with the conductive electrode (see Fig. 7.4.1). A thin layer is

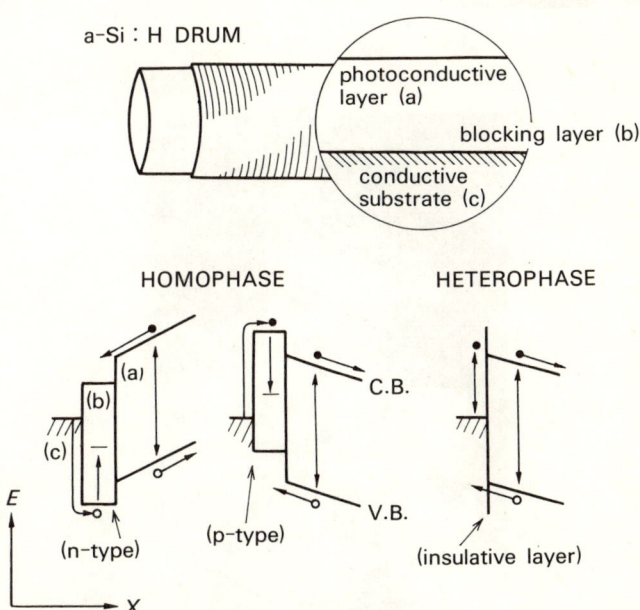

Fig. 7.4.1 Device structure of a-Si : H photoreceptor and its energy diagrams indicating blocking behaviors.
For homophase and heterophase blocking, a thin layer (n-type or p-type : 0.2~0.7 μm thick) and a very thin insulative film (30~300 Å thick) are provided between the photoconductive film and the conductive substrate respectively. [after E. Inoue and I. Shimizu.[2]]

Table 7.4.1 Typical performances of an a-Si : H photoreceptor of electrophotography.

acceptance voltage (by corona)	40~50 V/μm
photosensitivity	1 erg/cm^2
spectral sensitivity (peak)	670 nm
life	10^6 copies
image quality	superior
durability (heat, chemical, mechanical)	superior

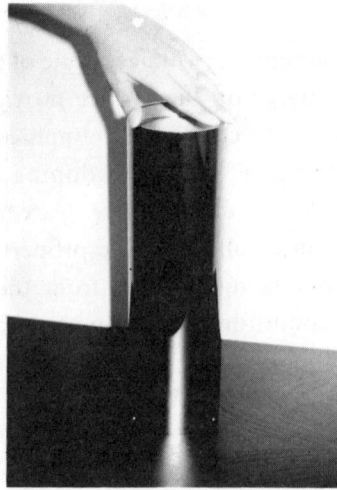
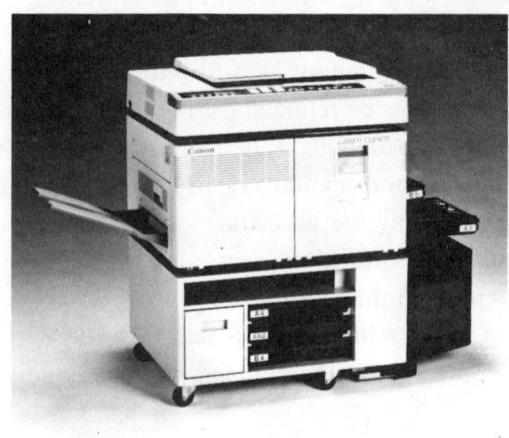

(a)

(b)

Fig. 7.4.2 A photograph of the a-Si : H drum furnished in the "Laser Copier®
(a)" and a copy of a half-tone image (b) modulated from continuous-tone image made by the "Integrated Electronic Filing System®". (courtesy of Canon, Inc.,)

furnished on the top to passivate the a-Si:H surface. Some insulative dielectric films, i. e., a-SiN$_x$, a-SiC$_x$ and a-SiO$_x$ were employed for this purpose. The device structure and typical performances are summarized in Fig. 7.4.1 and Table 7.4.1, respectively.

It is certainly noteworthy that the system performance of a copier, namely, sensitivity, operation velocity, image quality and durability are determined not only by the characteristics of the photoreceptor but also by those of the entire copymaking process.

As a result of systematic investigations from a technological point of view, a "Laser Copier®" furnished with a-Si photoreceptor has been demonstrated by Canon Inc. as a component of the new image-processing system termed "Integrated Electronic Filing System®". In this system, all information including images are digitalized and reconstructed by the laser printer. Consequently, a continuous tone image can be reproduced as a half tone image (dot image) as shown in the sample. A photograph of the a-Si drum as the photoreceptor is also shown (see Fig. 7.4.2). In this printer, the spectral sensitivity of the a-Si drum is tuned with light emitted from a diode laser. The spectral response of this photoreceptor is shown in Fig. 7.4.3.

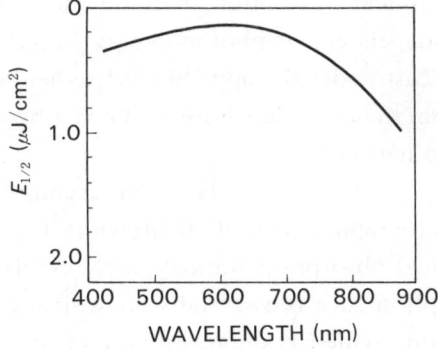

Fig. 7.4.3 Spectral photosensitivity of the a-Si:H drum in the "Laser Copier®". $E_{1/2}$ denotes the light energy illuminated to reduce the surface charges by to half the initial value. (courtesy of Canon, Inc.)

7.4.3 Enhancement of Long Wavelength Sensitivity

Neither the electrical nor the optical properties of a-Si:H are uniquely defined because its microstructure and even its chemical constituents as an inherent nature of amorphous materials depend greatly upon the preparation condition. Conversely a degree of freedom is conveniently available in making materials to conform with the requirements for application to devices. The optical gap E_g^0 (eV), for example, of a-Si:H is varied from 1.2 eV to 2.0 eV by changing preparation conditions.[9] Changes in the chemical constituents, i. e., the content of hydrogen C_H (%) or structural randomness are considered to be the causes of these changes in the optical properties. In a-Si:H prepared by glow discharge of SiH$_4$, the maximum quality from an optoelectric aspect is achieved in the film with $E_g^0 \sim$ 1.7 eV including hydrogen of 10~15 atm%, and the spectral response of the photoconduc-

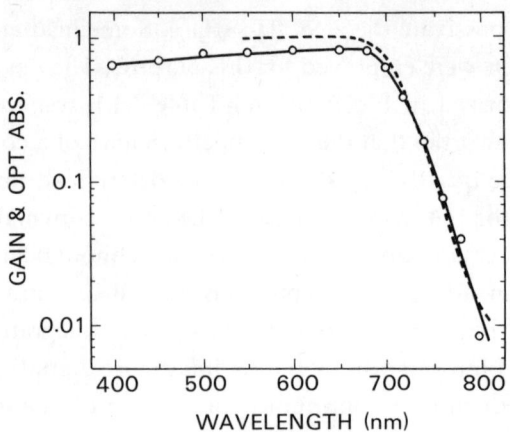

Fig. 7.4.4 Photoconductivity gain spectrum of an a-Si : H photoreceptor consisting of 10 μm thick film.
The optical absorption (1-transmittance) of an a-Si : H film 10 μm thick is illustrated by the dotted line. [after E. Inoue and I. Shimizu[2)]]

tivity is controlled by its optical absorption, because the generation efficiency η of the photoinduced carriers is independent of photon energy $h\nu$ (eV), temperature T (K) and electric field E (V/cm), at least under the applied electric field of $\sim 10^4$ V/cm. This characteristic is quite different from that of a-Se whose η is given by a function of $h\nu$, T and E due to the geminate recombination.[10)]

The spectral sensitivity of an a-Si : H photoreceptor with a film 10 μm thick measured by the electro-photographic method[2)] is shown in Fig. 7.4.4. A comparison of the gain spectrum and the optical absorption (dotted line) reveals that the spectral response corresponding to the absorption edge is essentially controlled by the optical absorption of the a-Si : H. According to this evidence, photoresponse of the a-Si : H photoreceptor falls markedly in the near infra-red region (700~800 nm in wavelength).

Several procedures are being considered to enhance the photosensitivity to near infra-red light, i. e.,
(1) using a thicker film ($> 20 \mu$m),
(2) preparing films exhibiting stronger optical absorption in the absorption tail, or
(3) adopting photoconductive films with narrower bandgaps.

The main factor in determining the thickness of the photoreceptor is the need to satisfy the demand for attaining sufficient surface potential for developing images with powdery toner. A thickness of 15~20 μm is commonly adopted in the a-Si : H drum. No dramatic improvement in optical absorption could be expected by increasing the film thickness, regardless of the amount of time and money spent. Therefore, other procedures, viz., items (2) or (3) are chosen to tune the spectral sensitivity with the light source (700 ~800 nm).

The Urbach tail of a-Si : H corresponding to the optical absorption $\alpha((\text{cm})^{-1})$, $5 \times 10^2 < \alpha(\text{cm})^{-1} < 5 \times 10^3$, tends to shift to the longer wavelengths as its structural and thermal disorder[9)] are increased. This is one of the most practical solutions because a slight shift of the Urbach tail brought on by changing the preparation conditions results in a fairly large increase in optical absorption. But this kind of solution is not essential since there is a fundamental "tradeoff" in a-Si:H between the optical absorption and photoelectric properties.

Item (3) must be considered an essential solution and thus further efforts must be made to prepare amorphous alloys and develop proper device structures.

7.4.4 Photoconductive a-SiGe Alloys

Preparation of photoconductive alloys with narrower gaps

Extensive efforts have been made to prepare amorphous silicon alloys, i. e., a-SiGe, a-SiSn, a-SiC$_x$ and a-SiN$_x$, for the purpose of excluding the defect density as well as the defect levels in a-Si:H. Although some improvement has been achieved,[11] the number of dangling bonds in these alloys, for example, still remains in the order of $10^{17} \sim 10^{18}$/cm^3 and is larger by two or three orders of magnitude than that of a-Si:H.

Among these alloys, a-Si$_x$Ge$_{1-x}$ and a-Si$_x$Sn$_{1-x}$ are the candidates for photoconductive films with a gap less than 1.7 eV. The band gaps of these alloys are mostly determined by their chemical constituents and decrease with an increase in the content of Ge or Sn, respectively. In the a-Si$_x$Ge$_{1-x}$:H prepared by glow discharge, the optical gap (E_g^0 eV) is given by a simple equation:

$$E_g^0 \text{ (eV)} = AX + B \tag{7.4.1}$$

Here x denotes the atomic ratio presented in a chemical formula of Si$_x$Ge$_{1-x}$. Both A and

Table 7.4.2 Relationship between optical gap (E_g^0 (eV)) and chemical constituency for (a) a-SiGe alloy and (b) a-SiSn.

(a) a-SiGe alloy whose E_g^0 value is given by the equation $E_g^0 = Ax + B$. Here, x is in the formula Si$_x$Ge$_{1-x}$.

A	B	films
0.70	1.02	GD a-SiGe:F:H[5]
0.70	0.95	GD a-SiGe:H[12]
0.75	1.10	GD a-Si:H[13]
0.45	0.40	Sputtered a-SiGe[11]

(b) a-SiSn:H prepared from (CH$_3$)$_4$Sn-SiH$_4$.[after S. Tsuda et al.[14]]

E (eV)	Sn content (atm %)
1.49	5
0.98	30
0.62	70

B are constants which are dependent to some extent on the preparation conditions (see Table 7.4.2).[12,13] In a-SiSn alloys made from a $(CH_3)_4$ Sn-SiH$_4$ gaseous mixture, on the other hand, virtually no linear relationship is recognized between E_g^0 and its consituents.[14] Furthermore, a small amount of Sn reduces the gap drastically as shown in Table 7.4.2 (b), reflecting the difference of the binding energy between Si-Ge and Si-Sn. However, there are a few open questions about the properties of a-Si$_x$Sn$_{1-x}$ due to a shortage of available information. Concerning a-Si$_x$Ge$_{1-x}$ alloys, on the other hand, several attempts have been made at improving their photoelectric properties besides the shift of the band edge towards long wavelengths.[15] Accordimg to the ESR study of a-Si$_x$Ge$_{1-x}$ alloy prepared from SiH$_4$-GeH$_4$, an increase in localized states attributed to germanium dangling bonds is considered to be the major cause of degradation of its optoelectric properties.[16]

Recently, we have succeeded in significantly reducing the defects in the a-SiGe alloy by adopting SiF$_4$-GeF$_4$-H$_2$ gases instead of hydrides. Because of the difference in the binding force between Ge-F and Ge-H, the number of dangling bonds has been dramatically reduced down to 8×10^{15}/cm^3 in an a-Si$_{0.7}$Ge$_{0.3}$: F : H film with $E_g^0 \sim 1.4$ eV. Electric conductivity either in the dark (σ_d) or under light illumination ($\Delta\sigma_p$) is plotted as a function of the optical gap in Fig. 7.4.5. The broken lines show the results of the samples prepared from hydrides, i. e., SiH$_4$ and GeH$_4$. Conspicuous differences are seen between these cases as follows:

(1) σ_d increases steadily as E_g^0 is decreased in the case of a-SiGe : F : H, and a minimum level is seen instead in the case of a-SiGe : H.

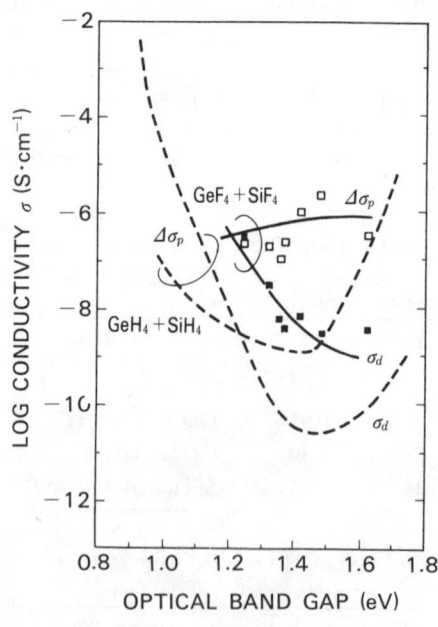

Fig. 7.4.5 Electric conductivity in the dark (σ_d(S·cm^{-1})) and under light illumination (4×10^{14} photons/cm^2·s)($\Delta\sigma_p$(S·cm^{-1})) plotted as a function of the optical gap (E_g^0(eV)) for a-SiGe alloys prepared from GeH$_4$-SiH$_4$ and GeF$_4$-SiF$_4$-H$_2$, respectively (All measurements were made at room temperature).

(2) $\Delta\sigma_p$ value measured under constant illumination decreases significantly for a-SiGe : H but is almost constant for a-SiGe : H : F as the E_g^0 value is decreased. With regard to the quality of the photoconductive film, we can obviously conclude that the a-SiGe : F : H is a more promising material for alloys with a narrow gap.

The drift mobility of μ_d (0.2~0.3cm²/V·s) was established by time-of-flight measurement for the electron transport of the a-Si$_{0.7}$Ge$_{0.3}$: F : H film. The doping feasibility has also verified a clear reduction of these dangling bonds. In a-Si$_x$Ge$_{1-x}$: F : H, doping a small amount of a foreign element (10~200ppm B or P) resulted in a marked change in the value (several orders of magnitude), in relation to the effective shift of its Fermi-level.[5] All this evidence thus gives support to the conclusion that the dangling bonds in a-SiGe : H : F can be largely eliminated, and brought down to the same level as in a-Si : H film.

A conspicuous difference was also found in the depositing behavior of a-SiGe alloy between the gaseous mixtures of SiH$_4$-GeH$_4$ and SiF$_4$-GeF$_4$-H$_2$. The chemical constituency of the film deposited in the former case is roughly determined by the mixing ratio of gases, whereas that in the latter case depends greatly upon the preparation conditions, i. e., rf power, substrate temperature and flow-rate in addition to the mixing ratio of the gases. The optical gap of the a-Si$_x$Ge$_{1-x}$ alloys plotted as a function of the mixing ratio, viz., GeX$_4$/SiX$_4$ + GeX$_4$ is shown in Fig. 7.4.6. These behaviors may be elucidated in terms of the difference in the chemical activities of the radicals made in the plasma, with respect to forming three-dimentional networks on the substrate. Virtually no a-Si networks are made from the radicals as a consequence of plasma-induced dissociation of SiF$_4$ whereas a-Si : H : F is easily prepared from a mixture of SiF$_4$ and H$_2$. Moreover, the evidence that the chemical constituency of a-SiGe alloy depends greatly upon the substrate temperature suggests the important role which the heterophase reaction on the substrate plays in the growth of the

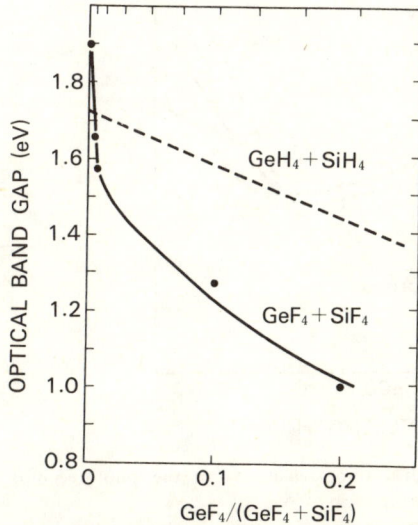

Fig. 7.4.6 The optical gap of a-SiGe alloys is plotted as a function of the mixing ratio of gases. The solid and the broken lines indicate respectively the results from GeF$_4$-SiF$_4$-H$_2$ and GeH$_4$-SiH$_4$.

silicon networks from the mixture of SiF_4-GeF_4-H_2.

Device structures

In some image devices, excellent photoresponse is attained by time-integration of the signals (photocurrent) during the processing time (T_p) under conditions of $T_p \ll \tau_r$. Here, τ_r denotes the dielectric relation time of the photoconductor used. On the other hand, there is a tradeoff in the semiconductive films between the optical absorption of light with long wavelengths and dielectric relaxation time because the narrower the gap is, the higher the electric conductivity is. Several successful attempts have been made to satisfy these demands by means of sophisticated device structures.[17] In the image pick-up device comprised of Se-As-Te and given the name of SATICON®, the sensitivity to red light is enhanced by adding a controlled amount of Te and As into a-Se without any accompanying reduction in the dielectric relaxation time.[18]

Device structures comprised of multi-layers have often been employed for photoreceptors made of organic coatings. Each layer in the device has an individual role, i. e., blocking carriers injected from the electrodes, carrier generation resulting from optical absorption and carrier-transport.

Recently this type of photoreceptor has been made of a-Si : H and a-$Si_x$$Ge_{1-x}$: H.[19] The a-$Si_x$$G_{1-x}$: H 1 μm thick ($E_g^0 \sim 1.3$ eV) was adopted to carrier generation at the position as illustrated schematically in Fig. 7.4.7. Though the spectral sensitivity was successfully expanded up to near the infra-red region as shown in the figure, an unfavourable reduction in the acceptance voltage ($20 \sim 30$ V/μm) accompanied the addition of the a-$Si_x$$Ge_{1-x}$: H

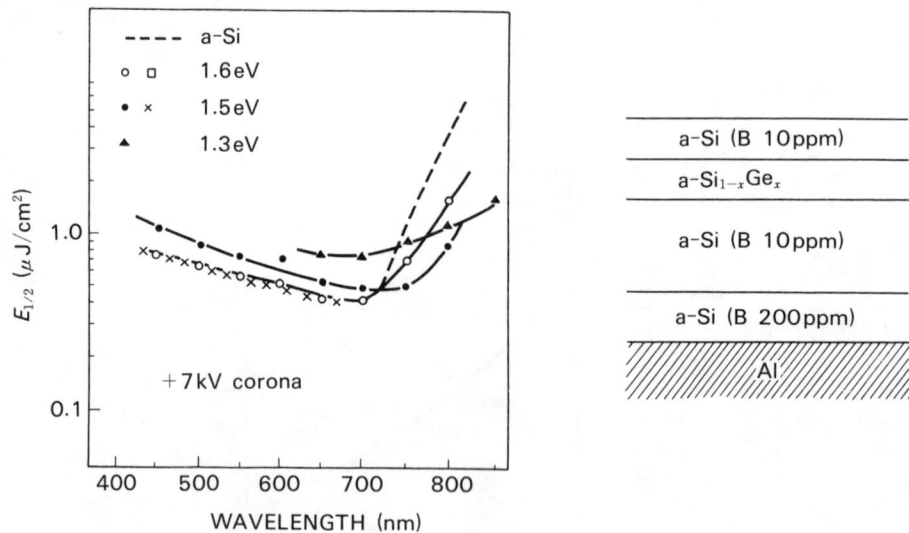

Fig. 7.4.7 Spectral photosensitivity of the photoreceptor with a stacked structure.
The structure is illustrated schematically on the right-hand side. We can clearly see an increase in photosensitivity at the infra-red region of the photoreceptor utilizing an a-SiGe alloy showing narrower gap energy. [after S. Nishikawa et al.[19]]

layer. A further reduction in the acceptance voltage or retention time of the surface charges may be brought about in a photoreceptor comprised of the a-Si$_x$Ge$_{1-x}$: F : H with a minimum number of defects because σ_d value of a-SiGe : F : H with the same E_g^0 value is higher by severel orders of magnitude than that of a-SiGe : H as a result of dangling bonds being removed. Consequently, further investigations are necessary to devise a photoconductor with narrower gap materials so as to provide a solution for this dilemma.

An excellent photoconductive gain was established in a CdSe/Se photoconductive film with thin layers (10~20 Å) of CdSe and Se deposited sequentially. By virtue of the strong optical absorption of the CdSe layer for red light, the sensitivity to long wavelength light was effectively enhanced with no loss of acceptance voltage.[20] Conversely, very poor acceptance voltage was found on a single layer of the same constituent. The suppression of the electric current in the dark of the multi-layer film may be attributed to the potential barriers formed at the interface of layers known as "barrier photoconductivity" shown in a polycrystalline PbS.[21]

The device with a multi-layer structure may be applicable to the a-Si : H/a-Si$_x$Ge$_{1-x}$ system in order to enhance photoresponse to long wavelength light with no loss in S/N value.

7.4.5 Conclusion

An a-Si : H photoreceptor has been successfully adopted to copiers and laser line printers with extremely favourable performance, i. e., excellent image quality and very high durability on an order of magnitude longer than the life of a conventional photoreceptor.

To tune the spectral sensitivity with the light source, viz., a diode laser emitting near infra-red light (700~800 nm), several attempts have been made as follows :
(1) using thick a-Si : H films ($>20\mu$m thick) with their Urbach tail modified to increase the optical adsorption as a tradeoff of the optoelectric properties, or
(2) using alloys (a-SiGe, or a-SiSn) with gap energy narrower than that of a-Si : H.

An a-SiGe film exhibiting desirable photoconductivity has been prepared by glow discharge from a gaseous mixture, GeF$_4$-SiF$_4$-H$_2$, by successfully removing the dangling bonds.

Some sophisticated device structures have been proposed for photoconductive devices comprised of a-chalcogenides with a wide spectral range. It is expected that by adopting these alloys a suitable structure of photoelectric devices exhibiting spectral sensitivity to near infra-red can be achieved with no deterioration in the S/N value of the devices.

Acknowledgements

We are gratetul to Mr. Eiichi KONDO and Mr. Tadaji FUKUDA of Canon Inc., for supplying the copying samples and the photographs of the a-Si : H drum.

References

1) Y. Hamakawa : J. Non-Cryst. Solids, *59 & 60* (1983) 1265.
2) E. Inoue and I. Shimizu : Phot. Sci. Eng., *26* (1982) 148.
3) T. Hamano, H. Ito, T. Nakamura, T. Ozawa, M. Fuse, and M. Takenouchi : Jpn. J. Appl. Phys., *21* (1982) Suppl. 21-1, 245.
4) S. Ishioka, Y. Imamura, Y. Takasaki, C. Kusano, T. Hirai and S. Nobutoki : Jpn. J. Appl. Phys., *22* (1983) Suppl. 22-1, 461.
5) K. Nozawa, Y. Yamaguchi, J. Hanna, and I. Shimizu : J. Non-Cryst. Solids, *59 & 60* (1983) 533.
6) I. Shimizu, T. Komatsu, K.Saito, and E. Inoue : J. Non-Cryst. Solids, *35 & 36* (1980) 773.
7) T. Kawamura, N. Yamamoto, and Y. Nakayama : JARECT *6* "Amorphous Semicond. Tech. & Devices" (1983) 325, Y. Hamakawa (ed.) OHMSHA & North-Holland.
8) Y. Imamura, S. Ataka, Y. Takasaki, C. Kusano, S. Ishioka, T. Hirai, and E. Maruyama : Proc.11th Conf. Solid State Devices, Tokyo (1979) 573.
9) G. D. Cody, T. Tiedje, B. Abeles, B. Brooks, and Y. Goldstein : Phys. Rev.Lett., *47* (1981) 1480.
10) D. M. Pai and R. C.Enck : Phys. Rev., *B 11* (1975) 5163.
11) G. Nakamura, K. Sato, Y. Yukimoto : J. de Phys., *C-4* (1981) Suppl. 10, C-4-483.
12) J.Chevallier, H. Wieder, A. Onton and C. R. Guarnieri : Solid State Comm., *24* (1977) 864.
13) Nguyen Van Dong, Tran Huu Danh and J. Y. Leny : J. Appl. Phys., *52* (1981) 338.
14) S. Tsuda, H. Tarui, M. Ohnishi, S. Sakai, K. Uchihashi, T. Matsuoka, S. Nakano, S. Kiyama, H. Kawata, and Y. Kuwano : J. Non-Cryst. Solids, *59 & 60* (1983) 1135.
15) Y. Yukimoto : JARECT *6* "Amorphous Semicond. Tech. & Devices" (1983) 136, Y. Hamakawa (ed.), OHMSHA & North-Holland.
16) A.Morimoto, T. Miura, M. Kumeda, and T. Shimizu : Jpn. J. Appl. Phys., *20* (1981) 833.
17) Y. Taniguchi, H. Yamamoto, S. Saito, and E. Maruyama : J. Appl. Phys., *52* (1981) 7261.
18) E. Maruyama, T. Hirai, T. Fujita, N. Goto, Y. Isozaki, and K. Shidara : Suppl. J. Jpn. Soc. Appl. Phys., *44* (1975) 97.
19) S. Nishikawa, H.Kamimura, T. Watanabe, and K. Kaminishi : J. Non-Cryst. Solids, *59 & 60* (1983) 1235.
20) Y. Yamaguchi and I. Shimizu : The 30 th JSAP spring meeting 4p-B10 (1983) 362 [in Japanese].
21) R. H. Bube : "Photoconductivity of Solids", Krieger Publishing Co., (1978) 359.

Authors' Profile

Editor

Yoshihiro Hamakawa was born in Kyoto, Japan, on July 12, 1932. He has been engaged with Osaka University since his completion of M. S. degree in 1958, and received Ph. D. degree in 1964 from Osaka University. He is now Professor of Electrical Engineering Science at Osaka University. He held the visiting posts at the University of Illinois (1965-1967) engaged as a Visiting Reseach Asistant Professor both at Department of Electrical Engineering and the Material Research Laboratory, Urbana, Illinois, USA.

Professor Hamakawa has performed research in the field of Semiconductor Physics, Optoelectronics and Solar Photovoltaic Conversion, particularly, on Optical Properties and Band Structure of Solids, Optoelectronic Devices, Solar Cells, Amorphous Semiconductor and Devices.

He has authored or coauthored eight books and more than 150 papers, including text books on "Semiconductor Physics and Devices, Volume 1 & 2" and an academic review volume; "Recent Advances in Modulation Spectroscopy" in OPTICAL PROPERTIES OF SOLIDS-New Developments (North-Holland Pub. Co., 1976) and has been serving as the Editor-in-Chief of "Amorphous Semiconductor Technologies & Devices" JARECT Vol. 2 (1982) and Vol. 6 (1983).

Committees/Boards: A member of Board of Japan Society for the promotion of Science No. 125, Optoelectronic Energy Conversion Committee (1970-79). The chairman of the Solar Photovoltaic Committee, Sunshine Project (1978), a neutral committee member (1974-79) and the chairman of Workshop for Amorphous Silicon Solar Cells (1979) in Sinshine Project, AIST, MITI. The chairman of the Amorphous Material R&D Survey Committee in Japan Society of Electronic Industry. A member of Board of "Japan Society of Applied Physics" (1979).

Academic Society Memberships: A member of IEEE, American Physics Society, ISES, SID Japan Society of Applied Physics, IEE of Japan, Physical Society of Japan. He was awarded the 1970 RCA Fundamental Research Grant for his work of "Electro-optical Effects and Band Structure in Mixed Compound Semiconductors" (1970), 1977 Yamada Science Foundation Grant for his research on "Valency Controls in Amorphous Semiconductors" (1977), and 1980 Hattori Hako-Syo for his great contributions to amorphous silicon solar cells.

1

Yoshihiro Hamakawa: aforementioned

2.1

Fumiko Yonezawa was born in Osaka Prefecture, Japan, on October 19, 1938.

In 1966, she obtained her D. Sci. degree from Department of Physics, Kyoto University, Kyoto, Japan.

Between 1966 and 1984, she worked at several places including (1) Research Institute for Fundamental Physics, Kyoto University, Kyoto, Japan, (2) Department of Applied Physics, Tokyo Institute of Technology, Tokyo, Japan, (3) Belfer Graduate School of Science, Yeshiva University, New York, USA, (4) Department of Physics, University of Bristol, Bristol, UK, (5) Department of Physics, City College of New York, New York, USA, and (6) at present full Professor of Physics, Keio University.

She has been doing research on theoretical physics and particularly interested in staistical physics of random system such as disordered alloys, doped semiconductors, liquid metals, amorphous metals and amorphous semiconductors. She is one of the four researchers who invented the coherent potential approximation for substitutionally disordered systems. Society of a splendid future for Female Scientists has awarde to her "the 4th Saruhashi Prize" for her excellent theoretical work on noncrystalline materials.

2.2

Tatsuo Shimizu was born in Osaka, Japan on July 12, 1936. He was graduated from Kanazawa University, Japan, in 1959, and received his M.S. and D. Sci. degrees from the University of Tokyo, Tokyo, in 1961 and 1967, respectively.

In 1961, he joined the Central Research Laboratory, Hitachi, Ltd., Tokyo, Japan. In 1962, he became a lecturer of electronics in Kanazawa University, and became an associate professor in 1964 and a professor in 1971. Since 1961, he has been working in the research of semiconductor physics. He is currently engaged in the research of amorphous semiconductors.

Dr. Shimizu is a member of the Physical Society of Japan, and the Japan Society of Applied Physics.

2.3

Toshikazu Shimada was born in Kyoto, Japan, on July 30, 1939. He received the B.E. degree from Ritsumeikan University, Kyoto, Japan in 1963, and the M.S. and Ph.D. degrees from Osaka University, Osaka, Japan, in 1965 and 1968, respectively.

In 1958, he joined Mitsubishi Electric Corp., Amagasaki, Japan, and in 1969, the Central Research Laboratory, Hitachi, Ltd., Tokyo, Japan, where he is a senior researcher. He was engaged in the research and development of ion implantation techniques, narrow gap semiconductors and amorphous semiconductors.

Dr. Shimada is a member of the Physical Society of Japan and the Japan Society of Applied Physics.

Yoshifumi Katayama was born in Kagawa Prefecture, Japan, on August 10, 1938. He received the B.E., M.E. and Ph.D. degrees from the University of Tokyo, Japan, in 1961, 1963, and 1966, respectively.

In 1966, he joined the Cenrtal Research Laboratory, Hitachi, Ltd., where he is a senior researcher. He was engaged in the fundamental research and development of narrow gap semiconductors, III-V and Si MOS devices, and molecular beam epitaxy. He is currently involved in the research of surface physics and amorphous semiconductors.

Dr. Katayama is a member of the Physical Society of Japan and the Japan Society of Applied Physics.

2.4

Kazuo Morigaki was born at Osaka, on February 10, 1932. He received his D. Sci. degree from Osaka University in 1959.

He is now a Professor of the University of Tokyo.

Dr. Morigaki is a member of the Physical Society of Japan.

3.1

Kazunobu Tanaka was born in Tokyo, Japan, on January 7, 1940, and received his B.S. degree in 1963, his D. Eng. degree in 1978 from the University of Tokyo, Tokyo, Japan.

In 1963, he joined the Matsushita Research Institute, Tokyo, Japan, worked in R & D of electroluminescent devices and new glass materials. In 1971, he moved to Electrotechnical Laboratory, Ibaraki, Japan. Since then, he has been primarily specializing in the field of amorphous semiconductors involving chalcogenide glasses as well as amorphous silicon, and in 1978, has joined the Sunshine Project as a leading staff of materials science.

Dr. Tanaka, a head of Amorphous Materials Section of ETL, is a member of the Japan Society of Applied Physics.

Nobuhiro Hata was born in Tokyo, Japan, on July 15, 1955. He received B.S. and M.S. degrees in Physics from the University of Tokyo, Tokyo, Japan in 1978 and 1980, respectively.

Since 1980, he has been a research staff at Electrotechnical Laboratory, Ibaraki, Japan, and primarily specializing in the field of laser diagnostics of plasma processing involving deposition of amorphous silicon films.

He is a member of the Japan Society of Applied Physics.

Akihisa Matsuda was born in Tokyo, Japan, on March 11, 1944, and received the B.S. and M.S. degrees in applied chemistry both from Waseda University, Tokyo, Japan in 1967 and 1969, respectively.

In 1969, he joined the research and development section of Japan Columbia (Denon) Co., Ltd., Kawasaki, Japan, where he was engaged in the research of new

AUTHORS' PROFILE

vidicon target materials. In 1972, he moved to Electrotechnical Laboratory, Ibaraki, Japan. Since then, he has been working in the field of amorphous materials.

He is a member of the Japan Society of Applied Physics.

--- 3.2 ---

Masataka Hirose was born in Gifu Prefecture, Japan, on September 30, 1939. He received the B. S. and the M. S. degrees in electronic engineering both from Nagoya University, Nagoya, Japan, in 1963 and 1967, respectively. He also received the Ph. D. degree in electronic engineering from Tohoku University, Sendai, Japan, in 1975.

From 1963 to 1964, he worked in Central Research Laboratory, Fuji Electric Co., Ltd. Since 1970, he has been with Hiroshima University, Hiroshima, Japan, and now is a Professor engaged in basic research on Si and GaAs MOS devices and amorphous silicon.

Dr. Hirose is a member of the Japan Society of Applied Physics and the Institute of Electronics and Communication Engineers of Japan.

--- 3.3 ---

Yukio Osaka was born in Hokkaido, Japan, on June 23, 1930. He recieved his B. S. degree in 1953, and in 1962 his Doctor of Science degree from Tohoku Univeresity.

He joined Faculty of Science, Tohoku University as lecturer from 1955 to 1962. He joined Faculty of Engineering, Tohoku University as Associate Professor from 1963 to 1971. Since 1972, he has been with Hiroshima University as Professor of the Department of Electrical Engineering where he is engaged in basic research on surface physics and amorphous semiconductor.

Dr. Osaka is a member of the Japanese Physical Society, the Japan Society of Aplied Physics and the Institute of Electronics and Communication Engineering of Japan.

Takeshi Imura was born in Toyama Prefecture, Japan, on September 29, 1940. He received his B. E. degree in 1963 in industrial chemistry from Kyoto University, his M. E. degree in 1969 in chemistry and his D. Eng. degree in 1974 in electrical engineering both from Osaka University.

Between 1963 and 1967 he worked at Wireless Research Laboratory, Matsushita Electric Industrial Co., Ltd. From 1970 to 1983 he was a faculty member of Osaka University at Department of Electrical Engineering. During this time, from 1975 to 1976, he was sent to Department of Chemistry, Cornell University, New York, U.S.A. In August 1983 he joined Faculty of Engineering, Hiroshima University as an associate professor, where he has been interested in physico-chemical and technological aspects of the semiconductor.

Dr. Imura is a member of the Japan Society of Applied Physics, Japanese Physical Society, and the Electrochemical Society of Japan.

--- 3.4 ---

Katsumi Aota was born in Sendai, Japan, on July 2, 1951. He received the M. Eng. degree in applied physics from Tohoku University, Sendai, Japan, 1978.

In 1978, he joined the Technical Laboratory, Citizen Watch Co. Ltd., Saitama, Japan, since then, he has been working in the field of thin-film and plasma technology. In April 1983 he has been sent to the department of electronic engineering, Tokyo University of Agriculture and Technology, Tokyo, Japan, as a research staff, and attends to the research and development of the amorphous semiconductor materials until now.

Mr. Aota is a member of the Japan Society of Applied Physics.

Yasuo Tarui was born in Tokyo, Japan, on June 4, 1929. He received the B. E. degree in electrical engineering from Waseda University, Tokyo, Japan, in 1951, and the Ph. D. degree in engineering from the University of Tokyo, Tokyo, Japan, in 1965.

After graduation in 1951, he joined the Electrotechnical Laboratory, Tokyo, Japan. He worked on solid-state devices and was Chief of the Semiconductor Device Section, Electrotechnical Laboratory. From 1962 to 1963, he was a Visiting Research Associate at the Solid-State Electronics Laboratory, Stanford University, Stanford, CA, USA. From 1976 to 1980, on leave of absence from the Electrotechnical Laboratory, he served as Director of the Cooperative Laboratories of the VLSI Technology Research Association, Kawasaki, Japan. Since 1981 he has been a professor of electronic engineering, Tokyo University of Agriculture and Technology, Tokyo, Japan.

Dr. Tarui is a member of the Institute of Electronics and Communications Engineers of Japan and the Japan Society of Applied Physics. He was Chairman of the Asian Committee of the International Solid-State

Circuits Conference from 1965 to 1966.

Tadashi Saitoh was born in Otaru, Japan, on April 10, 1940. He received the B. E. and M. E. degrees in applied chemistry from Hokkaido University, Sapporo, Japan, in 1962 and 1964. In 1978, he received the Dr. Eng. degree in electrical engineering from Osaka University, Osaka, Japan.

In 1964, he joined the Central Research Laboratory, Hitachi Ltd., Tokyo, Japan, where he was engaged in the preparation and characterization of semiconductor materials. Since 1973, he has been involved in researches on low cost silicon and indium phosphide materials, and ion implanted solar cells.

His current activities are centered around the development of highly-efficient, ion-implanted solar cells, and the research on amorphous silicon films prepared by photo-induced process.

Dr. Saitoh is a member of the Japan Society of Applied Physics.

---------- 3.5 ----------

Yukinori Kuwano was born in Fukuoka Prefecture, Japan, on February 14, 1941. He received the B. E. degree in 1963 from Kumamoto University, Kumamoto, Japan, and the Ph. D. degree in 1982 from Osaka University, Osaka, Japan.

He joined the Research Center of Sanyo Electric Co., Ltd. in 1963, where he was engaged in the research and development of amorphous devices. His present position is the manager of the 1st Department of the Research Center, and also the manager of Kuwano Laboratory of the Research Center.

Dr. Kuwano is a member of the Japan Society of Applied Physics, the Institute of Electrical Engineers of Japan. He is also a committee member of the Photovoltaic Science and Engineering Conference in Japan.

Dr. Kuwano was awarded the "Minister's Award for Scientific Technology" in 1980, and the "Richard M. Fulrath Award" in 1983.

Shinya Tsuda was born in Kyoto, Japan, on December 2, 1954. He received the B. E. and the M. E. degrees in electronic engineering both from Kyoto University, Kyoto, Japan in 1977 and 1979, respectively.

He then joined the Research Center, Sanyo Electric Co., Ltd. where he is engaged in research on amorphous semiconductor devices.

He is a member of the Institute of Electronics and Communication Engineers of Japan and the Japan Society of Applied Physics.

---------- 4.1 ----------

Hiroaki Okamoto was born in Hyogo Prefecture, Japan on December 21, 1951. He received his D. Eng. degree in 1980 from Osaka University, Osaka, Japan.

In 1981, he joined the Faculty of Engineering Science, Osaka University. Since then, he has been working in the basic and application fields of amorphous materials.

Dr. Okamoto is a member of the Japan Society of Applied Physics and the Physical Society of Japan.

---------- 4.2 ----------

Akio Hiraki was born on September 19, 1932. He received his B.S. degree in 1956, and in 1962 his Doctor of Science degree from Osaka University.

In 1963 he joined Faculty of Engineering, Osaka University. His main research area has been physics of semiconductor, especially its surface physics.

Dr. Hiraki is a member of the Japanese Physical Society and the Japan Society of Applied Physics.

---------- 4.3 ----------

Shumpei Yamazaki was born in Shizuoka, Japan, on July 8, 1942, and received his B. S. degree in 1965, his D. Engr. degree in 1970, from Doshisha University, Kyoto, Japan.

In 1970, he joined TDK Electronics Co. Ltd., he moved to TDK-Fairchild Co. Ltd.. He founded Semiconductor Energy Laboratory Co. Ltd., on July 1st 1980. Since then, he has been primarily specializing in the field of amorphous and crystalline Si and its application such as solar cells.

Dr. Yamazaki, president of Semiconductor Energy Laboratory Co. Ltd., is an active member of the Japan Society of Applied Physics, IEEE and Electorochemical Society.

AUTHORS' PROFILE

Satsuki Watabe was born in Kanagawa, Japan, on 1956, and recieved her B. S. degree in 1981 from Meisei University.

She joined Semiconductor Energy Laboratory Co. Ltd. in 1981. Since then, she has been majoring amorphous solar cells and Laser lithography for the integration.

Kenji Itoh was born in Tochigi, Japan, on 1958.

In 1977, he joined Nippon Precision Circuits Co. Led. and moved to Semiconductor Energy Laboratory Co. Ltd. in 1980. Since then, he has been majoring amorphous solar cells and Laser lithography for the integration.

--- 4.4 ---

Akira Yoshikawa was born in Mie Prefecture, in 1942. He received the B. S. and M. S. degrees in applied physics from Waseda University, Tokyo, in 1966 and 1968, respectively.

In 1968, he joined the Electrical Communication Laboratoty, Nippon Telegraph and Telephone Public Corporation, where he has been mainly working in the research and development of group VI and amorphous semiconductor materials. He is presently a head of Patterning Technology Section of Atsugi Electrical Communication Laboratory, where his responsibility is the research and development of microfabrication technologies of VLSI.

Mr. Yoshikawa is a member of the Japan Society of Applied Physics, the Physical Society of Japan, and the Institute of Electronics and Communication Engineers of Japan.

Yasushi Utsugi Member of staff engineer of Electrical Communication Laboratories in Atsugi, Japan. His activities presently include research on basic aspects of inorganic resist and new photolithographic systems. He received the B. S. degree in physics from Hokkaido University, Sappro, Japan, in 1967, and the Dr. Eng. degree in electrical engineering from the Tohoku Universty, Sendai, Japan, in 1980.

In 1971, he joined Electrical Communication Laboratories in Musashino, Tokyo, where he worked on studies on GaAs crystals and functional Gunn devices. In 1973, he got involved in amorphous semicondutor work to develop optical recording materials and systems. Since 1981, his main interests have been in the field of LSI processing technologies.

--- 4.5 ---

Toshio Ogino was born in kyoto, Japan, on June 3, 1951. He received the B. S., M. S. and Ph. D. degrees in electronic engineering from the University of Tokyo, in 1974, 1976 and 1979, respectively.

In 1979, he joined the Electircal Communication Laboratories, Nippon Telegraph and Telephone Public Corporation. He has been concerned with research on semiconductor devices.

Dr. Ogino is a member of the Japan Society of Applied Pnysics and the Institute of Electronics and Communication Engineers of Japan.

Yoshihiko Mizushima was born in Sendai, Japan, on January 13, 1925. He received the B. S. degree in 1946 from the University of Tokyo, the Dr. rer. nat. degree in 1968 from the Technische Hochschule Hanover, Hanover, Germany, and the D. Eng. degree in 1972 from the University of Tokyo.

In 1946, he joined the Electrotechnical Laboratory of the Japanese government as a research staff member and engaged in the study of the physical chemistry of electron tube materials. Since the establishment of the Electrical Communicaion Laboratory of the Nippon Telegraph and Telephone Public Corporation in 1948, he has been concerned primarily with research on thin films, surface physics, intermetallic semiconductors, and transport phenomena in solids. In 1983, he joined Hamamatsu Photonics Co., Ltd., Hamamatsu, Japan, as a scientist, vice president.

Dr. Mizushima is a member of the Institute of Electronics and Communication Engineers of Japan, the Physical Society of Japan, the Japan Society of Applied Physics, and the Institute of Electrical Engineers of Japan.

--- 5.1 ---

Yoshiyuki Uchida was born in Saitama Prefecture, Japan, on February 26, 1940, and recieved the B. S. degree in 1962 in elctronic engineering from Tohoku University, Sendai, Japan.

In 1962, he joined the Central Research Laboratory, Fuji Electric Co., Ltd., Kawasaki, Japan, where he was engaged in the research and development of silicon devices. Since 1980, he has been with Fuji Electric Corporate Research and Development Ltd., Yokosuka, Japan. He is now the deputy director of the Electron Device Laboratories.

He is a member of the Institute of Electrical Engineers of Japan, the Japan Society of Applied Physics, the American Physical Society and the Electrochemical Society.

---------- 5.2 ----------

Yoshihiro Hamakawa : aforementioned
Hiroaki Okamoto : aforementioned

---------- 5.3 ----------

Hiroshi Morimoto was born in Hyogo Prefecture, Japan, on May 26, 1951. He received the B. S., M. S. and D. Sci. degrees in physics from Osaka University, Osaka, Japan, in 1974, 1976 and 1980, respectively.

In 1980, he joined the Central Research Laboratory, Sharp Corporation, Nara, Japan, where he was engaged in the research and development of advanced mass production technology. His main interest is in the development of low-cost amorphous silicon solar cells.

Dr. Morimoto is a member of the Physical Society of Japan and the Japan Society of Applied Physics.

Masatsugu Izu attended Kyoto University (Dept. of Hydrocarbon Chemistry, School of Engineering) and received his Doctor's Degree from the same University.

After three years as a Research Instructor at Kyoto University, he joined Energy Conversion Devices, Inc. in 1972 where he is now Vice President, Photovoltaics.

---------- 5.4 ----------

Yukinori Kuwano : aforementioned

Shoichi Nakano was born in Hyogo Prefecture, Japan, on February 20, 1944. He received the B. E. degree in electric engineering from Kyoto University, Kyoto, Japan in 1968.

He then joined the Research Center, Sanyo Electric Co., Ltd., where he is engaged in research on electric components and amorphous semiconductor devices.

He is a member of the Japan Society of Applied Physics.

---------- 6.1 ----------

Masakiyo Matsumura was born in 1941. He recieved B. Eng., M. Eng. and Dr. Eng. degrees from the Tokyo Institute of Technology in 1964, 1966 and 1972, respectively.

In 1966, he joined Nippon Electric Company as a Researcher and was engaged in the developments of semiconductor devices and display devices. From 1976, he was Associate Professor of Physical Electronics Department at Tokyo Institute of Technology.

---------- 6.2 ----------

Katsumi Murase was born in Kyoto, Japan, on March 7, 1950. He received the B. E., M. E. and D. Eng. degrees from Kyoto University, in 1972, 1974 and 1984, respectively.

In 1974, he joined the Electrical Communication Laboratories, Nippon Telegraph and Telephone Pubilc Corporation. He has been concerned with research on MOS IC's, ion implantation, submicron technology and amorphous materials.

Dr. Murase is a member of the Institute of Electronics and Communication Engineers of Japan and the Japan Society of Applied Physics.

Yoshihiko Mizushima : aformentioned

---------- 6.3 ----------

Hideki Hasegawa was born in Tokyo, Japan, on June 22, 1941. He received the B. S., M. S., and Ph. D. degrees in electronic engineering from the University of Tokyo, Tokyo, Japan, in 1964, 1966, and 1970, respectively.

In 1979, he joined the faculty of the Department of Electrical Engineering, Hokkaido University, Sapporo, Japan, as a Lecturer. He became an Associate Professor in 1971 and a Professor, in 1980, both at the same department. From 1973 to 1974, he was on sabbatical leave at the Department of Electrical and Electronic Engineering, the University of Newcastle-upon-Tyne, Newcastle-upon-Tyne, England, as a Visiting Research Fellow. His current research interests include growth, characterization and processing of a-Si films, new devices using a-Si films, surface and interface properties of compound semiconductors, MBE and MOCVD growth and deep-level characterization of compound semiconductors, high-speed logic IC's and microwave IC's using GaAs and InP.

Dr. Hasegawa is a member of the Institute of Electronics and Communication Engineers of Japan, Institute of Electrical and Electronics Engineers (IEEE), the Japan Socity of Applied Physics, the Institute of Electrical Engineers of Japan and Japan Association of Crystal Growth.

Hidekazu Yamamoto was born in Hokkaido, Japan, on December 7, 1956. He received the B. S., and M. S. degrees in electrical engineering from Hokkaido University, Sapporo, Japan, in 1979, and 1981, respectively. He is working towards the Ph. D. degree at the same department.

His research interests include growth, characterization and processing of a-Si films, new devices using a-Si films, and solar cells using GaAs and InP.

Mr. Yamamoto is a member of the Institute of Electronics and Commuication Engineers of Japan, and the Japan Society of Applied Physics.

Satoshi Arimoto was born in Tokyo, Japan, on February 26, 1960. He received the B. S. degree in electrical engineering from Hokkaido University, Sapporo, Japan, in 1982. He is working towards the M. S. degree at the same department.

His recearch interests include growth, characyerization and processing of a-Si films, and new devices using a-Si films.

Mr. Arimoto is a member of the Japan Society of Applied Physics.

Hideo Ohno was born in Tokyo, Japan, on December 18, 1954. He received the B. S., M. S., and Ph. D. degrees in electronic engineering from the University of Tokyo, Tokyo, Japan, in 1977, 1979, and 1982, respctively. He spent 1979 to 1980 academic year at the School of Electrical Engineering, Cornell University, Ithaca, NY, USA, as a grantee of a scholarship by Japan Socity for the promotion of Science.

In 1982, he joined the faculty of the Department of Electrical Engineering, as a Lecturer, Hokkaido University, Sapporo, Japan, where he is now an Associate Professor, working on III-V compound semiconductor growths by MBE and MOCVD, and on various new electron devices based on III-Vs, as well as on a-Si : H based devices.

Dr. Ohno is a member of Institute of Electronics and Communication Electronics Engineers of Japan, the Japan Society of Applied Physics, the Institute of Electrical and Electronics Engineers (IEEE), and American Vacuum Society.

---------- 7.1 ----------

Mutsuo Takenaga was born in Kumamoto Prefecture, Japan, on October 9, 1954. He received his M. S. degree in 1970 from Kumamoto University and D. Eng. degree in applied chemistry from Osaka University in 1977.

He joined Matsushita Electric Industrial Co., Ltd., Osaka, in 1970, and was engaged in the study of thermo-luminescent materials for radiation dosimetry. Since 1973, he has been working in the research and development of optical memory materials.

Dr. Takenaga is a member of the Chemical Society of Japan, the Japan Society of Applied Physics.

Masanari Mikoda was born in Fukuoka, Japan, in 1933. He received his B. S. degree in 1955 from Kumamoto University and D. Eng. degree in applied chemistry from Kyoto University in 1981.

He joined Matsushita Electric Industrial Co., Ltd., Osaka, in 1959. He has been concerned with research and development of amorphous and ceramic materials for electronic component and electron devices.

Dr. Mikoda is a member of the American Ceramic Soc., the Society of Material Science, Japan, and a senior member of IEEE.

---------- 7.2 ----------

Nozomu Harada received his B. S. degree in 1968 and M. S. degree in 1970 from Shizuoka University, Japan. In April, 1970, he joined Toshiba Research and Development Center. He has been engaged in the development of image pickup devices, silicon vidicon, CCD image sensor. He is now in the VLSI Reseach Center. He is a member of the Institute of Television Engineers of Japan and the Japan Society of Applied Physics.

7.3

Toshihisa Tsukada received the B. S., M. S., and Ph. D. degrees, all from the University of Tokyo, Tokyo, in 1963, 1965, 1968, respectively.

In 1968, he joined the Central Research Laboratory, Hitachi, Ltd., Tokyo. He has been engaged in research and development of GaAsP light-emitting diodes, and GaAs-GaAlAs semiconductor lasers, and amorphous silicon thin film devices. He is now a chief researcher of the Central Research Laboratory, Hitachi, Ltd. During 1975-1976, he stayed at Cornell University, Ithaca, N. Y., as a faculty member.

Dr. Tsukada is a member of the Institute of Electronics and Communications Engineers of Japan, the Japan Society of Applied Physics, and the Institute of Electrical and Electronics Engineers (U.S.A). He received the 1984 Achievement Award from IECE, Japan.

7.4

Isamu Shimizu was born in Tokyo on August 17, 1939. He received the B. S. degree in printing engineering from Chiba University in 1962, and the M. S. and Ph. D. degrees in chemical engineering from Tokyo Institute of Technology in 1963 and 1968, respectively.

Since 1976, he has been with Tokyo Institute of Technogy as an associate profesor. His research interests include material design for imaging technology.

Dr. Shimizu is a member of the American Physical Society, Japan Society of Applied Physics, the Chemical Society of Japan, and the Society of Photographic Science and Technology.